*The Dependability
of Behavioral
Measurements*

# The Dependability of Behavioral Measurements:

*Theory of Generalizability for Scores and Profiles*

Lee J. Cronbach, *Stanford University*
Goldine C. Gleser, *University of Cincinnati*
Harinder Nanda, *A. N. Sinha Institute of Social Studies*
Nageswari Rajaratnam (*Deceased*)

John Wiley & Sons, Inc.

*New York    London    Sydney    Toronto*

Library of Congress Catalog Card Number: 70-180269

ISBN 0-471-18850-6

Printed in the United States of America.

10 9 8 7 6 5 4 3 2 1

# *Preface*

What we call generalizability theory has evolved slowly and circuitously. The tidy theory of error laid down for psychology at the start of the century by Spearman and Brown has always seemed just a little too tidy to describe the perverse behavior of real data. Nearly every specialist in behavioral measurement has tried to develop a less restrictive model. Each new formulation has had its virtues and its defects, but the combined thrusts of the proposals have propelled the profession a long way beyond the classical position.

In recent years a chief problem has been to assemble what is known into a comprehensible, intuitively appealing structure. Our own efforts date back more than twenty years, and the more intensive work that produced this monograph dates back to 1957. Time and again, we prepared what we thought was a comprehensive system. Each such reorganization, brushing away scraps, tying in loose ends, and exposing the central structure more clearly, has shown the structure to be unfinished and suggested ways to extend the argument. This monograph as well surely has a built-in obsolescence. The multivariate parts of the theory, in particular, began to come into focus only in 1967. Much further work remains to be done with them, and as that work is carried forward it is likely to alter the whole structure. It seems unlikely that further developments will displace our basic scheme, with which other investigators have also had considerable experience.

Many persons and agencies have assisted us. From 1957 to 1963, the work was aided by the Bureau of Educational Research of the University of Illinois and by grant M-1839 from the National Institute of Mental Health. Cronbach and Gleser were the principal investigators. Rajaratnam was full-time associate from 1957 to 1960, shared authorship of the basic technical reports, and contributed many of the key ideas. Many consultants and correspondents made valuable suggestions and comments; we are particularly indebted to Frederic Lord and Hubert Brogden. Assistants with the project for one year or more were Hiroshi Azuma, Milton Meux, Peter Schönemann, and James Terwilliger.

In 1960, with two main reports distributed and the third in draft form,

Rajaratnam took a position at the University of British Columbia and Gleser took on increased responsibilities at the University of Cincinnati. Cronbach continued to work on these and other measurement problems at the Institute for Advanced Study, Princeton. When he returned to Illinois in 1961, the project staff consisted of Victor McGee, associate, and Hiroshi Ikeda and T. Douglas McKie, assistants. From 1962 to 1963, Jean Cardinet was associate and McKie and R. A. Avner were assistants. John E. Hunter independently carried out pertinent mathematical studies. During this period, Cronbach, Gleser, and Rajaratnam revised the technical reports for journal publication. Cronbach served as a member of the committees that produced the 1966 *Test Standards;* this provided an opportunity for further discussion of our concepts with helpful colleagues. Support from the National Institute of Mental Health was terminated in 1963. Dr. Rajaratnam died in 1964, while serving on a research staff at the University of Minnesota.

There was no further systematic work on these matters until late 1965. In the interim, however, Cronbach and Gleser attempted to help several researchers use the theory in their substantive investigations. The experience showed that the original papers were not specific enough to guide the investigator; nearly every study required some adaptation of the basic methods and interpretations.

Cronbach had taken a position at Stanford University, and in 1965 applied for funds to permit preparation of a monograph. The Cooperative Research Branch of the U.S. Office of Education provided support for an assistant, Nanda, throughout 1966. Some supplementary support came from the Center for Research and Development in Teaching, a U.S.O.E.-funded agency at Stanford. Our original intention was simply to expand the published papers with more explicit advice on procedures and with worked examples. Such an expansion and clarification was distributed in a preliminary version in 1967. The 1967 report, with revisions, forms the heart of Chapters 1–7 of this volume. We are indebted to Haruo Yanai for suggesting the use of Venn diagrams to improve the exposition.

Meanwhile, Nanda's studies on interbattery reliability and studies by Kenneth W. Travers opened the way to a multivariate extension of the system. That part of this monograph has not appeared previously, save for a paper on difference scores in which Lita Furby collaborated. Michael Ravitch has served as assistant in preparing numerical examples and Leigh Burstein assisted with Chapter 8.

To spend more than ten years in producing a monograph seems on its face to require some apology. But even in retrospect it is hard to see how the work could have proceeded much faster. In such an endeavor one cannot head straight toward the final product. One casts about for useful paths, puzzles over trails that branch ambiguously, sometimes spends considerable time exploring an alternative pathway that, once traced out, can be omitted from the final map. Sometimes one pursues a line of development for a time as a major independent problem, only to have it shrink back into a minor role when the theory consolidates at a higher level of generality. (Our work on "stratified-parallel" tests is an example.) Perhaps the greatest amount of time goes into the detection and eradication of errors and blind spots, some due to the perversity of typewriters and computers, some arising from our own illogic, and some—the most elusive—being misconceptions built into habitual, "time-tested" concepts. Our chief process of work has been the ceaseless revision of each page of draft manuscript by each of the coauthors, to ensure the accuracy of everything that is said.

We owe a special debt to the project secretaries, whose duty has been to put illegible technical manuscript, revised by three writers in at least three colors, into immaculate form—so that the process of revision could begin again. The secretaries who served the project for a year or more are Gloria Block, Martha Francisco, Clara Hahne, Dorothy Humes, Thelma Wasson, and Jenny Cloudman.

We take this opportunity to thank the agencies, including our Universities, that have supported the work. We also thank the colleagues whose criticisms, encouragement, and loan of data have moved us forward, and the successive members of our research group.

*Lee J. Cronbach*
*Goldine C. Gleser*
*Harinder Nanda*

# Contents

CHAPTER 3

*Inferences From D-Study Data Regarding the Universe
Score $\mu_p$*                                                     73

CHAPTER 4

*Universes With Fixed Facets*                                      113

## CHAPTER 10

### *Multivariate Estimation of Universe Scores: Profiles, Composites, and Difference Scores*     309

## CHAPTER 11

### *Contributions and Controversies—a Summing-Up*     350

# Summary
# of Notation

The symbols used in this book are identified below. The page reference indicates initial use. Notation used at only one point will not appear in this summary.

### Lower-Case Letters

| | |
|---|---|
| $a, b, \ldots$ | Used to identify specific conditions of facet $j$ (p. 36). |
| $a$ | Used to represent an unknown or arbitrary quantity (p. 93). |
| $b$ | Slope of regression of $\mu_p$ on $X_{pi}$ (p. 143). |
| $c$ | As a prescript, identifies a criterion score (p. 325). |
| $d$ | Days, as a facet (p. 195). As a prescript, identifies a difference score (pp. 263, 330). |
| $e$ | A component of the score not identified with persons, classified conditions, or their interaction (p. 27). The "within-cell" component of an analysis of variance. |
| $f$ | Test forms, as a facet (p. 217). |
| $g, h, i, j, k, \ell$ | A facet, or a condition of the facet so labelled (pp. 26ff., 172, 265ff.). |
| $n$ | Number of conditions of a facet (e.g., $n_i$) employed in an experimental design (pp. 9, 34ff.). Also, number of person $n_p$, number of variables $n_v$. |
| $o$ | Occasions, as a facet (p. 176). |
| $p$ | Persons, as a basis for classifying observations (p. 26) |

| | |
|---|---|
| *r* | Raters or recorders, as a facet (p. 191).<br>A sample correlation coefficient (p. 78).<br>Residual, as in MS *r* (p. 83). |
| *s* | Schools, as a facet (p. 217).<br>Sample covariance (p. 234).<br>Sample standard deviation (p. 83). |
| *t* | Teachers, as a facet (p. 191).<br>Trials, as a facet (p. 238).<br>As a subscript, identifies a test score as distinct from an item score (rare) (p. 121). |
| *v* or **v** | Variable (pp. 265, 272). |
| *w* | Multiplier used in forming weighted composite scores (p. 265).<br>As a prescript, identifies a composite score (p. 327). |

*Capital Letters*

| | |
|---|---|
| *A, B, ...* | Used to identify specific conditions of facet *i* (p. 26).<br>Used to identify sums of components (pp. 48, 250) |
| $\mathscr{E}$ | Expected value. Limit approached by an average as the number of elements averaged increases (p. 26). For example,<br>$$\mathscr{E}_i X_{pi} = \text{Lim} \frac{1}{n_i} \sum_1^{n_i} X_{pi}.$$ |
| *E* | In classical theory, the error of measurement (p. 75). |
| EMS | Expected mean square (p. 43). |
| *G, H, I,* etc. | A set of conditions used to make an observation; e.g., several conditions of facet *g* make up the set *G* (p. 28). Any letter whose lowercase form identifies a facet may be used in upper-case form also. |
| *N* | $N_i$ is the number of conditions of facet *i* in the universe of admissible observations. $N_j$, etc. may be defined similarly (pp. 9, 58ff.). |
| *P* | A group of persons (p. 28).<br>As a subscript, indicates that a score is a sample mean (p. 103).<br>Probability (p. 50). |

$R$         Multiple correlation coefficient (p. 321).

$T$         In superscript position, indicates the transpose of a matrix or vector (p. 319).

As a subscript, identifies a total score as distinct from an average (p. 82).

$X$         Observed score. The score $_vX_{pi}$ is the observed score of person $p$ under condition $i$ (pp. 26, 265).

*Greek Letters*

$\alpha$         Intraclass correlation from a one-facet study where condition means are not regarded as a source of variance in scores (p. 82).

$\Delta$         Difference between observed score and universe score (pp. 24, 76).

$\delta$         Difference between observed deviation score and deviation score in the universe (pp. 24, 93ff).

$\varepsilon$         Difference between universe score and universe score estimated by means of a regression equation (p. 25).

$\in$         "Belongs to the set," as in $i \in I$ (p. 28).

$\mu$         Mean in population or universe (p. 26). In particular, a component of the observed score. The component $\mu_p$ is ordinarily the universe score of interest.

$\nu$         Portion of interaction uncorrelated with $\mu_p$ (p. 143).

$\rho$         Population value of a correlation coefficient (p. 75). The expression $\mathscr{E}\rho^2$ is an abbreviated notation for $\mathscr{E}\rho^2(X_{pi}, \mu_p)$ or more generally, the expected value of the squared correlation of observed score with universe score (p. 98).

$\sigma$         Population value of standard deviation or covariance. $\sigma^2(X_{pi})$ is the variance of $X_{pi}$ over all $p$ and $i$ (p. 27). $\sigma^2(pi)$ is the variance of the $pi$ "component" over all $p$ and $i$ (p. 27).

$\Sigma$         Sum of.

$\sum$         Variance–covariance matrix (pp. 267, 272).

### Numerals

**1, 2**        Two particular variables, as in $_1X$, $_2X$ (p. 265). Especially, pretest and posttest measures (p. 350).

**1, 2, . . .**   Persons whose scores are discussed (p. 36).

### Auxiliary Signs

$i, i'$        Two distinct conditions of facet $i$. Used similarly with $j$ (pp. 94, 231).

$p, p'$        Two distinct persons (p. 94).

$v, v'$        Two distinct variables (p. 265).

$n, n'$        Number of conditions employed in the G and D studies, respectively (p. 74).

$N'$          Number of conditions in the universe of generalization (p. 58).

$i^*, p^*$      A particular condition or person (p. 26). A double asterisk (as in $i^{**}$) is used where emphasis is on a condition or person fixed in the D study and the universe of generalization (pp. 114ff.).

As prescript, identifies a variable of particular concern (p. 313). Also in $v^*$.

Caret, Estimator. For example, $\hat{\sigma}(\mu_p)$ is an estimate of $\sigma(\mu_p)$. But $\sigma(\hat{\mu}_p)$ is the standard deviation of an estimate of $\mu_p$ (P. 47).

$\bullet, \circ$        Bullet and open circle.

$i \cdot g$        Condition $i$ for observing variable 1 is drawn simultaneously with condition $g$ for observing variable 2, causing $i$ and $g$ to be "linked" (p. 268).

$i \circ g$        Drawing of condition $g$ is independent of the drawing of condition $i$; i.e., $i$ and $g$ are independent (p. 270).

$^\bullet\sigma(_vX, _{v'}X)$      Covariance of observed scores $_vX$ and $_{v'}X$ when $i \bullet g$ (p. 271).

$^\circ\sigma(_vX, _{v'}X)$      Covariance of observed scores when $i \circ g$ (p. 311).

$\bar{e}$          Overline; average of a set, as in $\bar{e}$ (p. 59) or $\bar{X}$ (p. 75).

~ Tilde; functions as an ellipsis (p. 41). The score component for the *pi* interaction may be written $\mu_{pi}\sim$ instead of $\mu_{pi} - \mu_p - \mu_i + \mu$. Similarly with other components.

[1] As a subscript, the bracket indicates that the index it replaces takes on all values except 1 (p. 54).

( | ) The conditional symbol implies that any index to the right of the symbol is to be treated as fixed for the time being. Thus, $\sum (X_{pi} \mid i)$ has the same significance as $\sum_p X_{pi}$ (pp. 82, 83). In an experimental design e.g., $(i \times p) \mid J^*$, all *pi* combinations are observed under conditions $J^*$ (pp. 59, 114).

× The cross, colon, and comma are used in describing
: experimental designs, being read respectively as "crossed
, with," "nested within," and "joint with" (pp. 35ff., 286).

# CHAPTER 1

# The Multifacet Concept of Observational Procedures

The investigator who tests a person twice is likely to obtain scores that differ. Determination of the magnitude of such inconsistencies in measurement has been recognized as important since the time of Bessel and Gauss, as the investigator thereby learns how much confidence he can place in his data.

In psychology and education, a mountainous literature on "reliability" of measures has been built upon the foundation of Spearman's 1904 paper. In nearly all this literature, the observed score is seen as the sum of a "true score" and a purely random "error," the error being looked on as a sample from a single undifferentiated distribution. The classical procedure for reliability analysis estimates the standard deviation of this hypothesized distribution (the standard error of measurement) and the closely related reliability coefficient. Such a coefficient is interpreted as an estimate of the squared correlation of observed score with true score, or as the ratio of the variance of true scores to the variance of observed scores.

A generation ago, R. A. Fisher (1925) revolutionized statistical thinking with the concept of the factorial experiment in which the conditions of observation are classified in several respects. Investigators who adopt Fisher's line of thought must abandon the concept of undifferentiated error. The error formerly seen as amorphous is now attributed to multiple sources, and a suitable experiment can estimate how much variation arises from each controllable source. With the estimates of the several variance components in hand, the investigator can understand how unwanted variation arises, and he can plan an efficient design for collecting further data. This is demonstrated in the context of industrial statistics by Tippett (1950, pp. 124–133).

The behavioral scientist, like other investigators, can learn far more by

1

allocating variation to facets[1] than by carrying out the conventional reliability analysis. The principal methods for multifacet analysis of error were presented rather fully to behavioral scientists and educational researchers in Lindquist's text on experimental design (1953). These methods have been less extensively treated elsewhere. However, they have not been widely adopted. Indeed, we know of no instance where multifacet techniques were used to organize statistical evidence in a test manual, and only rarely have they appeared in publications on ratings and observation procedures. The tester's neglect of multifacet analysis probably reflects the fact that the design of experiments branched off as a specialty in itself, with the consequence that advances in variance analysis were not brought forcefully to the attention of students of behavioral measurement. The separation was encouraged by the fact that experimenters characteristically regard subjects (persons) as a source of "error" in their analyses, whereas the tester is interested chiefly in the person tested and only secondarily in the conditions of observation. Methodological statements directed to experimenters do not communicate well to students of measurement.

Among the compelling arguments for adopting multifacet analysis of error as a standard technique are these:

1. Explicit consideration of the several facets of a measuring operation dispels ambiguities that were present in, and concealed by, the classical model.
2. The multifacet study can appraise interactions inaccessible to the older methods, and so can improve one's understanding of the measure.
3. One multifacet study answers questions that formerly required several separate sets of data.
4. Multifacet information enables one to design more efficient procedures for collecting data, either for the measurement of individuals or for the determination of group means.

Concurrently with the multifacet conception of measuring operations, a tradition of multi*variate* analysis has evolved. In factor analysis, multiple correlation, profile interpretation, and a number of other techniques, conclusions are reached through the simultaneous consideration of diverse measures. Even though specialists in psychometric theory have played a large part in developing multivariate statistical methods and their applications, this work has been almost completely isolated from the theory of true scores and error of measurement. Better information on a true score can be obtained by combining a direct observation on that variable with

---

[1] Following Guttman, we speak of "facets" rather than "factors," because the latter term evokes, in the psychologist, associations with factor analysis.

observations on variables correlated with it than can be obtained from the direct observation alone.

Just a few scattered papers have suggested a multivariate approach to the estimation of true scores. When this line is pursued, it becomes evident that some of the most traditional procedures for test analysis are incorrect or open to misinterpretation. We shall introduce a systematic model for multi-variate theory of measurement and show how it joins hands with multifacet theory to make all measurement theory more logical and more useful. However, because of the novelty of both approaches, we combine them only to discuss a few simple cases. A completely integrated presentation would be too abstract to be assimilated at this time.

While the paper of Gleser, *et al.* (1965) presents the basic theory and formulas for multifacet analysis with one variable, that presentation is too compressed to be an adequate guide to the researcher. Endless variations of the basic problem are encountered in practice, requiring adaptation of formulas and interpretations. We have rarely undertaken a multifacet analysis of a new instrument without encountering surprises and paradoxes that required decisions for which the literature has provided no explicit guidance. Therefore, we have compiled in this monograph illustrative studies in sufficient variety and detail that the rationale, the computational procedures, and the interpretation can be shown at length. Most of the examples are taken from studies for which one or another of the authors served as con-sultant. Matters of technique and formal interpretation that puzzled us are emphasized here, rather than the substantive questions that gave rise to the studies.

This monograph goes beyond the 1965 paper in several respects, one of them being the multivariate extension mentioned previously. We have in-cluded some relevant theory of inference from the statistical literature, and a more complete discussion of mixed model analyses. Primary emphasis is now placed on variance components, where the earlier paper followed psycho-metric tradition in emphasizing ratios of variances (coefficients). We have altered details of the presentation, adopting, for instance, a new system of notation to describe experimental designs.

Despite our intent to display techniques clearly for the reader who wishes to use them, the book is complexly organized and by no means simple to follow. If we had been able to stop with a straightforward series of illus-trations of the well-worked-out part of the theory, the presentation might have adhered to a genuine textbook style. But it is necessary to bring in complex arguments, some of which have not yet reached a stable form. It appears important to draw attention to unfinished business within multifacet theory—for example, to the effect of sampling errors upon results of the analysis.

The questions that appear following the chapters must be seen, not as routine textbook exercises, but as opportunities for the reader to exercise his wits on some of the complex judgments that multifacet studies require. To be sure, a problem may simply provide the opportunity to apply an algorithm and check one's understanding of it. But another problem on the same page may pose a dilemma, and we offer our preferred answer only most tentatively. Many of the problems will prove useful as a basis for discussion in classes and seminars. While it is hoped that others will be able to confirm our numerical results, it is anticipated that they will often prefer alternative interpretations.

Because of the endless variety of the analyses generalizability studies call for, we do not regard most of what is presented as a set of procedures to be "mastered" and reduced to routines. Hence, there has been no systematic attempt to develop exercises on all topics. While it is hoped that the exercises will assist the reader to comprehend the theory, this volume is a theoretical monograph and not a textbook.

At some points it has been necessary to venture into metatheory—that is, to discuss *why* the person analyzing a measuring technique ought to ask certain kinds of questions. Issues forced to our attention by generalizability theory raise fundamental questions about the rationale and even the legitimacy of traditional approaches to the scoring, reporting, and evaluation of tests. Some readers will no doubt find multifacet theory cumbersome, and decide to return to simpler models. Even such a reader will find that multifacet theory has shed new light on his long familiar procedures, and has cast dark shadows upon the acceptability of some of them.

It is awkward to develop an argument at several levels of abstraction, especially when readers can be expected to vary in sophistication and technical background. The reader must proceed through the book in his own way; it would be unwise for him to struggle to comprehend each page as it comes. Perhaps the best strategy is to scan the whole volume to identify the kinds of material to be found, and then to follow a selected theme through the book. Occasionally, the reader may wish to skip ahead to locate a numerical example, or to turn to a further development of a theoretical issue. The reader is certain to gain far more from his third reading of most sections than from his first or second. By the time he returns to this or that puzzling matter for a fourth reading he is likely to discover implications in the topic that escaped *us*. In the course of writing and rewriting the book, we have repeatedly discovered further, sometimes dramatic, significance in topics that we thought had already been exhausted.

One reason for the continuing power of multifacet theory to provoke new thoughts is that it permits sensitive use of concepts that had a restricted or ambiguous meaning in the traditional framework. The hallowed Spearman–Brown formula, for instance, estimates the accuracy of the score that one can obtain by doubling the number of observations. But if a teacher is to be observed by $n_i$ observers on $n_j$ occasions, and it is proposed to "double the number of observations," the psychometric properties of the score will differ according to whether $n_i$ or $n_j$ is doubled. The properties also depend on such subtler points of design as whether or not the $n_i$ observers are the same on each occasion. The correlational language could be elaborated to cope with such complexities, but clarity is served by tracing how a change of design affects each component of observed-score variance.

Another example of ambiguity and misinterpretation arises in correction for attenuation. In all too many instances, an investigator has "corrected" an observed correlation by dividing it by a reliability index of an inappropriate type. Occasionally, for example, an investigator who desires to show that two tests reflect distinct traits correlates the two tests and then "corrects" the correlation by dividing it by an index of scorer agreement. A low "corrected" correlation is taken as evidence for his hypothesis. It is virtually impossible to explain why this is unsound and what the investigator should have done instead until we possess the conceptual apparatus of multifacet analysis. When we return to this topic, it will be evident that there are many different "corrected" coefficients for the same two sets of scores. Each coefficient allows the investigator to evaluate a different substantive proposition.

It is a straightforward matter to obtain unbiased estimates of components of variance and covariance, at least for the more regular experimental designs (Vaughn & Corballis, 1969). A certain variance component may contribute to "true," "error," or "observed score" variance or perhaps to none of these; the interpretation of the component will depend on how the measuring procedure is to be applied. The magnitude of errors of measurement depends on the type of decision to be made from the scores and on the experimental design by which scores are to be collected. A new estimate will usually be required for another type of decision or for an experiment differently designed. However, estimates of components remain useful to everyone testing subjects similar to those of the original study. Each user can derive from the components the estimates that pertain to *his* design and intended decision. It is expected that behavioral scientists will drift away from their present concern with coefficients, toward the reporting and interpreting of components of variance and covariance. This will bring their thinking more nearly in line with the theory of error used in other sciences, where correlation coefficients play little or no part.

## A. *Historical Notes*

The multifacet approach squarely faces the old criticism that reliability coefficients for a test are diverse and sometimes mutually contradictory. Under the classical theory, an investigator was expected to obtain two independent but interchangeable measurements on each individual, and to examine their agreement. As was forcefully pointed out by Goodenough (1936), the investigator who compares two administrations of the same list of spelling words asks a different question than the investigator who compares performance on two different lists. Inconsistency of observers, inconsistency in the subject's response to different stimulus lists, inconsistency of his response to the same stimulus on different occasions—all of these may be sources of error, but any one comparison will detect some inconsistencies and not others. This line of criticism has led various workers to classify the types of variance that can contribute to "error."

Thorndike (1947) classified variance into five categories:

1. Lasting and general. For example, level of ability, and general test-taking ability.
2. Lasting but specific. For example, knowledge or ignorance regarding a particular item that appears in one test form.
3. Temporary but general. For example, buoyancy or fatigue reflected in performance on every test given at a particular time.
4. Temporary and specific. For example, a mental set that affects success in dealing with a particular set of items.
5. Other, particularly chance success in "guessing."

The lasting–general variance is almost always "wanted" information about individual differences; the "other" or residual category is almost always "error." Temporary–general characteristics are significant for the investigator who is studying response to immediate conditions, but they are "error" for the investigator who wants to know the subject's typical level of response. Thus, an evaluator may wish to detect how much an adolescent's interest in reading about science here and now is aroused by the particular kind of stimulation a visiting lecturer provides. This evaluator is interested in the subject's temporary state. But a guidance counselor wants to measure the same adolescent's everyday, typical interest in science. He regards temporary departure from the student's norm as a source of error, since he is interested in a characteristic that transcends the stimulation of the moment. A split-half analysis treats the temporary variation as consistent information; the heightened interest raises scores on both halves of the measure. Therefore, from the viewpoint of the guidance counselor, the split-half index of agreement is falsely encouraging.

Thorndike's breakdown is multifacet in conception, recognizing occasions and stimuli (test forms) as logically distinct facets. However, it does not suggest how the investigator can estimate the magnitude of each type of variation separately. Cronbach's rather similar treatment (1947) went only a small step further in this direction.

The views to be developed here were foreshadowed in Guttman's (1953) review of Gulliksen's *Theory of Mental Tests* (1950). Guttman's remarks are worth quoting at length, because they exemplify how the experimental procedure for investigating generalizability derives from substantive considerations. Because a study of consistency among samples of behavior challenges or confirms the investigator's working concept of the variable, it is a part of instrument validation as well as a study of instrument precision. From Guttman's review:

> Current sampling theory by itself cannot solve many problems of prediction and external validity. Conventional sampling problems concern the selection of people from a large population. Mental test theory faces also another type of sampling problem—that of selecting items from one or more indefinitely large universes of content. This is a basic problem of item analysis. To this reviewer it appears that *there can be no solution without a structural theory* [p. 129].

. . . . . . . . . . . . . . . . . . . . . . . .

> Tests are parallel if they have common means, variances, and inter-correlation coefficients. It is not so easy to see, however, that the definition is unique. It seems to this reviewer that one could find the same test to belong to more than one set of parallel tests and thus in general to have more than one "reliability coefficient."
>
> Consider the following example of a series of "parallel" tests. Let test 1 consist of but a single item: "Write down all the words you can think of that begin with the letter *t*." For a given population, and a given time limit, the score for each person is the number of words he writes down beginning with *t*.
>
> There are at least two different directions in which one could go to construct tests parallel to this one. One direction is to vary *the letter involved*. For example, test 2 could be: "Write down all the words you can think of that begin with *p*," while test 3 could use instead the letter *d*, say. By adjusting the time limits, all three tests can be made to have the same mean. There seems no absolute barrier to their also having common variances and correlation coefficients. For our particular population, let us suppose the three tests are actually parallel, and that their common correlation coefficient is 0.70. Then, according to the book's theory, test 1 has reliability coefficient 0.70.
>
> Another direction in which we could have gone to construct tests parallel to test 1 is to vary the places of the letter, and not the letter itself. Thus, test 2 could be: "Write down all the words you can think of in which the second letter is *t*," and test 3 could ask for *t* as the *third* letter. Again, for our population, there is no physical bar to the tests turning out

to be parallel. But this time, let us assume that the mutual intercorrelations turn out to be equal to 0.60. Then test 1 has reliability 0.60.

Therefore, test 1 has reliabilities 0.70 and 0.60 simultaneously, according to the theory of parallelism [p. 125].

To go beyond the limited information contained in a simple correlation between two tests, one would collect data with at least four and preferably eight or more fluency tests. These tests would be designed to vary systematically with respect to the letter prescribed, the position in the word of the letter prescribed, and the length of word prescribed. Thus, one could compare, for example, a test asking the subject to list four-letter words beginning in $t$ with a test that asks for four-letter words beginning in $d$, then with one asking for four-letter words whose second letter is $t$, and with one asking for six-letter words beginning in $t$. With more than four tests the tester could introduce additional variations. For example, he might then discover that the "second-letter" tests consistently rank people differently from the "first-letter" tests. If so, they call upon different mental processes.

As it happens, Gulliksen (1936) had published perhaps the first formal multifacet analysis of test consistency in a paper that was summarized in the book Guttman criticized. Two forms of an essay test were administered, and each paper was scored by two graders. The cross-correlations of the four resulting scores answer three distinct questions about consistency of measurement having to do with different forms, different scorers, or both. Gulliksen's procedures could estimate some but not all of the components of variance that nowadays are determined by analysis of variance.

There has been a steady flow of concepts from Fisherian analysis of variance into educational and psychological statistics, but the presentations encountered by most students emphasize the testing of null hypotheses by means of the $F$ ratio. This reflects the earlier phase of factorial experimentation, during which the effects an experiment was designed to assess were regarded as fixed. An agricultural experimenter testing the effects on yield of three types of fertilizer, for example, is primarily concerned with those specific fertilizers. He is not studying fertilizers-in-general. The same is true in many psychological and educational studies of treatment effects. A study in which results of the PSSC physics course are compared with those of the course developed by Harvard Project Physics regards those treatments as fixed; there is no intention to formulate conclusions about any larger set of physics curricula.

In the late 1940's, statisticians came to distinguish among "fixed," "random," and "mixed" models for the analysis of variance. Attention to the components of variance followed. The random and mixed models recognize that sometimes the conditions used in an experiment are of little interest in themselves; rather, they represent a class of conditions. For example, the plot on which a crop is grown is a sample from a population of plots over which the experimenter intends to generalize. An experimenter who establishes that soil variation is significant is ordinarily not interested in the fields that happen to have been employed in the research. Rather, he wants to know to what extent yields vary from plot to plot, so that he can take this into account in his recommendations. The random model assumes that the experimental conditions are randomly selected from the set of possible conditions. The original (fixed) model assumes that the study has obtained data under every one of the conditions that is currently of interest. The mixed model allows some aspects of the experiment to be fixed and others to be determined by sampling (e.g., the set of fertilizers fixed, soils random).

In an especially significant theoretical paper on multifacet designs, Cornfield and Tukey (1956) embodied all three models in a single formulation. The paper considers the $n$ values of a facet used in an experiment to be samples from the $N$ values in the universe of conditions for that facet, where $N$ can take any value from $n$ to $\infty$. (For example, in the Guttman example of fluency tests, the initial letters $t$ and $d$ are presumably sampled from a set where $N = 26$.) There is a set of general formulas for estimating the expected magnitude of the effects (i.e., of the components of variance) for any $N$ between $n$ and $\infty$. But the intermediate possibilities are commonly ignored; analysis proceeds as if, for any facet, $N$ equals either $n$ or $\infty$. Where $N = n$, the conditions of the facet are fixed.

In the behavioral sciences, conditions of measurement or observation are commonly thought of as representative of a large set of conditions. In observational studies of teachers, the persons doing the observing and the occasions on which observations are made represent many other equally admissible observers and occasions. Similarly, where two forms of a test are used in a study, these two item-sets are considered to be samples from a universe of item-sets "like these." While *universe* and *population* are logically interchangeable terms, we shall reserve the word *population* for subjects, and apply the word *universe* to conditions under which the subjects might be observed.

The classical theory of reliability postulates strictly "parallel" measures such that test forms have equal means and variances and there is no interaction of subject with test form. Variance is considered to arise from "true" subject differences combined with random variation among observations ("error"). While this model is reasonable for carefully equated parallel forms of tests, it is less descriptive of other types of measures. For example, raters are likely to differ in the central tendency of the values they assign (producing a main effect for raters) in the spread of their ratings, and in the qualities they attend to (these latter producing a subject–rater interaction). Behavioral observations, such as of talkativeness, for instance, can also be expected to exhibit both situation (main effect) and interaction variances. Tests often lack second forms, and investigators turn to internal-consistency analyses. But half-tests are imperfectly parallel, and item scores do not conform at all to the classical model.

The theory presented here derives from amendments to the classical theory that, in the 1950's, proceeded along these lines:

1. It was formally recognized that conditions of observation are not necessarily parallel. (See Ebel, 1951, who made a place for inequality of condition means.)
2. Conditions (particularly test items) were thought of as sampled from a universe randomly or in accord with a stratified design. (See Lord, 1955a, 1955b; Tryon, 1957.)
3. Two or more facets were analyzed simultaneously.

The papers bearing on the third point are closely enough related to this monograph to be catalogued.

Applications of analysis of variance in psychology and education stem directly from the work of Fisher and others at the University of London in the 1930's. The senior educational psychologist at London at that time, Cyril Burt, translated Fisher's materials for the benefit of his students and applied them to the reliability problem, but his formulations reached print only in fragmentary form after World War II. Burt's 1955 paper was a comprehensive exposition of the application of analysis of variance to reliability problems, with particular attention to test forms and occasions as separable sources of variation. A companion paper by Mahmoud (1955) treated the same data factor-analytically, demonstrating some links between the two systems of analysis, such as, for example, the correspondence of person–form interactions to the specific-factor content of a test form. Burt dealt only with the completely crossed design where each test form is given to all subjects on two occasions. Other treatments of the reliability of tests by means of analysis of variance reflect the influence of Palmer Johnson, an associate of Jerzy Neyman in the mid-1930's at London. The list of associates and students of Johnson who have contributed to the literature on reliability includes R. W. B. Jackson (who worked with Neyman), Hoyt, Mitzel, and Medley. Reference should also be made to the continually developing thoughts of another Burt associate, R. B. Cattell. While the present monograph does not coincide fully with Cattell's views, his thinking has been oriented toward similar analyses for a long time (see Cattell & Warburton, 1967, p. 36 ff.).

Lindquist's extensive exposition of multifacet theory (1953) focused on reliability coefficients and treated components of variance incidentally. Lindquist only partially developed the methods of using one study to estimate the precision of measurements that could be collected with various alternative designs. However, he did make clear that a multifacet analysis allows for alternative definitions of error. Hence, several distinct coefficients can be obtained for any one measuring procedure. He demonstrated that increasing the number of observations has different effects, according to which facet the added observations sample.

Very likely the first report on reliability of measurement in terms of the analysis of variance components was Finlayson's (1951) study of grades assigned to essays where the student writes on more than one topic and the paper is graded by several readers. Pilliner, with whom Finlayson worked, published (1952) a theoretical exposition of the relations between intraclass correlations and analysis of variance.

In 1965 Pilliner assembled the thinking of many years into a doctoral dissertation. This coincides with our paper (Gleser, *et al.*, 1965) not only in time but in much of its thinking. Most of Pilliner's illustrative applications concentrate on one-facet studies of agreement among graders, but some studies treat pupils as nested within schools. The recognition that the school mean is at times the variable of interest takes the work into ground that has rarely been touched upon. (See also Pilliner, Sutherland, & Taylor, 1960, and a paper on two-facet studies by Maxwell & Pilliner, 1968).

Loveland (1952) carried out a doctoral dissertation under the direction of E. E. Cureton in which he computed components of variance to estimate the magnitude of variation from five sources: persons, person–occasion interaction, person–form interaction, a form–occasion effect, and a residual. The model appears to differ at least in minor particulars from those used subsequently, because of a preoccupation with individual differences characteristic of older test theory.

Another application of multifacet analysis to educational measurement was that of Medley, Mitzel, and Doi (1956). They carried out a three-way analysis of classroom observations of teachers to demonstrate the effect of conditions of observation, stressing the distinction between the mixed and random models for analysis. Medley and Mitzel (1963) presented the argument more completely, displaying a four-way analysis. This work is to be examined further in Chapter 7 (p. 189 ff.).

The intraclass correlation, originally developed by Pearson, was made a part of the theory of variance analysis by Fisher. Many well established reliability formulas, including those of Kuder and Richardson, are now recognized to be intraclass correlations, as are all the coefficients of generalizability that our procedures generate. Many persons discussed the intraclass formulas during the 1950's and 1960's, and connected them with reliability theory. Particularly extensive work on the variants of the intraclass formulas was done by Buros (1963). The authors' first publication on generalizability theory (Cronbach, *et al.*, 1963) emphasized the interpretation of intraclass correlations. However, the correlations considered came from data organized with respect to only a single facet: multiple ratings of each person on a single trait, for example, or a persons × items matrix of scores. An unpublished paper by Stanley (*ca.* 1955) emphasized the multifacet conception of the reliability problem, and showed that with a relatively complex design one can arrive at a number of intraclass correlations, each having its own meaning. This same conception was emphasized in the paper of Gleser, *et al.* (1965) on which the present monograph is based. With regard to univariate studies our chief additions to Stanley's formulation are the concept of the universe of generalization and the distinction between G and D studies; these considerably enrich interpretations.

In his 1954 *Psychometric Methods*, Guilford applied analysis of variance to ratings, discussing the effects for subjects, raters, traits, and their interactions. However, he did not extract variance components, nor did he relate the analysis to the reliability problem as conventionally stated. Stanley (1961) returned to the problem, and indicated the desirability of estimating and interpreting the several variance components. However, Stanley formulated the problem in terms of various types of covariances (e.g., the mean covariance between pairs of raters rating the same trait). The covariances can be estimated directly from the mean squares of the analysis of variance, and can be interpreted as composites of the variance components. The covariance formulation, while mathematically equivalent to the analysis of variance components, is probably less satisfactory. This is because the covariance formulation is less directly tied to conventional statistical procedures, and because it omits some information obtainable from the variance components. Stanley derived a number of recommendations for improving the design of the rating procedure, similar to those stemming from a generalizability analysis.

Where the literature reviewed above touches on internal consistency of tests, it usually regards items as randomly sampled. Tests are often constructed according to complex specifications regarding the distribution of content and perhaps of difficulty. This suggests (see Lord, 1955; and Tryon, 1957) that tests should be regarded as having items sampled within strata, which calls for a relatively complex analysis. We have explored such possibilities, particularly in a 1960 technical report, published with some revisions in 1965 (Rajaratnam, *et al.*, 1965; see also Cronbach, Schönemann, & McKie, 1965). Independently, Pilliner discussed similar applications of analysis of variance in his 1965 dissertation.

Guttman (1958) suggested the possibility of analyzing data from a three-way matrix of scores factor-analytically, in order to arrive at factors representing each of the (fixed) main effects and each of the interactions. This is directly pertinent to his conception of abilities as organized according to facets, and might be applied to the hypotheses implicit in the Guilford "structure of intellect." Unfortunately, no systematic procedure was offered and the method was neglected. Only very recently have studies of this character been carried out (Boruch & Wolins, 1970; and Merrifield, 1970). This kind of controlled factor analysis can carry much of the information that appears in the estimates of variance components, and adds information about the magnitude of the effect associated with each particular condition of the facet.

The three-mode factor analysis of Tucker (1964, 1966) (see also Snyder, 1968) approaches the problem somewhat differently. The aim is to describe the complex of variables in terms of a small number of factors. Whereas Guttman would investigate how much of the score variance is accounted for by the hypotheses represented in the facet structure, Tucker attends to common factors that may not have been hypothesized. Generalizability theory, like Guttman's analysis, examines the power of the gross facet structure to account for variance, whereas Tucker's method tends to suggest new structures. As LaForge (1965) pointed out, a factor analysis of correlations between conditions will serve purposes a generalizability study cannot.

Reliability theory and generalizability theory have hitherto looked at the accuracy of one score at a time. That is, they have been univariate in conception. Even for examining profiles of scores, the only special procedure invoked was the calculation of various difference scores, for each of which a univariate reliability study was made. In 1966 we stumbled into the realization that all the data in a profile may help one to estimate the universe score on any one of the variables. Travers, in an unpublished paper, developed for us the multivariate extension of the mathematics for one-facet generalizability studies. This has evolved into the theory discussed in Chapters 9 and 10.

There seem to have been almost no predecessors of multivariate error theory. The one general paper that has come to our attention is the proposal by Bock (1966) to evaluate the multivariate reliability of a battery by obtaining a coefficient for each canonical variate found in scores from parallel batteries. It can be seen now that work on "profile similarity" of the early 1950's (Cronbach & Gleser, 1953) needed only an additional twist to unlock the door to these psychometric riches, but the opportunity was missed. Nearly all of the current developments are implicit in Lord's first paper (1956) on the measurement of change, where multivariate methods much like ours are applied. Even though Lord's work stimulated many papers by leading psychometricians on the narrow problem of "change," the broader usefulness of the model escaped attention.

The possibilities of the multivariate model have recently been given significant publicity in a presentation by Novick (1971), who discusses it as part of a general defense of Bayesian methods in test analysis. The more technical reports that are to come will, we trust, be compatible with and so augment and clarify our theory.

### B. *Formulation*

The theory to be presented employs several interrelated concepts. While each of these will be elaborated in turn, a brief initial statement setting out all the main concepts should give a helpful perspective.

A measuring procedure is used as a basis for decisions or conclusions, and the accuracy of measurement must in principle be examined separately for each application of the procedure. At least four kinds of interpretation may be made:

1. Absolute decision. Where an individual is to be classified in some way, performance standards may be set which determine how he will be treated. For example, a test is given an applicant for a driver's license, with the predetermined rule that a score of 85% is considered adequate for licensing. In making the decision, this person is considered by himself; we shall call this an "absolute" decision, in contrast to comparative decisions (item 3 below). Another kind of absolute decision is that made in evaluation, where the performance of a group must reach a predetermined standard if the treatment is to be judged satisfactory. For example, an author of a programmed textbook determines how many errors students make after studying a lesson, proposing to revise all lessons where the percentage of errors exceeds some specified figure.

Instead of stating the standard in terms of the test performance itself, the decision maker may state what criterion performance is desired. Thus it may be decided that any student whose expected grade average is below C will not be admitted to a certain curriculum. Since the expected grade average is inferred from some pretest, any error of measurement on that test will affect the decision.

2. Comparison between two courses of action for an individual. Here, as in the first type of decision, each individual is considered separately. This type of decision is especially common in guidance, where the person chooses one curriculum rather than another on the basis of a difference between his scores on the abilities pertinent to each. The decision maker asks whether the difference between two measures, or the difference between two expected

outcomes, reaches a predetermined standard. (If there is a quota limiting the number of persons who can enter any curriculum, comparison of type 3 is also involved.)

3. Comparison between persons. An interpretation is made of a difference between scores of persons or groups. Examples: a selection test is used to choose among applicants; a dependent variable is used to compare groups in an experiment.

4. Conclusion about the relation between pairs of variables. For example, the relation between ability to solve Hidden Figures problems and ability to attain concepts is investigated.

The score on which the decision is to be based is only one of many scores that might serve the same purpose. The decision maker is almost never interested in the response given to the particular stimulus objects or questions, to the particular tester, at the particular moment of testing. Some, at least, of these conditions of measurement could be altered without making the score any less acceptable to the decision maker. That is to say, there is a universe of observations, any of which would have yielded a usable basis for the decision. The ideal datum on which to base the decision would be something like the person's mean score over all acceptable observations, which we shall call his "universe score." The investigator uses the observed score or some function of it as if it were the universe score. That is, he generalizes from sample to universe. *The question of "reliability" thus resolves into a question of accuracy of generalization, or generalizability.*

The universe of interest to the decision maker is defined when he tells us what observations would be equally acceptable for his purpose (i.e., would "give him the same information"). He must describe the acceptable set of observations in terms of the allowable conditions of measurement. This gives an operational definition of the class of procedures to be considered. The investigator may, for example, say that he would accept the score on any form of the Jones mental test, administered at any time during the Spring of the high-school student's senior year. In this way he defines the universe in terms of two facets: test form and occasion. The investigator may fix the condition of a certain facet; e.g., he may specify that only Form A of the Jones test is acceptable. He invariably leaves out of the description certain aspects of the conditions of observation. The investigator in our example appears to be willing to accept the result from any tester, obtained in any room, etc. The facets that are mentioned in his specifications will be explicitly represented in the experimental design when scores for decision making are collected.

Knowing that observed score and universe score are not identical, the decision maker will want to take the discrepancy into account. One way to do this is to accompany each report by an expression of uncertainty, in the way that a physical scientist reports a value as 0.065 ± 0.003. Another possibility is to "correct" the observed score in some manner so that it better approximates the universe score; this corrected value also will have an uncertainty.

A confidence interval—a band within which it is reasonable to suppose that the true measurement falls—is often reported when an absolute decision is contemplated. Suppose that a standard of 88 has been set, so that persons known to have universe scores above 88 are to be treated in one way and those below are to be treated in another way. If it can be said that a certain person's universe score very likely falls in the interval 91–97, the decision about him will be made with great confidence. If the interval for another person is 85–90, the decision about him cannot be made with any confidence until further evidence is collected. Another frequent use of such intervals is to examine whether a difference between two scores can confidently be regarded as greater than zero. Suppose one wishes to advise a student that his interest in scientific activities is greater than his interest in mechanical activities. The two scores are presumed to be expressed on comparable scales, and bands are established for the two scores. If the upper end of the mechanical score-band does not reach as high as the lower end of the scientific score-band, the statement that his scientific interest is greater would very likely be confirmed by further testing. If the bands overlap, however, one has to entertain the possibility that the universe-score difference is in the reverse direction from the observed-score difference. When the person is a member of some group (e.g., a high-school class) whose score distribution is known, another option becomes available. One can estimate the person's universe score by a regression equation that describes the relation between universe scores and observed scores in the reference group.

We distinguish decision (D) studies from generalizability (G) studies. A G study collects data from which estimates can be made of the components of variance for measurements made by a certain procedure; a D study collects data for the purpose of making decisions or drawing conclusions. For example, the published estimates of reliability for a college aptitude test are based on a G study. College personnel officers employ these estimates to judge the accuracy of data they collect on their own applicants (D study). The G data may be analyzed to determine the generalizability of D data that will be collected under other designs. Sometimes, of course, the same data serve for both G and D studies.

In a G study, one obtains two or more scores for the person by observing him under different conditions, and examines the consistency of the scores. The analysis estimates components of variance, each attributable to one facet or combination of facets represented in the experimental design. These estimates may show, for example, that one set of test stimuli (within the universe) elicits about the same behavior as the next, but that variation in behavior from occasion to occasion is substantial. In the light of this, one proposes a suitable design for collecting D data and estimates how well one can generalize from the scores that will be obtained.

Where individual differences are the primary concern, the variance of universe scores for the population being studied, and also the variance of observed scores likely to be obtained in the D study will be of special interest. One can estimate the universe-score variance from the components of variance and can also estimate the "expected" observed-score variance (i.e., the variance likely to be obtained under a certain experimental plan).

The plan for collecting data is often evaluated by estimating the *coefficient of generalizability*, the counterpart of the traditional "reliability coefficient." This is defined as the ratio of universe-score variance to the expected observed-score variance. It expresses, on a 0-to-1 scale, how well the observation is likely to locate individuals, relative to other members of the population.

The size of the coefficient depends on the experimental design used for the decision study, as well as on the population of persons considered. The coefficient is employed in interpreting the correlation of this variable with other variables. It is also calculated as an intermediate step in obtaining the formula for making a point estimate of the universe score.

### G and D studies

Our separation of G and D studies formalizes and extends an idea implied in the Spearman–Brown "prophecy" formula. When one 20-item test has been correlated with another, the prophecy formula estimates the reliability of a 40-item test. The study with the 20-item tests was a G study, an investigation of the instrument; the prophecy is made because a 40-item test might be used in collecting subsequent data for decision making (D study). The idea is also present in the customary correction of a reliability coefficient to fit a new range of ability. The scores in the G study have a certain standard deviation, but one can forecast the reliability in a D study where the sample has a different standard deviation.

Rajaratnam (1960) introduced the distinction between G and D studies to clarify analyses of ratings. Often, in a G study, a certain set of raters is asked to judge all subjects. An intraclass correlation among raters is calculated that ignores differences in rater means (p. 79 ff.). This coefficient is pertinent if whatever raters are used in the D study will rate all the subjects. But if the raters in a subsequent D study differ from subject to subject, one needs to know the intraclass correlation that treats rater leniency or severity as a source of error (p. 77). This correlation can be estimated from the ḓinal G data. In general, the plan and purpose of the D study determine ṭ questions to be asked of the G data.

The distinction between G and D studies is no more than a recognition that certain studies are carried out during the development of a measuring procedure, and then the procedure is put to use in other studies. When an investigator factor analyzes questionnaire items in order to decide how to organize them into dimensions and how many items of each kind to use, he is making a preliminary study analogous to our G study. An even closer parallel is the research on optimal design of a battery for predictive purposes (Horst, 1949, among others). The data for the design study are collected by means of a trial version of the battery. On the basis of intercorrelations, correlations with criteria, and reliability coefficients in these data, a new battery is designed that is used to collect D data.

Generalizability studies ought to be regarded as a part of instrument development, and therefore G studies should take place prior to collection of the D data. To be sure, one will occasionally use the actual D data for an analysis of generalizability. But since it is then too late to take advantage of the information to improve the D data, this is a weak use of the method.

Distinguishing between G and D studies is especially valuable in multifacet investigations, because separation of facets makes possible a great variety of experimental designs. Different designs may be and usually should be used for the G and D studies. Even if the investigator who conducts the original G study knows that he will adopt that same design in his D study, another investigator may choose an alternative design for collecting similar observations. *Information from the G study should therefore be reported in such a form that each new investigator can plan his D study and estimate the error of generalization arising under that plan.*

### Universes

A behavioral measurement is a sample from the collection of measurements that might have been made, and interest attaches to the obtained score only because it is representative of the whole collection or *universe*. If the decision maker could, he would measure the person exhaustively and take the average over all the measurements.

Educators and psychologists have traditionally referred to the average reached via exhaustive measurement as "the true score" for the person. We speak instead of a *universe score*. This emphasizes that the investigator is making an inference from a sample of observed data, and also that there is more than one universe to which he might generalize. Any person fits within many different populations. John Doe may be considered a sample from any of several sets: residents of California, electricians, persons with a $15,000 income, Republicans, etc. Any observation likewise fits within a variety of universes. "The universe score is estimated to be 75" is without meaning until we answer the question, "Which universe?" This ambiguity is concealed in the statement "The estimated true score is 75," for no one thinks to inquire, "Which truth?"[2]

An observation is described in terms of *conditions:* the task or stimulus presented, the day and hour, the setting in which the observation is made, the observer, and possibly additional features of the operations performed. The general term referring to conditions of a certain kind is *facet.* Thus, observations may be classified with respect to the facet of tasks presented, the facet of days of testing, the facet of observers, etc.

The facets, alone or in combination, define universes. A child is asked to draw a cowboy on Friday, May 12. This drawing belongs to a universe of cowboy drawings made on various days, to a universe of drawings on various themes that might have been made on May 12, to a universe of cowboy drawings that might have been solicited by various testers, etc. To ask which universe is relevant is to ask how the investigator proposes to interpret the measure.

A universe of observations will be characterized with respect to one, two, or more facets. Almost everywhere we shall assume that joining one condition of the first facet with a condition of every other facet defines a possible observation. For example, in a drawing task the facets may be themes for drawing and testers. Any tester might ask the child to draw a picture on any of the admissible themes, hence, any pairing of theme and tester defines an observation that may, in principle, be made. If occasions constitute a third facet, the argument is extended: any combination of theme, tester, and occasion defines a possible observation.

The universe to which an observation is generalized depends on the practical or theoretical concern of the decision maker. Consider a supervisor's rating of an employee. This rating differs from what would be recorded on another occasion, since the supervisor's mood at the time of rating and his recent experience with the employee have some transient effect. The investigator concerned with employee effectiveness surely wants to generalize over the class of ratings the supervisor might have given at other moments. The investigator will generalize over a time period of perhaps a month if the

[2] Another difficulty with the term *true score* is that the statistical concept of a limiting value approached through extensive observation is readily confused with some underlying in-the-eye-of-God reality. Sutcliffe (1965) referred to the way things "really are" as a "Platonic" concept of the true score. The Platonic measure of the mean income of Americans, for example, might be determined by some all-seeing and impartial accountant. It would be quite unlike the operationally defined mean produced by compiling the incomes reported to the Internal Revenue Service. Any bias in the class of measuring procedures adds a corresponding bias to the universe score, which is in that sense "untrue."

rating is taken as an end-of-year report of the employee's qualities. Any of the moments within that month would presumably have been a suitable time for the inquiry. In another study, where the rating is a datum for an intensive study of week-to-week changes in supervisor attitudes during a human-relations course, the investigator will generalize over only a single day. If the rating is a criterion against which he will validate an ability test, he needs to generalize over supervisors as well as occasions. But if the sole concern is whether the employee is getting along with *this* supervisor, the universe of possible supervisors is irrelevant. Since it is impossible for the developer of a procedure to anticipate all its uses, his G study can at best report data for prospective users to assemble in the light of their own decisions and designs.

At times it is necessary for us to distinguish among universes that perform different functions. The test developer or other investigator who carries out a G study takes certain facets into consideration and, with respect to each facet, considers a certain range of conditions. The observations encompassed by the possible combinations of conditions that the G study represents is called the *universe of admissible observations*. We may also speak of the universe of admissible conditions of a certain facet. A decision maker, applying essentially the same measuring technique, proposes to generalize to some universe of conditions all of which he sees as eliciting samples of the same information. We refer to that as the *universe of generalization*. The G study can serve this decision maker only if its universe of admissible conditions is identical to or includes the proposed universe of generalization. Different decision makers may propose different universes of generalization. A G study that defines the universe of admissible observations broadly, encompassing all the likely universes of generalization, will be useful to various decision makers.

The universe of admissible observations and the universe of generalization may be identical; then the decision maker may simply think of "the universe" in taking the G study into account. Some decision makers, however, will generalize less broadly, taking as universe of generalization a subset of the universe on which the G study was based. The decision maker may propose to generalize over only one facet, for example, where the G study took several facets into account. In such a case, a careful selection among formulas is required to make proper use of the G study.

The third possibility is that the decision maker will propose to generalize beyond the universe of admissible observations. His universe of generalization includes conditions not present in the universe from which the G study samples conditions. The G study then does not give him the information he needs, though it may give him some rough ideas as to the accuracy of his proposed generalization. *We shall rule this case out of consideration.*

The universe of generalization is necessarily determined by the decision maker. The instrument developer, carrying out a G study to guide users of his instrument, will, in the design of that study, treat systematically the facets that are likely to enter into generalizations of various users. Sometimes he will examine a facet over which a particular user does not care to generalize. The extra information does no harm, so long as the user properly interprets the report of the G study.

A G study treats a facet in one of three ways:

1. The facet is systematically represented by sampling two or more conditions of the facet; or
2. A single constant condition of the facet is employed in all G-study observations; or
3. Conditions of the facet vary in the G study without direct experimental control.

In a study of ratings, the investigator might treat occasions and supervisors as facets of the first kind, collecting ratings from two or more supervisors on two or more occasions. The investigator might regard his pencil-paper rating form as fixed, that is, as a condition of the second kind. He would collect all G data with this form; then he gets no information as to what would happen if the wording of the items, their content, or format were changed. Such a G study will not serve someone who needs to generalize over contents or formats. Inevitably, a great number of potential facets remain uncontrolled, falling into the third category. For the study of ratings it is inconceivable that the investigator would control (for example) the number of minutes since the supervisor has interacted with the worker. Decision makers presumably intend to generalize over uncontrolled facets; such facets ordinarily contribute to the undifferentiated residual variance in the G study. The investigator designs the G study in terms of facets of types 1 and 2 (i.e., those to be controlled through multiple representation and through single representation, respectively). He does not attempt the impossible task of listing the many uncontrolled facets, though he and his readers must be aware of their existence. A future, more elaborate G study may profitably turn its attention to one of these presently unanalyzed sources of variance.

With regard to any facet of the first type, admissible conditions must be defined. For example, what category of supervisors is the sample to represent? It is one thing to investigate the agreement among supervisors who have just been handed the form for the first time, and quite another to investigate the agreement among supervisors specially trained by the investigator. The category may sometimes be limited to the conditions actually used in the G study, but it will ordinarily include a much larger number of conditions.

Ideally, the investigator carrying out a G study would formally define the universe of admissible conditions corresponding to each facet of the first type, would sample in a strictly random or stratified-random fashion, and would combine at random conditions drawn from the several facets. That is, he would pair a random rater with a random occasion, etc. Practice never conforms to this idealized model. Even if the investigator were formally to specify some collection of raters and sample from it for his G study, one can be sure that the raters in subsequent D studies will not be a truly random sample from the same collection. In this as in any use of models, practice departs from the ideal and results are indicative rather than definitive. The investigator should describe how conditions were selected for his G study, considering each facet that is systematically varied. The class of conditions should be described with sufficient clarity that a reader of his report will know whether the conditions (e.g., raters) are like those he plans to employ in the D study. Similarly, any fixed condition in the G study must be clearly described. If the D study brings in essentially different conditions of any facet, the G-study results convey no more than a hint about the variation to be expected from that facet in the D study.

It is not reasonable to regard the time limit set for a test as sampled from an array of time limits. The definition of the universe has to include the time limit as a fixed condition, for two reasons. The first is that tests similar in all respects save working time collect different amounts of information, and hence do not have the same degree of generalizability. Second, altering the time limit is likely to alter what the procedure measures, so it is unwise to consider procedures with different time limits as acceptable for the same purpose; that is, as members of the same universe.

While we mention occasions as a facet worthy of investigation, sampling-from-an-aggregation is a dubious model for occasions, because occasions occur in a time sequence. In the traditional investigation of "retest reliability" the psychometric model ignores the time interval between the two testings. Interpreters, however, regard the correlation obtained as an index of stability over a specified time interval, rather than as an index of accuracy of measurement alone. It is reasonable to think of occasions as randomly sampled from a certain time span, whenever we regard the behavior observed as being in a "steady state" for the subjects (i.e., as not undergoing systematic change due to learning, fatigue, etc.).

A model that takes the sequence of conditions into account can surely be developed. Instead of generating an overall index of agreement between the observed value and the expected value over all conditions, one would describe degree of agreement between observations as a function of their separation in time. Within the present model we have no alternative but to treat occasions as an unordered facet, and to speak of generalization from the observed score to the mean of possible scores during a reasonable period of time—perhaps, during the same month.

A so-called facet may involve two or more entangled effects. Indeed, the facet on which observations are classified is usually impure. Ratings are usually classified with respect to the persons who do the rating; but the ratings reflect the information available to each rater as well as rater bias, etc. Because, in principle, data might be classified with respect to both the situation where behavior is observed and the observer, whenever the situation is not controlled the two effects are confounded in the "observer" facet.

## Estimating a universe score

The heart of traditional measurement theory is the so-called reliability coefficient, the ratio of "true score" variance to observed-score variance. This concept is altered in the theory we are about to present. First, because with several possible universes of generalization, there are correspondingly many variance ratios. Second, because each alternative for the D-study design generates a different variance of observed scores, and this alters the ratio. (This has been ignored in reliability theory because of a tacit assumption that the design of the D study will be essentially like that of the G study, or, indeed, that the D data are themselves the data of the G study.)

Our theory tends to subordinate coefficients in reporting a G study. The end point of measurement is a decision. The decision about a person is in principle based on his estimated universe score or his estimated criterion score. The primary question is: how may his score best be estimated? The secondary question is: how large is the error arising from incomplete observation?

Coefficients of generalizability bear directly on the original problem for which Spearman invented reliability theory. Where a study is intended to determine the correlation between two variables, it is valuable to find out how much the observed correlation is reduced (attenuated) by errors of measurement. Coefficients of generalizability can be employed in estimating correlations of universe scores, much as in classical theory, but our theory offers a more direct way of dealing with this question, and poses the question in a more complex form than has been traditional.

There are several ways of arriving at an individual score for decision making. The first is simply to use the raw score as an estimate of the person's average score over all observations in the universe of generalization. Most testers do this, though this may not be their conscious intention. Interpreting the raw score, one reaches a decision about the individual without considering scores of other persons. Such reasoning is commonplace in physical measurement, where the observed weight of a chemical sample, for example, is taken as the best estimate of its true weight. The statistician considers the experiment to be one of a population of possible experiments when reasoning about the dependability of the conclusion from it. But information about other specimens is not used in estimating the weight of this specimen.

Sometimes an observed score is expressed in terms of percentiles or standard scores or grade equivalents, and this value is taken as an estimate of a norm-referenced universe score. Data on other cases define the scale but are not used to adjust an individual's score. This is a procedure unique to behavioral science, and one that has serious faults (see Chapter 5).

When the decision is based on the person's rank within the sample tested or on his deviation from the mean, further use is made of reference-group data. Ranking is at the base of many quota-controlled decisions such as hiring, though it is rarely discussed explicitly in test theory. Test theory has traditionally focussed on individual differences, and in our formulation it is the statements about deviation scores that come closest to matching the conventional theorems.

The regression equation for estimating the universe score from the individual's observed score makes still more substantial use of information from other cases. While the regression technique is recognized in classical theory, it has not been very prominent. Perhaps this is because the regressed scores of individuals are perfectly correlated with observed score, so long as the same regression equation applies to the whole sample. Regression estimates do alter comparative decisions when the persons in the sample can be identified with subpopulations, since regressing each person toward the mean of his own group does alter ranks. Decisions based on an absolute standard also are changed when regression estimates replace observed scores. Multiple-regression methods alter the estimates still further.

We shall confine attention to linear estimating equations. Complex kinds of estimation such as Lord is currently investigating (see Lord, 1969; Ross & Lumsden, 1968) may ultimately be of practical value, but they cannot be accommodated within the present techniques of generalizability analysis.

For each of the kinds of universe-score estimate, there is a corresponding error. The following distinctions will be necessary:

1. Decision based directly on the observed score: error $\Delta$. The observed score is taken as an estimate of a universe score for the person, the universe score being expressed on the same numerical scale as the observed scores. (Thus, if the observed score is a standard score on a scale with a mean of 50 and a standard deviation of 10, the universe scores will also be expressed on that scale. The universe scores will, however, have a standard deviation smaller than 10.) The symbol $\Delta$ will be used to identify the error in such an interpretation; that is, $\Delta$ is the discrepancy between observed score and universe score.

2. Decision based on the observed deviation from the sample mean: error $\delta$. Interpretations that are entirely concerned with individual differences,

implicitly or explicitly, rest on deviation scores. The observed deviation score is, in effect, treated as an estimate of the deviation of the person's universe score from the group mean. Again, the universe score and the error are expressed on the scale of the observed score, except that a translation has made the mean of the score distribution under any condition equal to zero. The error $\delta$ is the discrepancy between the observed deviation score and the universe score expressed in deviation form. (For a formal statement, see p. 93.)

3. Decision based on a regression estimate: error $\varepsilon$. A third approach is to use a linear regression equation that combines information about the individual with information about the group mean. The equation may take other variables into account along with the observed score. This kind of estimation is only a specialized application of the ordinary technique for forecasting criterion scores. Consequently, the error $\varepsilon$ is an error of estimate, in the usual statistical terminology.

In this volume, the scale of observed scores is used for universe scores and for $\Delta$, $\delta$, and $\varepsilon$. Distinguishing the three kinds of estimation and their associated errors is an important step beyond classical theory.[3] Classical theory, being concerned primarily with individual differences, assuming uniform means for all conditions, and in most developments assuming that a single undifferentiated population is under consideration, does not need this distinction. Observed scores and the corresponding deviation scores differ only by a constant, and hence are perfectly correlated. Moreover, within an intact population, the regression estimate of the universe score described in classical theory is perfectly correlated with the observed score. With our greater interest in absolute scores, our weaker assumptions, our recognition of alternative universes of generalization, and our interest in regression estimates of universe scores that make use of multiple predictors and subgroup means, we find no such consistency from one method of estimation to another. In the typical study, the three errors have different variances; methods for estimating these variances will be discussed in Chapter 3.

### Score components and components of variance

The analysis of a G study generates estimated "components of variance." These are variances of hypothesized components of an observed score. We

---

[3] Gulliksen (1950, pp. 39–45) and Lord and Novick (1968, p. 66) distinguish several kinds of error. The "error of measurement" arising when the observed score is substituted for the true score is our $\Delta$. The "error in estimating true score" from a regression equation is our $\varepsilon$. The "error of prediction" is the difference between the observed score on one of two parallel forms and the estimate of that score made from the score obtained on the other form; this does not enter our discussion. Gulliksen mentions a fourth error that we can also ignore: the simple difference between observed scores on the two parallel tests.

start with the model for a one-facet study.[4] The universe of admissible observations is classified with respect to a facet $i$. We will also use $i$ as a general label for various conditions of that facet ($i = A,B,C, \ldots$). In the one-facet model the number of conditions of $i$ in the universe of admissible observations is ordinarily taken to be indefinitely large, and we assume that the universe of generalization is identical to the universe of admissible observations. For every person $p$ a score $X_{pi} (= X_{pA}, X_{pB}, \ldots)$ can in principle be observed for each condition of facet $i$. At times attention is directed to a particular condition or person by a notation such as $i^*$ or $p^*$. (See p. 113ff.)

The investigator wishes to generalize over all conditions of facet $i$; he would like to know $\mu_p$, the universe score of $p$. $\mu_p = \mathcal{E}_i X_{pi}$. We also define the mean $\mu_i$ for each $i$ and a general mean $\mu$ over persons and conditions. $\mu_i = \mathcal{E}_p X_{pi}$ and $\mu = \mathcal{E}_{p,i} X_{pi}$.

For the observation corresponding to a particular $p$ and $i$, we have the identity:

$$
\begin{array}{lll}
(1.1) & X_{pi} = \mu & \text{(general mean)} \\[2mm]
& + \mu_p - \mu & \text{(person effect)} \\[2mm]
& + \mu_i - \mu & \text{(condition effect)} \\[2mm]
& + X_{pi} - \mu_p - \mu_i + \mu & \text{(residual)}
\end{array}
$$

This equation divides the observed score into components representing hypothesized effects. Mathematically, it is no more than a tautology.

The observed value $X_{pi}$ will be larger or smaller than the sum of the first three components because of uncontrolled variation. Suppose rater $i$ happens to see the subject at a time when he is performing unusually well; in that event, the residual will be positive. The residual will also include any systematic effect of $p$ and $i$ in combination (an interaction). Perhaps, for example, rater $i^*$ is more favorable than other raters to introverts, In this case, if $p^*$ is an introvert, $X_{p^* i^*} - \mu$ will be greater than the sum of the two main effects: the rater's constant error over all subjects exhibited in $\mu_{i^*} - \mu$, and the person's general rating $\mu_{p^*} - \mu$. If some raters spread out their ratings

---

[4] All notation is summarized on p. xv following the Table of Contents. The reader's attention is directed to a mathematically strict development for the one-facet case offered by Hunter (1968). The model for this case is also treated by Lord and Novick (1968, pp. 154–165).

more than other raters ("use more of the scale"), this also contributes to the residual term.

In principle, the number of possible observations on $p$ *under any single condition i* is indefinitely large; within this set of observations, conditions of any facet other than $i$ vary in an uncontrolled manner. Hence, we can define a mean $\mu_{pi}$ over these varying conditions of other facets, and can resolve the residual into $(\mu_{pi} - \mu_p - \mu_i + \mu) + e_{pi}$. (There will almost never be a direct correspondence of $e$ with the error $\Delta$, $\delta$, or $\varepsilon$.)

Each score component has a distribution. Considering all the conditions $i$ in the universe, there is a distribution of $\mu_i - \mu$, whose mean is zero. The variance of these values $\sigma^2(i)$ $[= \underset{i}{\mathscr{E}}(\mu_i - \mu)^2]$ is called the "variance component for $i$" or the "$i$ component of variance." Over persons, there is a variance of $\mu_p - \mu$, symbolized by $\sigma^2(p)$. Over $pi$ combinations there is a variance of $X_{pi} - \mu_p - \mu_i + \mu$ [i.e., of $(\mu_{pi} - \mu_p - \mu_i + \mu) + e_{pi}$]. This is the residual component of variance, $\sigma^2(pi,e)$, which combines $\sigma^2(\mu_{pi} - \mu_p - \mu_i + \mu)$ with $\sigma^2(e_{pi})$.

The collection of $X_{pi}$ for all persons and conditions has a variance $\sigma^2(X_{pi}) = \underset{p,i}{\mathscr{E}}(X_{pi} - \mu)^2$. The variance here is defined over all admissible observations, and is analogous to the "total" sum of squares in analysis of variance. This variance equals the sum of the variance components:

$$(1.2) \qquad \sigma^2(X_{pi}) = \sigma^2(p) \qquad \text{(person component)}$$

$$+ \; \sigma^2(i) \qquad \text{(condition component)}$$

$$+ \; \sigma^2(pi,e) \qquad \text{(residual component)}$$

There is no variance component for $\mu$, because $\mu$ is constant for the population and universe. The covariance terms [e.g., $\underset{p,i}{\mathscr{E}}(\mu_p - \mu)(\mu_i - \mu) = \underset{p}{\mathscr{E}}(\mu_p - \mu)\underset{i}{\mathscr{E}}(\mu_i - \mu)$] vanish because expressions such as $\underset{p}{\mathscr{E}}(\mu_p - \mu)$ reduce to zero. It will later become clear that $\sigma^2(X_{pi})$ is different from what is traditionally called "the observed-score variance."

The component $\sigma^2(p)$ resembles the "true-score variance" of classical theory. The component $\sigma^2(i)$ is the variance of constant errors associated with various conditions—for example, the varying difficulty of test forms or the varying leniency of raters. The residual variation $\sigma^2(pi,e)$, which equals $\sigma^2(\mu_{pi} - \mu_p - \mu_i + \mu) + \sigma^2(e)$, combines the person–condition interaction with variation from unidentified sources. The two parts could be separated only by a two-facet study with more than one observation on each $p,i$ pair.

If conditions are classified with respect to two facets $i$ and $j$, seven components of the score and seven variance components are identified:

$$
\begin{array}{lll}
(1.3) & X_{pij} = \mu & \sigma^2(X_{pij}) = \\
\text{persons} \quad p & + \mu_p - \mu & + \sigma^2(p) \\
\text{conditions} \quad i & + \mu_i - \mu & + \sigma^2(i) \\
\text{conditions} \quad j & + \mu_j - \mu & + \sigma^2(j) \\
\text{interactions} & & \\
\quad\quad pi & + \mu_{pi} - \mu_p - \mu_i + \mu & + \sigma^2(pi) \\
\quad\quad pj & + \mu_{pj} - \mu_p - \mu_j + \mu & + \sigma^2(pj) \\
\quad\quad ij & + \mu_{ij} - \mu_i - \mu_j + \mu & + \sigma^2(ij) \\
\text{residual} \quad pij,e & + X_{pij} - \mu_{pi} - \mu_{pj} - \mu_{ij} & + \sigma^2(pij,e) \\
& + \mu_p + \mu_i + \mu_j - \mu &
\end{array}
$$

This model takes the number of conditions of $i$ and $j$ in the universe of admissible conditions to be indefinitely large. Formulas for estimating components of variance are considered in Chapter 2. The pattern of (1.3) generalizes to 15 definable components for a three-facet study, to 31 for a four-facet study, etc.

More often than not, a number of separate observations $X_{pi}$ are averaged (or added) to form a score. A test score, though it is a single observation representative of a universe of tests, is also a composite of observations on a set of items. At times $i$ will refer to a test, and at times to a half-test or an item. Which point of view is taken in any given analysis will be clear from context. We shall use a capital letter $I$ (as in $X_{pI}$) where it is necessary to speak of a set of conditions $i$. The score $X_{pI}$ is the average of some number of values of $X_{pi}$. The score $X_{pI}$ divides into components just as $X_{pi}$ does; the meanings of $\mu_I - \mu$, $\sigma(\mu_I)$, etc., will be obvious. So long as the $i$ are randomly assembled into sets of size $n_i$, $\mu_I = \dfrac{1}{n_i} \sum_{i \in I} \mu_i$, etc. Over sets of randomly assembled $I$, $\sigma^2(\mu_I) = \dfrac{1}{n_i} \sigma^2(\mu_i)$. Similarly for other components. Note that $i$ is simply a special case of $I$ in which $n_i = 1$. For a second facet, $J$ will be used to denote a set of $j$; for a third facet, we use symbols $K$ and $k$. A group of persons is denoted by $P$; thus, a sample mean of $X_{pi}$ is $X_{Pi}$.

In the G study, a set of observations is made on each person. Several tests may be given; alternatively, parts of a test, or items, may be treated as separate observations. The finest subdivision of a facet that yields a separate score in the G study will be assigned the letter $i$ (or $j$, etc.). Variance components are estimated for elements of that size, e.g., $\sigma^2(i)$, $\sigma^2(pi)$. These components of variance can be divided by the appropriate number of observations whenever the variances of components of a score $X_{pI}$ are needed.

The one-facet model considers the universe of generalization to be the set of $X_{pi}$ for the person under all admissible conditions. The universe mean to which the decision maker generalizes is $\mu_p$. A two-facet model allows the decision maker a choice between three types of universe of generalization. There is the grand universe that includes the full range of both $i$ and $j$, and is the same as the universe of admissible observations. There are two types of restricted universe, one in which an $I$ is fixed and one in which a $J$ is fixed. There are corresponding universe scores $\mu_p$, $\mu_{pI}$, and $\mu_{pJ}$. A universe with $I$ fixed at, say, $I^*$ is a subset of the grand universe. In it, $j$ ranges over all its possible values, each $j$ being paired with each of the conditions $i \in I^*$ in turn.

This implies that the label $\mu_p$ in the one-facet study and the corresponding concept of true score in classical theory are deceptively simple, since stating that generalization is over $i$ does not indicate whether other aspects of the procedure such as $j$ are held constant or are allowed to vary along with $i$. No matter how many facets are taken into account in the experimental design of the G study, there are additional potential facets to be considered in interpretation (see p. 122).

The conception of alternative universes of generalization is closely related to the test theory developed by Lord and Novick (1968, Chapter 8) for "imperfectly parallel" measurements. They consider the possibility that there are a number of distinguishable tests, and that each test may be administered more than once. One might then generalize over performances on the same form at different times or generalize over both forms and occasions. The expected value of the person's observed score over trials on a particular form is called a "specific true score" by Lord and Novick (1968, p. 43); it is comparable to our $\mu_{pi*}$. The expected value over forms and occasions, our $\mu_p$, they call a "generic true score." Lord and Novick do not distinguish our third possibility: the scores hypothetically obtainable by administering all the forms at essentially the same time. For this universe, the universe score is $\mu_{pj*}$—a true score corresponding to the person's temporary–general state rather than to a lasting trait. Our model goes beyond Lord and Novick also in considering the possibility of additional facets (e.g., testers).

## EXERCISES

*The exercises offered for Chapter 1 are similar to those of the usual textbook, allowing the reader to test his understanding of basic concepts. For some of these exercises, however, more than one answer can be defended. In later chapters the exercises often serve to extend the text, presenting problematic cases for which the body of the chapter does not provide a model solution. The reader will often be well advised to think briefly about the challenge an exercise poses and then to turn at once to our suggested answer, instead of trying to attack the problem independently from the outset. In an exercise with several parts, he should compare his answer to ours as each part is completed.*

**E. 1.** An "in-basket" test is devised to assess the judgment of an administrator (a factory manager, a school principal, etc.). In the test, a file of correspondence, memos, etc. is placed before the man and he is allowed two hours to work through the pile and indicate appropriate actions. There are various scores—for example, the number of items on which action is taken, and the number of items delegated to a subordinate.

Use this problem to illustrate, from the viewpoint of the person developing the procedure, the following concepts or distinctions:

    a. G study vs D study
    b. universe; facet; condition of a facet
    c. the *pi* component of the score (for a facet mentioned in *b*)

**E.2.** What facets should be investigated systematically in a G study of each of the following measures?

    a. A trait of impulsiveness is postulated. To measure it in nine-year-old children, items are prepared in which a main drawing is followed by six possible choices. All but one are exactly like the main drawing, and the subject is to find the variant. When a subject marks a choice identical to the main drawing, the response is regarded as an indication of impulsiveness.

    b. It is supposed that children's interests can be described in terms of emphasis on "people" or "things." As a test procedure, the child is shown a brief movie covering six incidents in the park. He is then asked two or three questions: "Tell me what you saw." "If we make another movie like this what would you like us to show?" etc. The response is recorded on tape and later scored for number of references to people and to things.

    c. Speech samples are recorded during therapeutic interviews. The investigator proposes to have experts rate each sample on "affectivity"—free expression of feelings. He wants to examine changes from month to month during the course of therapy.

    d. A test of proficiency in proofreading asks the job applicant to circle every error in spelling, printing, etc. on a page, doing as much as he can in three minutes.

**E.3.** The following exercise asks for careful application of terminology. For each of the lettered phrases *a* to *l*, apply one or more of the following labels. There is room for uncertainty in making some of the responses.

    A. Would not be termed a facet, or a condition of a facet.
    B. Can be termed a facet, or a condition of a facet.
    C. Likely to be fixed, in the universe of generalization.

D. Likely to be seen as representing a facet over which one wishes to generalize.

E. A facet which in the universe of generalization has a different condition(s) for different persons.

A high-school orchestra is to be formed in a suburban community where a reasonable pool of young people have had previous instrumental training and can go directly into ensemble work. To choose among candidates, the music director holds tryouts. Each pupil comes at a scheduled time, and has half an hour for warm-up and study. He may bring his own instrument or may borrow a school-owned instrument. When he enters the tryout room, after his warm-up, he plays two selections (or excerpts) of his own choice, each five minutes long. He is also to play a piece for which the sheet music was handed him at the start of the warm-up period, to test ability to read unfamiliar music.

a. The conductor (teacher) of the orchestra (who rates each candidate).

b. Music teacher Smith, who sits in on many of the tryouts and helps judge candidates.

c. The instrument used by the candidate who has chosen to bring his own instrument.

d. The instrument used by the candidate who borrows a school instrument.

e. The number of years the candidate has been playing the instrument.

f. The sex of the candidate.

g. The piece of music chosen by the candidate as a tryout piece.

h. The piece of music the candidate is asked to play after short study.

i. The room in which the tryout is held.

j. The duration of the tryout piece. (Set at five minutes in the procedure described above.)

k. The composer or period of the music the candidate chooses to play (e.g., baroque).

l. The number of candidates applying for the percussion section.

**E.4.** Classify the following according to whether the decision is absolute, or requires comparison of persons. (If you think "it depends," on what does it depend?)

a. A state licensing examination for lawyers.

b. As part of a medical checkup, vision is tested.

c. An experimenter wants to know whether learning in statistics is improved when students have access to a computer.

d. An experimenter wants to know how faint a sound signal sonarmen are likely to detect, late in a two-hour shift.

e. A counselor wants to know whether a student likes outdoor work better than indoor work.

**E.5.** List reasons one might have for carrying out a G study separate from a D study, instead of determining generalizability from the decision data.

**E.6.** "Three types of facets" in a G study are discussed on page 2. Illustrate the three types with reference to the following study.

To test proficiency of pilots in making instrument landings, the pilot is asked to land the plane with the cockpit hooded. The plane is equipped with recording instruments from whose records the quality of the performance can be judged. To get data on generalizability, each pilot is tested on several days.

**E.7.** Write the full algebraic expression for each of the following components of the observed score $X_{pijk}$ in a three-facet study.
a. person    b. *pj*    c. *pjk*

### Answers

**A.1.** a. In the G study some group of subjects would be tested with two or more separate in-baskets. Analysis will indicate how many separate baskets, and how many decisions regarding each basket, are needed to reach adequately precise scores. This information will be the basis for designing the procedure to be used on a large scale to investigate characteristics of principals trained in different ways. The latter inquiry is a D study.
b. There is a universe of possible memos, letters, etc., that could be included in the in-basket for the person playing the role of principal of Central High School. There is also a universe of schools, any of which might be considered. One may wish to generalize over schools, items of material within schools, and occasions of testing. Each of these three is a facet. Central High is a condition of the school facet. The letter asking that funds for the Red Cross be solicited is a condition of the facet items-of-material.
c. Principals who are generally similar but who differ in attitudes toward co-operation in community affairs will act differently on the Red Cross request; this is a *pi* effect.

**A.2.** a. Drawings, occasions of testing, possibly tester, etc. (Here and elsewhere where illustrations are called for, other answers might be added, or one might argue that an answer we suggest is relatively unimportant.)
b. Incidents portrayed, interviewers, scorers, occasions of viewing (trials), and probably probe questions.
c. Raters, occasions during a limited interval (one week?), topics of conversation. (The interviewer is likely to be fixed within the person. It will be noted that topics cannot ordinarily be assigned and that the topics that enter the conversation are very likely not a random sample of the person's concerns.)
d. Pages of text, sampled from diverse kinds of material; occasions or trials.

**A.3.** a. B, C. It seems that one wants to know whether the candidate suits this teacher's requirements, since this teacher is a part of the criterion situation.
b. B, D. Smith is only one of many teachers who might equally well help in choosing candidates.
c. B, C, E. The instrument will presumably also be used by him when the orchestra is formed, and so is part of the criterion task. If he does better on this than he would on instruments generally, this is not a source of "error" in the selection process. It is unlikely that he will do worse with a familiar instrument. (Admittedly, it is unfair to give the pupil who owns and has become familiar with

a good instrument a better chance for the orchestral experience than others; but that is how such orchestras are usually managed.)

d. **B, D.** Presumably this child will borrow or rent an instrument from a universe of instruments, if he enters the orchestra.

e. **A.** Subject characteristics are not called "facets" in our theory.

f. **A.**

g. **B, D, E.** The candidate has his own repertoire on which he has practiced, and the observers look on what he does as a sample of what he can do with pieces on which he has had plenty of practice.

h. **B, D.**

i. **B, D.** (While the tryout room may be the hall the orchestra will use for rehearsal and concert, there is surely no intent to build a one-room orchestra.)

j. Two points of view might be taken.

**B, D** appears to be the best answer. One certainly is interested in ability to play pieces of any duration, and one looks on the five-minute excerpt as a sample of what other excerpts would show. One could define a universe of selections that vary in length, but restrict the sample to five-minute selections. This is not the strictly random sampling called for by our model.

On the other hand, one might answer **B, C**, since the rules of this procedure define a universe of five-minute elements.

k. **B, D.** The judge surely wants to evaluate ability to play a fairly wide repertoire, and if time were not limited would perhaps call for several musical styles.

l. **A.**

**A.4.** a. Absolute, assuming that there is a predetermined passing level.

b. Absolute.

c. Comparative; scores in one group will be compared with those in another.

d. Absolute.

e. Absolute decision based on a difference score. Presumably scores are not expressed in terms of norms.

**A.5.** A G study carried out in advance can make the D study more efficient.

A single elaborate G study can give information pertinent to a variety of D studies for different purposes, calling for different designs.

A G study must employ two or more conditions of each facet; this may be impractical to do with a large sample of persons, yet the D study may require a large sample. A publisher has responsibility for providing G-study information to guide persons who may later carry out D studies. Data for decisions may come in only gradually over a long period of time.

**A.6.** The variable, controlled facets may be: Days. Approaches within a day. Fixed: Plane. Recording instruments. (Likely to be constant because of cost of duplicating.) Airfield.

Variable, uncontrolled: Winds. Pilot's physiological state, etc.

**A.7.** a. $\mu_p - \mu$

b. $\mu_{pj} - \mu_p - \mu_j + \mu$

c. $\mu_{pjk} - \mu_{pj} - \mu_{pk} - \mu_{jk} + \mu_p + \mu_j + \mu_k - \mu$

# CHAPTER 2

# Experimental Designs and Estimates of Variance Components

## A. Varieties of Experimental Design

With multiple facets, a great variety of experimental designs are possible. Each design, applied in the G study, calls for formulas to estimate variance components. These estimation formulas are the main topic of this chapter. There is also a choice to be made among experimental designs for the D study, and that choice determines how the information on components is assembled to evaluate generalizability. This last is the topic of Chapter 3.

### Terminology and notation

We refer to "facets" and "conditions" where the literature on experimental design speaks of "factors" and "levels." If two aspects of a measuring procedure vary systematically in a study, we speak of a two-facet study, though persons constitute a third basis for classification of data. Similarly, we speak of a conventional two-way matrix of several ratings for several persons as a "one-facet" study, raters being a "facet" of the observing procedure. (While our scheme of counting departs from statistical convention, referring to persons as a facet would lead us into extremely awkward locutions.)

The number of persons entering into a G study is denoted by $n_p$. We consider only designs in which each person is observed under the same number of conditions, and also restrict ourselves to designs in which there are $n_i$ observations per person or, with two or three facets, $n_i n_j$ or $n_i n_j n_k$, etc. Every such G study produces, in effect, a box-like array of observations; the dimensions of the box are $n_p$, $n_i$, $n_j$, . . . .

The terms *crossed* and *nested* are commonly used by statisticians to describe designs. If the design provides for observing every subject under every condition *i*, we say that *p* is crossed with *i*. Following Millman and Glass (1967), we denote this by $i \times p$. Similarly, $i \times j \times p$ is a crossing that produces a score for every subject *p* under every pairing of conditions *ij*. An example is the study in which $n_i$ raters observe the same pupils simultaneously on $n_j$ occasions.

Raters might visit the classroom, not simultaneously, but at different times. Then occasions are said to be nested within the rater; for each rater *i* there is a different set of $n_j$ occasions. For *j* nested within *i*, we write *j:i*. The crossing relation is commutative: $i \times j = j \times i$. But *j:i* is not the same as *i:j*. Nesting is not commutative.

These symbols may be combined in various ways. There might be nesting such that a rater observes all pupils during each of his $n_j$ occasions of observation, with occasions differing from rater to rater. The study is described as $(j:i) \times p$: pupils crossed with raters *i* and with occasions *j*, occasions nested within raters.

In learning this system, schematic diagrams are helpful. Figure 2.1 shows layouts for two one-facet designs. Each cell represents a different *p,i* combination. In the crossed design, every person is observed under every condition. In the nested design, there is a different set of conditions for each person.

The reader can sketch for himself the design *p:i*, which may be encountered (for instance) if there are several raters, each giving information on one subgroup of subjects. We shall give no more than incidental attention to designs in which subjects are nested. In measurement of individuals, and in generalizability studies, designs rarely have subjects nested. In this monograph persons appear as nested within schools in some illustrative studies where the measurement problem is to estimate the mean for the school or to estimate a population mean. However, in this context the "subject" of the

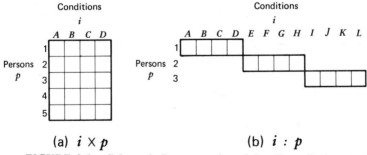

(a) $i \times p$          (b) $i : p$

**FIGURE 2.1.   *Schematic Representation of One-Facet Designs.***

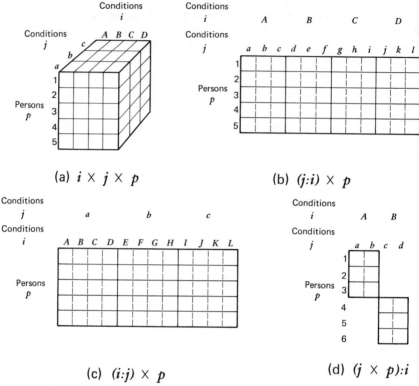

**(a)** $i \times j \times p$        **(b)** $(j{:}i) \times p$

**(c)** $(i{:}j) \times p$        **(d)** $(j \times p){:}i$

**FIGURE 2.2.    Schematic Representation of Some Two-Facet Designs.**

inquiry is the school or the population, not the person. It may be sometimes necessary to nest subjects in the G study. Fleiss (1970) points out that under some circumstances it is impossible to have a subject interviewed twice, where one wishes to appraise the extent to which the interviewer is a source of variance. For this purpose Fleiss designs a G study where persons are nested within interviewers, though this design does not permit one to disentangle person from person–interviewer effects. Excellent use of a persons-nested design is seen in an elegant but intricate study by Coffman and Kurfman (1968). Assigning essays to graders in a counterbalanced design, they were able to show large effects that implied shifts in the graders' standards over time.

A few of the possible two-facet designs are illustrated in Figure 2.2. The crossed $i \times j \times p$ design (diagram a) has an egg-crate structure. In the second design (diagram b), $(j{:}i) \times p$, $p$ is observed under each $i$ and under each $j$, but each $j$ is paired with only one $i$. The third sketch (diagram c) shows $(i{:}j) \times p$. The reader can sketch $i \times (j{:}p)$ for himself. The final sketch (diagram d) is for $(j \times p){:}i$. For each $i$, there is a $j \times p$ design, but different *p* and *j* are associated with each *i*. An example would be a study in which teachers are rated by a team of judges, but a different set of judges (and teachers) is used in each school *i*.

One further symbol will be helpful. Sometimes each rater observes on one and only one occasion, different for every rater. This is nesting ($j:i$) with $n_j = 1$. We employ the special symbol *i, j* for complete confounding of this type. We shall speak of the *i, j* pattern as having "*i* joint with *j*."

### Possible two-facet designs

In this section we present a great deal of information on two-facet designs, recognizing that there is too much detail to be readily comprehended. The compilation will serve primarily for reference. The patterns exhibited here have counterparts for designs with three or more facets, which the reader can trace for himself once he understands the approach.

Figures 2.3 and 2.4 present Venn diagrams of one- and two-facet designs. These diagrams make it possible to determine which components are confounded in nested designs. There is one circle for persons (solid line) and one for each facet (broken or dotted line). In diagram (*a*) of Figure 2.3, the

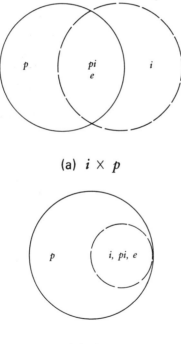

(a) *i* × *p*

(b) *i:p*

*FIGURE 2.3. Schematic Representation of Components of Variance for One-Facet Designs.*

| Design number | Structure | Structure within the person | Pattern of components |
|---|---|---|---|

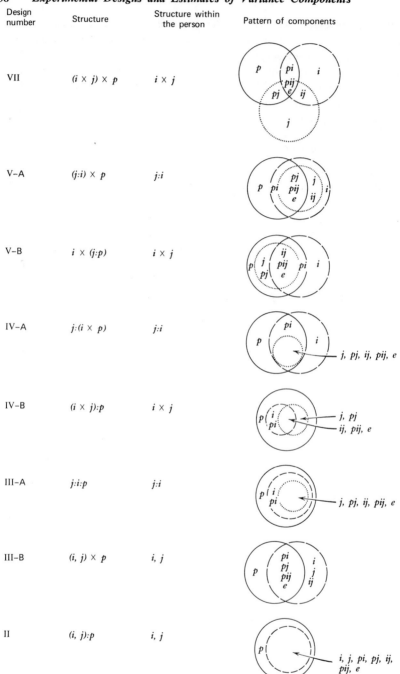

**FIGURE 2.4.  Schematic Representation of Components of Variance for Two-Facet Designs.**

crossed design, there are three areas, corresponding to the components of variance for $p$, $i$, and residual ($pi,e$) given by the two-way analysis of variance. In diagram (b), the nested design, the $i$ circle is within the $p$ circle. The $pi,e$ area is not separated off; the linking of $i,pi,e$ represents the fact that the $i$ component is confounded with $pi,e$. The reader is warned not to think of circle size as representing the number of observations of each kind. In the nested design, the number of different $i$ will be greater, not smaller, than the number of $p$.

When this scheme is extended to two facets, there are circles for $p$, $i$, and $j$ and there are seven components of variance to be accounted for, as shown in Figure 2.4.[1] (With three or more facets, it is sometimes possible to make similar diagrams, but many higher order designs cannot be represented in a plane figure.)

The first line of Figure 2.4 refers to part (a) of Figure 2.2. In this $i \times j \times p$ design, seven components of variance can be separately estimated; we therefore call it Design VII. The analysis of variance generates mean squares for "main effects" $p$, $i$, $j$, plus mean squares for interactions $pi$, $pj$, and $ij$, plus a residual mean square; for each mean square there is a component of variance. An example would be a study in which a checklist of $n_i$ symptoms of tension is filled out by a set of $n_j$ raters, each of whom examines all $n_p$ subjects. Illustrative G studies of this type will be presented in Chapter 6.

We shall need the within-person variance to estimate the magnitude of errors $X_{pij} - \mu_p$. The within-person design is a slice of the grand design made up of observations on a single person. As shown in Figure 2.2, the within-person pattern in Design VII is $i \times j$. These patterns are listed in Figure 2.4 for all designs.

Further information on Design VII appears in Table 2.1. The design samples each score component the indicated number of times. To see how these values are derived, consider a study with two persons (1, 2), two conditions of $i$ ($A,B$) and three of $j$ ($a,b,c$). Then we may express each score in terms of components according to (1.3):

$$X_{1Aa} = \mu + (\mu_1 - \mu) + (\mu_A - \mu) + (\mu_a - \mu) + (\mu_{1A} - \mu_1 - \mu_A + \mu)$$
$$+ (\mu_{1a} - \mu_1 - \mu_a + \mu) + (\mu_{Aa} - \mu_A - \mu_a + \mu) + e_{1Aa}$$

[1] The list of designs displayed in the figure is not exhaustive. One can develop a number of complicated designs by organizing conditions or persons into blocks. For example, to measure the teaching ability of teachers $t$, one might organize lessons $i$ into two groups $I_1$ and $I_2$, for instance, and assign pupils $j$ to the $tI$ combinations. This is a design with $j:(I \times t)$, but it gives more information than Design IV-A because it has $i$ nested in $I$. Such designs have apparently not been used in generalizability studies, and they appear in this monograph only in connection with an analysis of Wechsler scores in Chapter 8.

**TABLE 2.1.** *Number of Observations of Each Score Component in Various G Studies Having $n_i n_j$ Observations per Person*

| Design of G study | | Within-person design | Number of observations[a] | | | | | | |
|---|---|---|---|---|---|---|---|---|---|
| | | | $\mu_p \sim$ | $\mu_i \sim$ | $\mu_j \sim$ | $\mu_{pi} \sim$ | $\mu_{pj} \sim$ | $\mu_{ij} \sim$ | $\mu_{pij,e} \sim$ |
| VII | $i \times j \times p$ | $i \times j$ | $n_p$ | $n_i$ | $n_j$ | $n_p n_i$ | $n_p n_j$ | $n_i n_j$ | $n_p n_i n_j$ |
| V-A | $(j{:}i) \times p$ | $j{:}i$ | $n_p$ | $n_i$ | $n_i n_j$ | $n_p n_i$ | $n_p n_i n_j$ | $n_i n_j$ | $n_p n_i n_j$ |
| V-B | $i \times (j{:}p)$ | $i \times j$ | $n_p$ | $n_i$ | $n_p n_j$ | $n_p n_i$ | $n_p n_j$ | $n_p n_i n_j$ | $n_p n_i n_j$ |
| IV-A | $j{:}(i \times p)$ | $j{:}i$ | $n_p$ | $n_i$ | $n_p n_i n_j$ | $n_p n_i$ | $n_p n_i n_j$ | $n_p n_i n_j$ | $n_p n_i n_j$ |
| IV-B | $(i \times j){:}p$ | $i \times j$ | $n_p$ | $n_p n_i$ | $n_p n_j$ | $n_p n_i$ | $n_p n_j$ | $n_p n_i n_j$ | $n_p n_i n_j$ |
| III-A | $j{:}i{:}p$ | $j{:}i$ | $n_p$ | $n_p n_i$ | $n_p n_i n_j$ | $n_p n_i$ | $n_p n_i n_j$ | $n_p n_i n_j$ | $n_p n_i n_j$ |
| III-B[b] | $(i,j) \times p$ | $i, j$ | $n_p$ | $n_i n_j$ | $n_i n_j$ | $n_p n_i n_j$ | $n_p n_i n_j$ | $n_i n_j$ | $n_p n_i n_j$ |
| II[b] | $(i,j){:}p$ | $i, j$ | $n_p$ | $n_p n_i n_j$ | $n_p n_i n_j$ | $n_p n_i n_j$ | $n_p n_i n_j$ | $n_p n_i n_j$ | $n_p n_i n_j$ |

[a] Components with similar underscores are estimated as part of a single confounded variance, corresponding to an area in Figure 2.4.
[b] Number of observations is fixed at $n_i n_j$ per person for comparability to other designs; that is, $n_i n_j$ values of $i$ are paired with an equal number of $j$.

To save space and focus attention on the crucial symbols, let us write $\mu_1 \sim$ for $\mu_1 - \mu$, $\mu_{1A} \sim$ for $\mu_{1A} - \mu_1 - \mu_A + \mu$, etc. Then for the design under discussion we have

$$X_{1Aa} = \mu + \mu_1 \sim + \mu_A \sim + \mu_a \sim + \mu_{1A} \sim + \mu_{1a} \sim + \mu_{Aa} \sim + e_{1Aa}$$
$$X_{1Ab} = \mu + \mu_1 \sim + \mu_A \sim + \mu_b \sim + \mu_{1A} \sim + \mu_{1b} \sim + \mu_{Ab} \sim + e_{1Ab}$$
$$(2.1) \quad X_{1Ac} = \mu + \mu_1 \sim + \mu_A \sim + \mu_c \sim + \mu_{1A} \sim + \mu_{1c} \sim + \mu_{Ac} \sim + e_{1Ac}$$
$$X_{2Aa} = \mu + \mu_2 \sim + \mu_A \sim + \mu_a \sim + \mu_{2A} \sim + \mu_{2a} \sim + \mu_{Aa} \sim + e_{2Aa}$$

$$\cdots$$

$$X_{2Bc} = \mu + \mu_2 \sim + \mu_B \sim + \mu_c \sim + \mu_{2B} \sim + \mu_{2c} \sim + \mu_{Bc} \sim + e_{2Bc}$$

Evidently, $n_p = 2$ *different* values of the person component are sampled, $n_i = 2$ *different* values of the $i$ component, and $n_j = 3$ *different* values of the $j$ component. There are $n_p n_i = 4$ different $pi$ components ($1A, 2A, 1B, 2B$); likewise there are 6 $pj$ and 6 $ij$ components. Finally, each observation generates a different residual value, and there are $n_p n_i n_j$ of these. These frequencies are entered in Table 2.1.

The next design to be considered, $(j{:}i) \times p$, is shown as part (b) in Figure 2.2. The example given earlier will be recalled: each of the $n_i$ raters observes all $n_p$ subjects on $n_j$ occasions, but the raters do not make their observations simultaneously; there is a different set of $n_j$ occasions for each rater. Because a given $j$ is present in connection with only one of the $i$, one can never observe the $j$ main effect independent of the $ij$ interaction. The $j$ and the $ij$ score components are confounded, and appear together in the analysis as a "within $i$" mean square. Likewise, $pj$ is confounded with $pij,e$. Because of this confounding only five components of variance, two of them composites, can be estimated, and hence we call this a Design V. The Venn diagram in the last column of Figure 2.4 displays this regrouping of components.

For this design, suppose there are persons 1 and 2, and two conditions of $i$ $(A,B)$. If $n_j = 3$, under our convention there will be *six* conditions of $j$ ($a$, $b$, $c$ within $A$ and $d$, $e$, $f$ within $B$). Each observed score breaks into components as before.

$$X_{1Aa} = \mu + \mu_1\sim + \mu_A\sim + \mu_a\sim + \mu_{1A}\sim + \mu_{1a}\sim + \mu_{Aa}\sim$$
$$+ \mu_{1Aa}\sim + e_{1Aa}$$

(2.2)    $$X_{1Ab} = \mu + \mu_1\sim + \mu_A\sim + \mu_b\sim + \mu_{1A}\sim + \mu_{1b}\sim + \cdots$$

$$X_{1Ac} = \mu + \mu_1\sim + \mu_A\sim + \mu_c\sim + \mu_{1A}\sim + \mu_{1c}\sim + \cdots$$

$$\cdots$$

$$X_{2Bf} = \mu + \mu_2\sim + \mu_B\sim + \mu_f\sim + \mu_{2B}\sim + \mu_{2f}\sim + \cdots$$

There are 12 such rows. Just $n_p = 2$ components of the type $\mu_p\sim$ have been sampled, and $n_i = 2$ components of the $\mu_i\sim$ type. There are 6 different $j$ components ($6 = 2 \times 3 = n_i \times n_j$). The reader can work out the rationale for the four remaining entries in the V-A row of Table 2.1.

The reader who writes out the full set of equations above will see that $\mu_a\sim$ is present only when $\mu_{Aa}\sim$ is also present. Likewise for $\mu_b\sim$ and $\mu_{Ab}\sim$, etc. This is confounding of $j$ with $ij$. Variances for components so tied together cannot be separately estimated. Confounding is indicated in Table 2.1 by underscores. Thus, in the row for Design V-A, single under-scores appear in the $j$ and $ij$ columns, repeating in another code the indication of confounding that appears in the Venn diagram of Figure 2.4.

The design $(i:j) \times p$ is formally like $(j:i) \times p$, and the entries for the former can be obtained by simply transposing $i$ and $j$ wherever they appear in the V-A row of Table 2.1.

In Design V-B, with $i \times (j:p)$, the within-person design is $i \times j$. There is a symmetric design $j \times (i:p)$. Designs V-A, V-B, and their two transposes are basically similar; the analysis of variance is essentially the same for each of them.

Figure 2.4 or Table 2.1 can be used to trace relations among designs. For example, IV-B is like V-B except that in IV-B, $i$ is tied to $pi$. A study of these relations shows that V-B gives whatever information about components IV-B gives. We may say that Designs IV are weaker than the corresponding Designs V because, other things being equal, a study with Design

V-A or V-B yields more information about components of variance. Design III-A is like Design IV-A and IV-B save for additional confounding.

Designs III-B and II have $i$ joint with $j$, hence these two facets are completely confounded. An example is the study where every rater sees the subject on a different occasion, one occasion per rater. For this to occur, $n_i$ must equal $n_j$, so that in effect we are dealing with just one facet of $i,j$ pairs.

Design III-B can be seen as a one-facet design in which $p$ is crossed with the variable $i,j$. The three-way analysis of variance degenerates to a two-way analysis. Design II is the weakest of all these designs; the analysis degenerates to a one-way analysis yielding components for $p$ and "within $p$." There is no point in listing separately the degenerate design $i,j,p$ which results from setting $n_i = n_j = 1$ in Design II.

A stronger design may be preferred for a G study, as it separates the components more completely. But a weaker design is often appropriate. For example, an investigator who is fairly sure *a priori* that the $ij$ interaction is small may be quite content to leave $ij$ confounded with the residual, if this makes his G study easier to carry out, or less expensive. When the purpose of a G study is limited, one of the weaker designs often gives all the information required. For a D study also, the choice of design is dependent upon too many considerations for any rule of thumb to apply. Problems of design are illustrated concretely in Chapters 6 and 7.

Apart from its general significance as one of a set of alternative designs, Design IV-A has special interest because it embodies the Lord–Novick conception of "specific" reliability, which is a step away from classical theory in the direction of a multifacet model. They envision tests $i$ crossed with persons $p$ in a G study and in the universe, and they envision the possibility of "replications" of observations under condition $i$. These replications appear formally as the nested $j$ in Design IV-A (see p. 29).

### B. *Analysis of Design VII under the Random Model*

The first G study to be considered is that with Design VII, the completely crossed design that yields $n_p \times n_i \times n_j$ elemental scores. The first step in analysis is to perform the usual analysis of variance (McNemar, 1969, p. 359 ff.). This produces the familiar table of sums of squares, degrees of freedom, and mean squares. In our presentation the final row of the table is labelled "residual" rather than "error" so that "error" can be given other meanings later. The example in Table 2.2 comes from a study in which the facets are scorers and items. A sample of 30 patients took an individual test of 10 items; 3 qualified scorers simultaneously observed the performance, each recording a set of item scores. Thus, the data form a 30 × 10 × 3 array.

After having obtained the mean squares, the next step is to estimate the

**TABLE 2.2.** *Analysis of Variance for a Study with 30 Patients Crossed with 10 Test Items Crossed with 3 Scorers*[a]

| Source of variance | Sum of squares | Degrees of freedom | Mean square |
|---|---|---|---|
| Patients $p$ | 5300.24 | 29 | 182.7670 |
| Items $i$ | 1168.65 | 9 | 129.8494 |
| Scorers $j$ | 65.35 | 2 | 32.6744 |
| $pi$ | 2421.11 | 261 | 9.2763 |
| $pj$ | 214.38 | 58 | 3.6963 |
| $ij$ | 67.65 | 18 | 3.7584 |
| Residual | 817.26 | 522 | 1.5656 |

[a] Analysis of PICA Subtest III. See Chapter 6 for descriptive information.

seven components of variance. The equations used here assume that the population of persons and the universes of $i$ and $j$ are all infinite.

Any one study, carried out with random samples of $p$, $i$, and $j$, yields a mean square for a certain effect. Another study carried out in exactly the same way generates another such mean square. The average of mean squares for this particular effect, over all the possible studies applying the same design to the same universe and population, is the "expected" mean square for that effect. As any one study is considered to be a random sample of the possible studies, an obtained mean square can be taken as an unbiased estimate of the expected mean square (EMS).

The expected mean squares can be shown (Cornfield & Tukey, 1956) to be weighted sums of the components of variance defined in Chapter 1 (pp. 27, 28).

$$\text{EMS } p = \sigma^2(pij,e) + n_i\sigma^2(pj) + n_j\sigma^2(pi) + n_in_j\sigma^2(p)$$
$$\text{EMS } i = \sigma^2(pij,e) + n_p\sigma^2(ij) + n_j\sigma^2(pi) + n_pn_j\sigma^2(i)$$
$$\text{EMS } j = \sigma^2(pij,e) + n_i\sigma^2(pj) + n_p\sigma^2(ij) + n_pn_i\sigma^2(j)$$
(2.3)
$$\text{EMS } pi = \sigma^2(pij,e) + n_j\sigma^2(pi)$$
$$\text{EMS } pj = \sigma^2(pij,e) + n_i\sigma^2(pj)$$
$$\text{EMS } ij = \sigma^2(pij,e) + n_p\sigma^2(ij)$$
$$\text{EMS res} = \sigma^2(pij,e)$$

The structure of these equations is related to the Venn diagrams of Figure 2.4. The $p$ circle for Design VII contains four segments that correspond to the four terms in the equation for EMS $p$. Similarly, the $pi$ area defined by the overlap of $p$ and $i$ circles has two segments which correspond to the two terms in the equation for EMS *pi*. The EMS equations can be written from the Venn diagram, using $n_p$ as multiplier if a component does not include *p* in its label, $n_i$ if *i* does not enter the label, etc.

To solve for the unknown components, the actual mean squares from the analysis of variance are written into the equations in place of the EMS which they estimate. The equations are then solved for the components, starting from the bottom. With the use of the data in Table 2.2, we first note that the mean square for residual is 1.57 and take this as $\widehat{\sigma^2}(pij,e)$. Then $3.76 = 1.57 + 30\widehat{\sigma^2}(ij)$, hence $\widehat{\sigma^2}(ij) = 0.07$. Continuing, we find that $\widehat{\sigma^2}(pj) = 0.21$, $\widehat{\sigma^2}(pi) = 2.57$, $\widehat{\sigma^2}(j) = 0.09$, $\widehat{\sigma^2}(i) = 1.32$, and $\widehat{\sigma^2}(p) = 5.71$. The "hat" symbol ($\frown$) signifies "estimate of." These estimates are the main results of the G study. Their interpretation and use will be taken up in subsequent chapters.

We urge the reader to fix in mind the difference between a variance component and an ordinary variance. The observed-score variance is the variance, over persons, of scores $X_{pIJ}$. In Design VII, $X_{pIJ}$ is the average of $n_i n_j$ values of $X_{pij}$. The sample variance of observed scores is

$$\frac{1}{n_p - 1} \sum_p (X_{pIJ} - X_{PIJ})^2$$

The variance component for persons is the population variance of universe scores, i.e.,

$$\lim_{n_p \to \infty} \frac{1}{n_p - 1} \sum (\mu_p - \mu)^2$$

where $\mu_p$ is the average of all admissible values of $X_{pij}$ (any *i*, any *j*).

This type of variance analysis is related to correlational analysis, but we postpone discussion of the relationship to Chapter 8.

### Extension to simpler and more complex crossed designs

The equations given above for the two-facet study have simpler counterparts for the one-facet crossed study (design $i \times p$):

$$\text{EMS } p = \sigma^2(pi,e) + n_i \sigma^2(p)$$

(2.4)    $$\text{EMS } i = \sigma^2(pi,e) + n_p \sigma^2(i)$$

$$\text{EMS } pi,e = \sigma^2(pi,e)$$

The symmetry in this set of equations and in set (2.3) suggests how the equations would look for studies with a greater number of facets. Representative formulas for the three-facet crossed design are given in Table 2.3.

**TABLE 2.3.**   *Selected Equations*[a] *for Expected Mean Squares in a Crossed Three-Facet Design* $(i \times j \times k \times p)$

---

$$\text{EMS } p = \sigma^2(pijk,e) + n_i\sigma^2(pjk) + n_j\sigma^2(pik) + n_k\sigma^2(pij) + n_jn_i\sigma^2(pk)$$
$$+ n_in_k\sigma^2(pj) + n_jn_k\sigma^2(pi) + n_in_jn_k\sigma^2(p)$$

$$\text{EMS } pi = \sigma^2(pijk,e) + n_j\sigma^2(pik) + n_k\sigma^2(pij) + n_jn_k\sigma^2(pi)$$

$$\text{EMS } pij = \sigma^2(pijk,e) + n_k\sigma^2(pij)$$

$$\text{EMS } pijk,e = \sigma^2(pijk,e)$$

---

[a] Other equations may be written by symmetry.

## C. Analysis of Partially Nested Designs

A design that involves nesting or joint sampling confounds two or more of the components. Therefore, the G data do not allow us to estimate these components separately. Sometimes practical constraints make such a design necessary even though the crossed design would be more informative. On the other hand, a design in which there is some confounding may be entirely satisfactory or even preferable to a crossed design generating the same number of observations per person. In particular, where certain components are to be confounded in the D study, it may be better to use a G study where there is similar confounding, because more precise estimates of the confounded components are obtained than would be obtained from the fully crossed design.

We identify a compound component of variance by entering in its label all the arguments used in labelling the underlying components. Thus, a random-model component of variance in which the $pi$ and $pij,e$ effects are combined is labelled $\sigma^2(pi,pij,e)$. In Design III-B, $(i,j) \times p$, Figure 2.4 indicates that the available components are $\sigma^2(p)$, $\sigma^2(i,j,ij)$, and $\sigma^2(pi,pj,pij,e)$.

### Determining mean squares

Where the analysis of variance is to be made by computer, the investigator will often be able to locate a specific program for analyzing whatever design he has employed, and if a ready-made program is not available he can always prepare one. It is often more convenient, however, to follow an all-purpose procedure that generates the mean squares for any design where there are $n_in_j$ (or $n_in_jn_k$, etc.) observations per person. The scores $X_{pij}$ can be treated as if the design were completely crossed by arbitrarily assigning the nested $i$ and/or $j$ to rows and columns. Then one carries out the analysis of variance by the formulas that give sums of squares for Design VII. Any of our two-facet designs can be rearranged into a data box of size $n_p \times n_i \times n_j$. For example, the $(j:i) \times p$ design (part b of Figure 2.2) can be

**TABLE 2.4.**   *Recombination of Sums of Squares and Degrees of Freedom to Recognize Confounding in an $i \times (j:p)$ Design*[a]

| Analysis of variance as if crossed[b] | | | Analysis as actually nested | | | |
|---|---|---|---|---|---|---|
| Source of variance | Sum of squares | Degrees of freedom | Sources as confounded | Combined sums of squares | Combined degrees of freedom | Mean square |
| $p$ | 154.37 | 28 | $p$ | 154.37 | 28 | 5.51 |
| within $p$ | | | | | | |
| $j$ | 26.72 | 18⎫ | $j, pj$ | 643.02 | 522 | 1.24 |
| $pj$ | 616.30 | 504⎭ | | | | |
| $i$ | 0.01[c] | 1 | $i$ | 0.01 | 1 | 0.01 |
| $pi$ | 39.89 | 28 | $pi$ | 39.89 | 28 | 1.42 |
| within $pi$ | | | | | | |
| $ij$ | 14.45 | 18⎫ | $ij, pij, e$ | 275.85 | 522 | 0.53 |
| $pij, e$ | 261.40 | 504⎭ | | | | |

[a] Analysis of data from the study of Belgard and others discussed in Chapter 7. In this study $n_p = 29$, $n_i = 2$, and $n_j = 19$.
[b] Rows are ordered to correspond to the combination of effects in the design.
[c] This value is small because data had been put into standard-score form.

rearranged into a $5 \times 4 \times 3$ box, ignoring the fact that the entries in the slice representing the "first value" of $j$ actually come from a different $j$ for each $i$. The analysis of variance is performed with degrees of freedom determined from $n_p$, $n_i$, and $n_j$—the dimensions of the box, not the actual number of different $i$ or $j$ in the study. *The mean squares coming out of this program are discarded;* only the sums of squares and degrees of freedom are retained. These are recombined to obtain the appropriate mean squares.

First the confounded components are identified, perhaps with the aid of Table 2.1 or Figure 2.4. The sums of squares and degrees of freedom from the crossed analysis are pooled to derive mean squares for the confounded components, as shown in Table 2.4 for an example of Design V-B: $i \times (j:p)$. Here $j$ is confounded with $pj$, and $ij$ with $pij, e$.

### Estimating random-model variance components

The random-model equations relating expected mean squares to variance components are different for each design, but they follow a pattern defined by the following rules:

1. For each source of variation shown as a separate area in the Venn diagram for the design (see Figure 2.4), there is an equation; the left side of the equation is the corresponding EMS.

2. For a particular expected mean square, one lists every component of variance whose argument contains the letter or letters that identify that mean square. Thus the equation for EMS $p$ contains any components in which $p$ appears. In Design VII, EMS $p$ contains the components for $p$, $pi$, $pj$, and $pij,e$. In Design IV-A, EMS $p$ contains three components: $p$; $pi$; and $j$, $pj$, $ij$, $pij,e$. These are the areas appearing within the $p$ circle of the Venn diagram for each design. For EMS $j$, $pj$ of Design IV-B the equation contains the two components in whose labels $j$ or $pj$ appears (but not the component for $p$ or $pi$). The Venn diagram can be used to identify the terms of any equation.

3. Each component is multiplied by $n_i$ if $i$ does not appear within the argument of the component, by $n_j$ if $j$ does not appear, etc. Hence in a two-facet study $\sigma^2(p)$ is multiplied by $n_i n_j$ in any equation where it appears; likewise, the weight for $\sigma^2(pi)$ or $\sigma^2(p,pi)$ is always $n_j$.

For Design V-B, $i \times (j{:}p)$, the equations are as given below. This set of equations is written in the order in which components are estimated, which is the reverse of the order of (2.3) and (2.4).

$$\text{EMS } ij, pij,e = \sigma^2(ij,pij,e)$$

$$\text{EMS } pi = \sigma^2(ij,pij,e) + n_j\sigma^2(pi)$$

(2.5) $$\text{EMS } j, pj = \sigma^2(ij,pij,e) + n_i\sigma^2(j,pj)$$

$$\text{EMS } i = \sigma^2(ij,pij,e) + n_j\sigma^2(pi) + n_p n_j\sigma^2(i)$$

$$\text{EMS } p = \sigma^2(ij,pij,e) + n_i\sigma^2(j,pj) + n_j\sigma^2(pi) + n_i n_j\sigma^2(p)$$

These equations allow us to estimate components from the data of Table 2.4 by substituting a mean square for each expected mean square.

One may also identify simple computing algorithms from the Venn diagram, as in Figure 2.5. For instance, the $i$ circle (diagram e) can be seen as the sum of the $p,i$ intersection ($A + C$) and $D$. Hence to obtain $D$, which equals $n_p n_j\sigma^2(i)$, one need only subtract EMS $pi$ from EMS $i$. One word of caution is required. If an estimate of $\sigma^2$ is negative, the simplified algorithm cannot be used to estimate succeeding components into whose expected mean square equation that variance enters (see p. 57). Starting with diagram (b) of Figure 2.5,

$$\text{MS } ij,pij,e = 0.53 = \widehat{\sigma^2}(ij,pij,e)$$

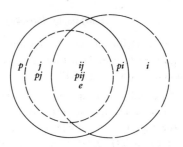

(a)   Components in the design

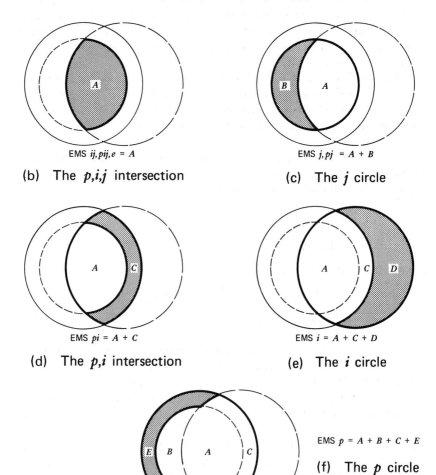

EMS $ij, pij, e = A$

(b)   The $p,i,j$ intersection

EMS $j, pj = A + B$

(c)   The $j$ circle

EMS $pi = A + C$

(d)   The $p,i$ intersection

EMS $i = A + C + D$

(e)   The $i$ circle

EMS $p = A + B + C + E$

(f)   The $p$ circle

*FIGURE 2.5.  Schematic Analysis of $i \times (j{:}p)$ Design to Show Composition of Expected Mean Squares.*

Then,

$$1.24 = 0.53 + n_i\widehat{\sigma^2}(j,pj); \quad \text{and} \quad \widehat{\sigma^2}(j,pj) = 0.36.$$

$$1.42 = 0.53 + n_j\widehat{\sigma^2}(pi); \quad \text{and} \quad \widehat{\sigma^2}(pi) = 0.05.$$

$$0.01 = 1.42 + n_pn_j\widehat{\sigma^2}(i); \quad \text{and} \quad \widehat{\sigma^2}(i) \text{ is taken to be zero.}$$

$$5.51 = 1.42 + 0.71 + n_in_j\widehat{\sigma^2}(p); \quad \text{and} \quad \widehat{\sigma^2}(p) = 0.09.$$

As a further example, consider equations for Design IV-A, $j:(i \times p)$.

(2.6)
$$\begin{aligned}
\text{EMS within } pi &= \sigma^2(j,pj,ij,pij,e) \\
\text{EMS } pi &= \sigma^2(j,pj,ij,pij,e) + n_j\sigma^2(pi) \\
\text{EMS } i &= \sigma^2(j,pj,ij,pij,e) + n_j\sigma^2(pi) + n_pn_j\sigma^2(i) \\
\text{EMS } p &= \sigma^2(j,pj,ij,pij,e) + n_j\sigma^2(pi) + n_in_j\sigma^2(p)
\end{aligned}$$

The estimation of components from G studies with Designs V-B and III-A, and also from certain three-facet designs, is discussed in Chapter 7.

## D. *Sampling Errors of Estimates of Variance Components*

The estimates of variance components obtained in one G study are not numerically identical to those from a second G study employing the identical design. Sampling of persons and conditions causes the estimate in any study to depart somewhat from the value for the population and universe. One would like some idea regarding the extent to which an estimate of a variance component reached in a G study departs from the true value. Adequate study of this matter will ultimately provide much-needed guidance to the person designing G studies. The literature reviewed below leads us to think that the behavioral scientist is on dangerous ground when he employs estimates of components and coefficients from a G study with the usual modest values of $n_i$ and $n_j$, unless he can confidently make assumptions of equivalence, homoscedasticity, and normality. In this monograph we have done what can be done with available techniques and samples of customary size, but all these G studies are primitive. Far more effort will need to be given to the collection of G data in the future, and far more subtlety of design will be needed if that effort is to be deployed economically.

Mathematical statisticians have studied the estimation of variance components in various ways. We cannot review that literature adequately, and any attempt to recapitulate the mathematical reasoning would have little meaning for the majority of our readers. More work is needed on the mathematical properties of components constrained by weak assumptions and on the numerical properties of estimates. From that work, advice for the behavioral scientist planning and carrying out G studies should emerge.

## Monte Carlo studies

One intellectually simple approach is to generate hypothetical data and to calculate statistics for one sample after another. The accumulation of sample values locates the parameter of interest with increasing precision and indicates how the sample statistics distribute themselves around that value. This Monte Carlo method has been applied to the statistics of reliability and generalizability, but only in relatively elementary problems (mostly one-facet).

For example, a person interested in studying the error of estimating the universe-score variance, i.e., the distribution from sample to sample of $\sigma^2(p)$, might proceed as follows:

1. Assume that an error-free ability measure T is normally distributed with arbitrary mean zero and variance one.

2. There are an indefinitely large number of items; every item can be characterized on this same scale by a scale value $\theta_i$. Specify the distribution of $\theta$. Usually, this distribution would be made normal or rectangular, and two parameters would have to be specified to define the distribution.

3. Assume that every response is scored 1 or 0, and that the probability $P$ that person $p$ will earn a score of 1 on item $i$ is an ogival function of $T_p - \theta_i$. Let $P$ approach 0 as this difference becomes large and negative, approach 1.00 when the difference is large and positive, and equal 0.50 when the difference is 0. This item-characteristic curve has a single parameter $\phi$, which is inversely related to the steepness of the curve. Assume this to be uniform for all items.

4. Choose values of the parameter $\phi$ and of the two parameters of the distribution of $\theta$. Specify that the G study will have the design $i:p$. Specify $n_p$ and $n_i$. This completes the definition of the problem.

5. Now draw at random one value from the distribution of T. Draw at random one value from the distribution of $\theta$. Enter the ogival function with this $T_p - \theta_i$ and read off $P$. Then, from an aggregation of zeros and ones mixed in the ratio $(1 - P):P$, draw one value. Call this the score $X_{pi}$.

6. Retaining the same $T_p$, draw another $\theta_i$ and determine another score. Repeat until there are $n_i$ scores for the first person, each on an independently sampled item.

7. Select a second $T_p$, and generate $n_i$ scores for that person as in steps 5 and 6. Repeat until sets of scores have been generated for $n_p$ persons. One has now simulated the collection of data for a single G study. Calculate $\widehat{\sigma^2}(p)$.

8. Repeat steps 5, 6, and 7. This gives a second estimate of $\sigma^2(p)$. Continue until a sufficient number of estimates have been assembled.

9. Summarize the distribution of $\widehat{\sigma^2}(p)$, over studies, in terms of a confidence interval for $\sigma^2(p)$, a standard deviation of the estimates, or some other statistic.

10. One may repeat the process, starting with step 3, to learn the effect of altering any one parameter or any combination of parameters on the sampling error.

The process is laborious and expensive even in the simple case detailed above. The labor rises exponentially when two- and three-facet models are considered. Some economies are possible. For example, in the course of carrying out the study with one value of $n_i$, it is easy to treat subsets of the data to obtain results for smaller values of $n_i$. With some elaboration of technique one can examine sampling errors of $\widehat{\sigma^2}(p)$, $\widehat{\sigma^2}(i,pi,e)$ and $\widehat{\rho^2}(X,\mu_p)$ at the same time. The basic plan described above can be amended to accommodate any experimental design, to accommodate continuously scored observations, etc.

This technique is illustrated in studies of the intraclass correlation coefficient $\mathscr{E}\rho^2(X,\mu_p)$ arising under the $i \times p$ design, with randomly sampled items (Cronbach & Azuma, 1962) and with stratified-sampling plans (Cronbach, Schönemann, & McKie, 1965). These studies were illuminating, but they fall far short of answering our present questions about sampling error for variance components in multifacet designs.

We strongly recommend further Monte Carlo work. In particular, work is needed on two-facet designs where scores are continuous (or vary over a wide range of integers). Comparison needs to be made of various G-study designs all of which involve the same total number of observations $n_p \times n_i \times n_j$.

### Procedures based on statistical theory

The majority of papers on sampling error of variance components in the statistical literature assume score components to be normally distributed. Furthermore, they either assume a very weak design such as $i:j:p$ or they make strong equivalence assumptions (e.g., that the population variance of scores under one condition is the same as the variance under any other condition).

In every analysis of variance the residual sum of squares leads directly to the estimate of a variance component. Depending on the design of the study, this component may be identified as $\sigma^2(pi,e)$, $\sigma^2(i,pi,e)$,

$$\sigma^2(i,j,pi,pj,ij,pij,e),$$

or something else. For convenience here, we shall omit the designation and simply speak of sums of squares (SS), degrees of freedom (d.f.), and $\sigma^2$, recognizing that the point estimate of $\sigma^2$ is given by sum-of-squares/degrees of freedom. Then under certain assumptions,

$$(2.7) \qquad \text{Sum of squares} = \sigma^2 \chi^2_{\text{d.f.}}$$

If this formula is applicable, and one wishes a 95% confidence interval for $\sigma^2$, one need only turn to the $\chi^2$ table for the number of degrees of freedom, look up $\chi^2_{0.025;\text{d.f.}}$ and $\chi^2_{0.975;\text{d.f.}}$ and divide these into the sum of squares. That is,

$$(2.8) \qquad P\left(\frac{\text{SS}}{\chi^2_{0.975;\text{d.f.}}} \leqslant \sigma^2 \leqslant \frac{\text{SS}}{\chi^2_{0.025;\text{d.f.}}}\right) = 0.95$$

The assumption is made that the score component is normally distributed and has the same variance for each person (Scheffé, 1959, pp. 226–229). In a one-facet $i:p$ study, (2.8) would give limits for $\sigma^2(i,pi,e)$. This use is warranted if the within-person variance over all conditions in the universe is the same for every person. This, however, is untrue if some persons are more variable, from task to task or occasion to occasion, than others. If the G study is $i \times p$, (2.8) might be applied to estimate $\sigma^2(pi,e)$. Additional assumptions are now required: that all within-condition distributions have the same variance, and that the correlation between pairs of conditions is uniform for all pairs. That is, observations must be equivalent in the sense of classical test theory. Scheffé (1959, p. 345) discusses the effect of violations of assumptions upon the trustworthiness of the confidence interval, and finds departures from normality to be a source of serious difficulty. His exploration of the effects of nonequivalence is less complete.

A somewhat more complicated formula of the same general character is given by Scheffé (1959, p. 231 ff.), following a development by Bulmer (1957). This applies, not to the case where $\widehat{\sigma^2} = \text{MS}$, but to the case where $\widehat{\sigma^2}$ is given by $(\text{MS}_1 - \text{MS}_2)/n$ and $\text{MS}_2$ satisfies the conditions of (2.7). Thus, in an $i:p$ design, the Bulmer–Scheffé procedure applies to $\sigma^2(p)$, because $\widehat{\sigma^2}(p) = (\text{MS } p - \text{MS res})/n_i$. The formula would not apply to the $p$ component in an $i \times j \times p$ design, because in that case $\sigma^2(p)$ is not proportional to a difference between two EMS. The reader should consult Scheffé for details of the argument and formulas. A representative formula has the form

$$(2.9) \qquad g_U = F_1 \frac{\text{MS}_1}{\text{MS}_2} - 1 + \frac{\text{MS}_2}{F_2 \text{MS}_1}\left(1 - \frac{F_1}{F_2}\right)$$

The product $\text{MS}_2 g_U$ gives the upper limit of the confidence interval for the variance component. The value $F_1$ is from the $F$-table for $\infty$, d.f.$_1$ degrees of freedom, while $F_2$ is the value for $\text{d.f.}_2$, $\text{d.f.}_1$. Again Scheffé warns of serious consequences of nonnormality.

A mathematical development by Welch (1956), reduced to a textbook form by Graybill (1961, pp. 368–374), applies to any of the estimated components in the kinds of balanced design our G studies employ. It assumes a normal distribution for each score component, but appears not to require homoscedasticity and equivalence of conditions.

Any of the components may be estimated by a linear composite of observed mean squares. To describe the basic procedure we may write general expressions in the Welch–Graybill notation. For the moment, then let $i$ symbolize any source of variance including persons or residual. Any one study generates several MS $i$. The equation for expressing the variance to be estimated in terms of EMS (i.e., of variances of certain scores or marginals) can be written:

$$(2.10) \qquad \sigma^2 = g_1 \sigma_1^2 + g_2 \sigma_2^2 + \cdots = \sum_1 g_i \sigma_i^2$$

where all but one $g$ may be zero or negative. For simplicity, write

$$x_1, x_2, \ldots, x_i, \ldots$$

for the mean squares, and write $n_1, n_2, \ldots, n_i, \ldots$ for the corresponding degrees of freedom. We now define an integer $n_0$ by

$$(2.11) \qquad n_0 \doteq \frac{(\sum g_i x_i)^2}{\sum (g_i^2 x_i^2 / n_i)}$$

The calculation on the right is carried out and rounded. If $n_0$ is less than 10, one is advised not to proceed, because the interval formed will be undependable.

$$(2.12) \qquad C = \tfrac{2}{3}(2z^2 + 1)\left(\left\{ \sum g_i x_i \left[ \sum \frac{(g_i x_i)^2}{n_i} \right]^{-2} \sum \frac{g_i^3 x_i^3}{n_i^2} \right\} - 1 \right)$$

Here $z$ is the normal deviate corresponding to the desired risk (e.g., 1.96 for a 95% confidence interval). The corresponding percentage points of the $\chi^2$ distribution with $n_0$ degrees of freedom are obtained from a table. Then the confidence interval for $\sigma^2$ takes the form

$$(2.13) \qquad P\left( \frac{n_0 \sum g_i x_i}{\chi^2_{0.975;n_0} - C} \leqslant \sigma^2 \leqslant \frac{n_0 \sum g_i x_i}{\chi^2_{0.025;n_0} - C} \right) = 0.95$$

The foregoing methods of establishing confidence intervals are closely related to the methods for obtaining unbiased estimates of the variance components. The statistical literature increasingly discusses maximum likelihood estimators of components (Hartley & Rao, 1967). The procedures developed include a method of establishing a confidence region for the set of variance components. For the most elementary one-way analysis, Wang (1967) and Klotz, Milton, and Zacks (1969) demonstrate that the unbiased estimator has a larger sampling error than a maximum likelihood estimator, and suggest still other estimation procedures that might improve results.

There is one more line of theoretical work to be mentioned: the use of *polykays*. Hooke (1956) and Dayhoff (1966) show how one may estimate the second and higher moments of the distribution of estimates of variance components. Thus, in principle one can obtain an interval estimate without relying on normal assumptions. The procedure is laborious and appears to require extensive data.

### A "jackknife" procedure[2]

Whereas all the approaches of the preceding section attempt to derive a formula appropriate to a particular theoretical model, the "jackknife" method makes only the random-sampling assumption. It offers a general procedure that will apply to virtually any investigation of sampling error (Mosteller & Tukey, 1968, p. 133 ff.; Miller, 1968). It requires judgment at various points, and it may give rather crude information where data are not well-behaved.

*A one-facet, mixed-model study.* By way of introduction we discuss the jackknife analysis of a study in which 5 raters rate 10 persons, and raters are regarded as fixed. Interest attaches to the variance of scores $X_{pI}$ in the population of persons. This study is of course not a G study, because no generalization over raters is attempted, but the observed score $X_{pI}$ is a universe score, for the five-rater universe. For the sake of analogy to our second example, $\widehat{\sigma^2}(p)$ will be written for the variance of $X_{pI}$ in the sample. The data are arrayed in a matrix with 10 rows and 5 columns. The procedure is as follows:

1. Label the usual mixed-model estimate of the person component $\widehat{\sigma^2}(p)_{[all]}$, to indicate use of all data in the 10 × 5 matrix. The calculation of this component, and all subsequent calculations, should be carried to more decimal places than would ordinarily be necessary.
2. Eliminate row 1 (person 1) from the data. Now define the symbol [1] to mean "not 1" (i.e., use of all data but those from 1). Treating the 9 × 5 matrix we have just formed by the analysis of variance produces an estimate we may call $\widehat{\sigma^2}(p)_{[1]}$.
3. Repeat for every row in turn.

---

[2] The advice of J. W. Tukey on this section is gratefully acknowledged.

4. For each row, compute a "pseudovalue" by an equation of

$$(2.14) \qquad y_{*1} = 10 \log \widehat{\sigma}^2(p)_{[\text{all}]} - 9 \log \widehat{\sigma}^2(p)_{[1]}$$

{Variance components sometimes have negative estimates, but one cannot take the logarithm of a negative number. Where a negative estimate seems likely with a given set of data, Tukey suggests that in place of the logarithm some other function of $\widehat{\sigma}^2(p)$ be used; one possibility, likely to give reasonable results, is $[\widehat{\sigma}^2(p)]^{1/5}$}.

5. The pseudovalues $y_{*1}, \ldots, y_{*10}$ are a new type of estimate that can usually be treated satisfactorily (as if independently sampled from a distribution of estimates) by the Student $t$ procedure. Compute an $s^2$ for the 10 pseudovalues as if they were a sample of 10 observations. Use the $t$ distribution as a basis for establishing a confidence interval [expressed in terms of $\log \sigma^2(p)$] around the mean pseudovalue. Take antilogs to return to the scale of $\sigma^2(p)$.

This slightly rough-and-ready procedure can be understood best by thinking of a split-half study. We might have divided the original $10 \times 5$ matrix into two $5 \times 5$ segments, computed $\widehat{\sigma}^2(p)$ for each segment, and compared the two values to get a rough indication of the adequacy of the estimate. A correction would be needed to take into account that the estimate in hand is based on 10, not 5, persons. The jackknife procedure is a version of this that tends to minimize bias and to use the full power of the data. The analysis outlined above could be made with $n_p$ as small as 3, though skimpy data can be expected to yield a disappointingly wide confidence interval.

*A one-facet study with random conditions.* Matters become a bit more complicated when sampling of both persons and raters must be considered. The procedure now takes this form:

1. Compute $\widehat{\sigma}^2(p)_{[\text{all}]}$ by the random-model equations.

2. Eliminate row 1 and column 1 and compute $\widehat{\sigma}^2(p)_{[1][1]}$ from the resulting $9 \times 4$ matrix.

3. Repeat for each pair of rows and columns, to get other values of $\widehat{\sigma}^2(p)_{[\text{row}][\text{col}]}$.

4. Repeat eliminating row 1 but not eliminating a column, to obtain $\widehat{\sigma}^2(p)_{[1][-]}$ [which was labelled $\widehat{\sigma}^2(p)_{[1]}$ in the earlier analysis]. Continue, eliminating all rows in turn. Repeat, eliminating columns and not rows, to get five values of $\widehat{\sigma}^2(p)_{[-][\text{col}]}$.

5. Calculate pseudovalues by letting row and col take on the various possible values in

$$(2.15) \quad y_{*\text{row,col}} = 50 \log \widehat{\sigma^2}(p)_{[\text{all}]} - 45 \log \widehat{\sigma^2}(p)_{[\text{row}][-]}$$
$$- 40 \log \widehat{\sigma^2}(p)_{[-][\text{col}]} + 36 \log \widehat{\sigma^2}(p)_{[\text{row}][\text{col}]}$$

6. From the $10 \times 5$ matrix of pseudovalues, estimate components of variance in the usual manner (random model) to get $\widehat{\sigma^2}(y_{\text{rows}})$, $\widehat{\sigma^2}(y_{\text{cols}})$, and $\widehat{\sigma^2}(y_{\text{res}})$. Then the required variance for $\log \widehat{\sigma^2}(p)$ is estimated as if we had a simple two-way sample in which both the columns and rows were sampled randomly and independently by

$$(2.16) \qquad \frac{1}{n_p} \sigma^2(y_{\text{rows}}) + \frac{1}{n_i} \sigma^2(y_{\text{cols}}) + \frac{1}{n_p n_i} \sigma^2(y_{\text{res}})$$

7. From the square root of this variance, the square root of the number of $y$, and the $t$ distribution, establish a confidence interval symmetric around the mean of all the $y$. This interval will be in terms of $\log \sigma^2(p)$. Take antilogs.

A similar procedure applies to other components, and in principle can be extended to more complex designs.

The computational labor involved in a jackknife analysis becomes very great as the total number of observations increases, and it is substantially greater for a two-facet study than a one-facet study. There are various possibilities for reducing the labor without serious loss of information. The most practical device is to form random groups of persons and random groups of conditions. Thus, if a one-facet study had data for 20 persons and 12 conditions, one might randomly assign persons to 5 groups of 4 persons, and randomly assign conditions to 4 groups of 3 conditions. One would form an average score for each block of the data. The resulting $5 \times 4$ matrix of $X_{PI}$ would be treated as the matrix of $X_{pi}$ was treated above. Since the variance component for $P$ is one-fourth as large as the component for $p$, the component for $I$ one-third as large as the component for $i$, etc., simple rescaling of the calculated confidence interval gives the desired intervals for $\sigma^2(p)$, $\sigma^2(i)$, and $\sigma^2(pi,e)$.

An extensive exploration of the application of the jackknife procedure to generalizability studies appears in a recent doctoral dissertation by Collins (1970). He concludes that the method can indeed be used with increasingly complex multifacet designs to establish rough confidence intervals for variance components. This is true provided that score components are normally distributed, that the number of conditions of each facet is appreciable, and that the error components are relatively small. He recommends against use of the technique for the coefficient of generalizability, and against its use with nonnormal data. He recognizes that the basic jackknife procedure can be altered in many ways. Hence, with further research it may be adapted for data that presently generate an interval in disagreement with the accurate interval determined by Monte Carlo techniques.

*Treatment of negative values*

Estimates of variance components and composites of components are the statistics most often reported in generalizability studies. Variances must be non-negative, but estimates obtained by the procedures given above are not infrequently negative because of sampling errors (Leone & Nelson, 1966). Interpretation of negative estimates is a problem, a problem much like the interpretation of *F* ratios that are less than 1.00. A plausible solution is to substitute zero for the negative estimate, and carry this zero forward as the estimate of the component when it enters any equation higher in the table of mean squares. This is the method we shall use. When a zero is substituted for a negative value, the short-cut equations that call for subtracting the corresponding mean square from a mean square higher in the table should not be used. The calculation must use the several $\hat{\sigma}^2$ already computed, including the zero value, to avoid error.

Scheffé (1959) recommends against substituting zero values for negative values on grounds that have to do primarily with formal statistical inference. The sampling distribution of estimates so modified is much more complicated than that for direct estimates. The simple formulas for the sampling variance of estimates under the normal assumption are no longer valid, and the modified estimates are biased.

Nelder (1954) notes that a negative result is a warning that the random-effects model may be invalid. For example, if there is an opportunity for one set of observations to influence another, this may violate the model. Hill (1965, 1970) takes a similar position. Hill (and also Tiao & Tan, 1965) considers the estimation of variance components from a Bayesian point of view. Hill concludes that a large negative unbiased estimate of a between-conditions component indicates that an uninformative experiment has been conducted in which the likelihood function for that variance component is extremely flat. The study is not to be taken as strong evidence, then, that the variance component is near zero. Hill would recommend that the investigator suspend judgment about the component, and consider alternative, weaker models. This work is extended by Novick, Jackson, and Thayer (1971), who find the Bayesian approach particularly valuable when measurement errors are large. The designs to which Bayesian methods have been applied are, to this point, quite restricted. The Bayesian estimates are never negative, and it may ultimately be possible to employ them in place of the more conventional estimates derived from equations such as (2.3).

### E. Components of Variance Where the Universe is Finite

To this point we have assumed that the G study samples $n_i$ conditions of $i$ from an indefinitely large number of admissible conditions of that facet. (Similar statements apply to facets $j$, $k$, etc.) It is possible, however, for the number of conditions of a certain facet to be limited, either in the universe of admissible observations or in the universe of generalization. To discuss the possibilities, we follow Cornfield and Tukey by introducing the symbol $N$. Let $N_i$ be the number of conditions of $i$ in the universe of admissible observations, with $n_i \leqslant N_i \leqslant \infty$. All observations in the G study are admissible; when the conditions of $i$ employed exhaust the admissible conditions, $n_i = N_i$.

Cornfield and Tukey offer equations expressing expected mean squares as a function of components of variance which take $N_i$, etc. into account. Earlier equations like (2.3) are the forms approached by the Cornfield–Tukey equations as $N_i$, $N_j$, etc. all become large. The limiting case where some $N$ are indefinitely large and others are equal to the corresponding $n$ is the set of equations fitting the so-called mixed model for variance components.

We shall not devote attention to the equations for intermediate cases where for some facet $n < N < \infty$. The Cornfield–Tukey equations then multiply certain components by $(1 - n/N)$. If $n = N$ these components vanish; as $N$ becomes large, the multiplier approaches unity. Ordinarily, when $n < N$, $N$ is sufficiently large that a multiplier of 1.00 is accurate enough.

We must go beyond Cornfield and Tukey to take account of the distinction between the universe of admissible observations and the universe of generalization. When $N_i$ is large, the decision maker may propose not to generalize over all conditions of $i$. Let $N_i'$ represent the number of conditions of $i$ defining the universe of generalization; $n' \leqslant N_i' \leqslant N_i$. The constraint $N_i' \leqslant N_i$ is required for the G study to be applicable, and $n_i' \leqslant N_i'$ is required to make the D study sensible. To simplify matters we shall not consider intermediate values of $N'$ nor the case $n_i' < N_i' = N_i < \infty$.

With respect to any facet, these possibilities are to be considered:

1. $N \to \infty$, $N' \to \infty$; $n$, $n'$ are unrestricted.

2. $N \to \infty$, $n' = N'$ takes on any value from 1 upward.

3. $N$ takes on any value greater than 1, and $n' = n = N' = N$.

The same conditions of the facet enter the G and D study. (If $n = 1$, the G study has a "hidden facet," which creates problems of interpretation that

Chapter 5 will discuss.) Case 1 is what we have discussed in previous sections. Case 3 is like the conventional mixed model. Case 2 arises only in generalizability theory. Throughout the following technical section, discussion will be limited to alternative interpretations regarding the universe of $j$, keeping $N_i$ large but letting $N_j$ and $N'_j$ vary. The argument can readily be applied to alternative interpretations of facet $i$, and to studies with more than two facets.

### Modified notation for components

The usual treatment of variance components writes $\sigma^2(p)$ for "the person component" regardless of the value of $N_j$ assumed. This, however, is in general equal to $\mathscr{E}(\mu_{pJ} - \mu_J)^2$, where $J$ is the set of $N_j$ conditions that defines the universe. The tautology

$$(2.17) \qquad \mu_{pJ} - \mu_J = (\mu_p - \mu) + (\mu_{pJ} - \mu_p - \mu_J + \mu)$$

makes it evident that the $\sigma^2(p)$ of the random model is the limiting case as $N_j$ becomes large, and that for each $N_j$, $\sigma^2(p)$ takes on a different definition. This can generate considerable confusion for us and we therefore develop a special notation for the finite universe where the fixed facet is crossed with all other facets.

For case 3, there is a fixed set of $N_j$ conditions of $j$ that appears in the G and D study. Here[3] we shall call this $J^*$. The universe score over all admissible observations is the expected value over $i$ and $j$ of $X_{pij}$, but since we are limited to $j \in J^*$, the universe score is $\mu_{pJ^*}$. In place of (1.3), where $N_j$ was taken to be indefinitely large, we write a new expression for the decomposition of the score. We make the preliminary tautological statement

$$(2.18) \qquad X_{pij} = X_{piJ^*} + (X_{pij} - X_{piJ^*})$$

and then decompose these two parts separately. Note that $\sum_j (X_{pij} - X_{piJ^*})$ equals zero. Decomposing the first term we have

$$(2.19) \qquad \begin{aligned} X_{piJ^*} &= \mu_{J^*} + \mu_{pJ^*} - \mu_{J^*} \\ &\quad + \mu_{iJ^*} - \mu_{J^*} \\ &\quad + X_{piJ^*} - \mu_{pJ^*} - \mu_{iJ^*} + \mu_{J^*} \end{aligned} \qquad \begin{aligned} \sigma^2(X_{piJ^*}) &= \sigma^2(p \,|\, J^*) \\ &\quad + \sigma^2(i \,|\, J^*) \\ &\quad + \sigma^2(pi, \bar{e} \,|\, J^*) \end{aligned}$$

The conditional notation for the variance is a way of emphasizing that the variance applies to scores obtained under the set of conditions $J^*$. The variance in the right-hand column is the mean square over all $p$ and/or all $i$ of the component at left. We write $\bar{e}$ for $e_{piJ^*}$ to distinguish it from $e_{pij}$.

Components of the second term in (2.18) contain "within $J^*$" information,

---

[3] A more complicated convention is used in Chapter 4. (See p. 114.)

that is, effects associated with the separate $j$ in the set $J^*$. The breakdown is as follows:

$$
\begin{aligned}
(2.20) \quad X_{pij} - X_{piJ*} &= \mu_j - \mu_{J*} & \sigma^2(X_{pij} - X_{piJ*}) &= \sigma^2(j \,|\, J^*) \\
&+ \mu_{pj} - \mu_{pJ*} - \mu_j + \mu_{J*} & &+ \sigma^2(pj \,|\, J^*) \\
&+ \mu_{ij} - \mu_{iJ*} - \mu_j + \mu_{J*} & &+ \sigma^2(ij \,|\, J^*) \\
&+ X_{pij} - \mu_{pj} - \mu_{ij} - \mu_{piJ*} & &+ \sigma^2(pij,e \,|\, J^*) \\
&+ \mu_{pJ*} + \mu_{iJ*} + \mu_j - \mu_{J*}
\end{aligned}
$$

The score components of (2.19) and (2.20) combined make up the whole of $X_{pij}$ and the variance components make up $\sigma^2(X_{pij})$.

## Analysis of crossed designs

The G study with Design VII ($i \times j \times p$) is analyzed as before to obtain mean squares and, if desired, the mixed-model equations for expected mean squares may be employed. These fall into two groups:

$$
\begin{aligned}
&\text{EMS } p = n_j\sigma^2(pi,\bar{e} \,|\, J^*) + n_i n_j \sigma^2(p \,|\, J^*) \\
(2.21) \quad &\text{EMS } i = n_j\sigma^2(pi,\bar{e} \,|\, J^*) + n_p n_j \sigma^2(i \,|\, J^*) \\
&\text{EMS } pi,e = n_j\sigma^2(pi,\bar{e} \,|\, J^*)
\end{aligned}
$$

For within-$J^*$ components there are the further equations:

$$
\begin{aligned}
&\text{EMS } j = \sigma^2(pij,e \,|\, J^*) + n_p\sigma^2(ij \,|\, J^*) \\
&\qquad\qquad + n_i\sigma^2(pj \,|\, J^*) + n_p n_i \sigma^2(j \,|\, J^*) & [j \in J^*] \\
(2.22) \quad &\text{EMS } pj = \sigma^2(pij,e \,|\, J^*) + n_i\sigma^2(pj \,|\, J^*) & [j \in J^*] \\
&\text{EMS } ij = \sigma^2(pij,e \,|\, J^*) + n_p\sigma^2(ij \,|\, J^*) & [j \in J^*] \\
&\text{EMS res} = \sigma^2(pij,e \,|\, J^*) & [j \in J^*]
\end{aligned}
$$

Considering (2.21) and (2.22) together, and comparing them with (2.3), we see as differences the confounding of $\bar{e}$ with $pi$, the absence of $\sigma^2(pij,e)$ from the first three expected mean squares, the absence of $\sigma^2(pj)$ from EMS $p$, and the absence of $\sigma^2(ij)$ from EMS $i$. All these changes are a consequence of the change in definitions of the components to be estimated.

So long as generalization in the D study is over facet $i$ from scores $X_{pIJ*}$ obtained using the fixed conditions $J^*$, there is no need for the within-$J^*$ components, and hence no need to solve equation (2.22). A two-way analysis of variance of the $X_{piJ*}$ yields the necessary mean squares. These, substituted for the expected mean squares in (2.4), give $\widehat{\sigma^2}(p \,|\, J^*)$, $\widehat{\sigma^2}(i \,|\, J^*)$ and $\widehat{\sigma^2}(pi,\bar{e} \,|\, J^*)$. The expected mean squares of (2.21) are $n_j$ times those of (2.4).

*Analysis of nested two-facet designs*

If $J^*$ defines the universe of $j$ but there are indefinitely many $p$ and $i$, it is not possible to have $j$ nested within $p$ or $i$. The design for the G study may have $p$ crossed with $J^*$, and conditions of $i$ nested in $p$, $j$, or $pj$. These are Designs V-B, V-A, and IV-A, respectively.

Considering first the V-B design, $j \times (i:p)$, one obtains equations (2.23) for the expected mean squares from a three-way analysis of variance.

$$\text{EMS } p = n_j\sigma^2(i,pi,\bar{e} \mid J^*) + n_i n_j \sigma^2(p \mid J^*)$$

$$\text{EMS } i, pi, \bar{e} = n_j\sigma^2(i,pi,\bar{e} \mid J^*)$$

(2.23) $\quad\text{EMS } j = \sigma^2(ij,pij,e \mid J^*) + n_i\sigma^2(pj \mid J^*) + n_i n_p \sigma^2(j \mid J^*)$

$$\text{EMS } pj = \sigma^2(ij,pij,e \mid J^*) + n_i\sigma^2(pj \mid J^*)$$

$$\text{EMS res} = \sigma^2(ij,pij,e \mid J^*)$$

Here, the expected mean for $p$ and for $i$ within $p$ contain no components involving $j$, $pj$, $ij$, or $pij$. Components for $j$, $pj$, and $ij$, $pij,e$ are estimated by the last three equations. However, as we noted for Design VII, it is unnecessary to estimate these components when $J^*$ is to be used throughout. A one-way analysis of variance of $X_{piJ^*}$ provides MS $p$ and MS $i$ within $p$. These have to be multiplied by $n_j$ to estimate the EMS $p$ and EMS $i$ within $p$ of (2.23).

In Designs V-A and IV-A, since $pi$ and $pij$ are confounded, the variance components are best obtained from an appropriately nested three-way analysis. In both of these cases $\widehat{\sigma^2}(p \mid J^*)$ is given by

(2.24) $$\frac{\text{MS } p - \text{MS res}}{n_i n_j}$$

Estimates of all other variance components with these designs are numerically identical to those obtained using the random model.

*Relationship between variance components obtained*
*under fixed and random assumptions*

It is always possible to obtain estimates of variance components for a universe with a fixed facet from estimates computed under the random model. Consider Design VII for the moment. Because the values of mean squares calculated from a set of data are the same no matter what is assumed about the universe, the same set of values would be entered on the left-hand sides of (2.21) and (2.22) as are entered in (2.3). Equating the right-hand terms from the two sets of equations we obtain

$$\widehat{\sigma}^2(pij,e \mid J^*) = \widehat{\sigma}^2(pij,e)$$

$$\widehat{\sigma}^2(ij \mid J^*) = \widehat{\sigma}^2(ij)$$

$$\widehat{\sigma}^2(pj \mid J^*) = \widehat{\sigma}^2(pj)$$

(2.25)
$$\widehat{\sigma}^2(j \mid J^*) = \widehat{\sigma}^2(j)$$

$$\widehat{\sigma}^2(pi,\bar{e} \mid J^*) = \widehat{\sigma}^2(pi) + \frac{1}{n_j} \widehat{\sigma}^2(pij,e)$$

$$\widehat{\sigma}^2(i \mid J^*) = \widehat{\sigma}^2(i) + \frac{1}{n_j} \widehat{\sigma}^2(ij)$$

$$\widehat{\sigma}^2(p \mid J^*) = \widehat{\sigma}^2(p) + \frac{1}{n_j} \widehat{\sigma}^2(pj)$$

Estimates of components of variance defined by the random model can be combined in this way to determine estimates for components defined in terms of a fixed facet.

These same equations can be used to arrive at the relationship between the two sets of components in any permissible confounded design. In Design V-B, for example, $i$ and $pi$ are confounded. Amalgamating the equations for $i$ and $pi,\bar{e}$ above, one obtains:

(2.26)
$$\widehat{\sigma}^2(i,pi,\bar{e} \mid J^*) = \widehat{\sigma}^2(i,pi) + \frac{1}{n_j} \widehat{\sigma}^2(ij,pij,e)$$

Similarly,

(2.27)
$$\widehat{\sigma}^2(ij,pij,e \mid J^*) = \widehat{\sigma}^2(ij,pij,e)$$

In V-A we have $i$ and $ij$ confounded, and

(2.28)
$$\widehat{\sigma}^2(pi,pij,e \mid J^*) = \widehat{\sigma}^2(pi,pij,e)$$
$$\widehat{\sigma}^2(i,ij \mid J^*) = \widehat{\sigma}^2(i,ij)$$

To give some sense of the effect of the shift in universe definition, we reanalyze the data of Table 2.2 under the assumption of fixed scorers, admitting that this is not a particularly likely interpretation.

In Table 2.2, MS $pi$ is 9.28, $n_j = 3$. Then from (2.21) $\widehat{\sigma^2}(pi,\bar{e} \mid J^*) = 3.09$. Continuing, $\widehat{\sigma^2}(i \mid J^*) = 1.34$ and $\widehat{\sigma^2}(p \mid J^*) = 5.78$. These may be compared with $\widehat{\sigma^2}(pi) = 2.57$, $\widehat{\sigma^2}(i) = 1.32$, and $\widehat{\sigma^2}(p) = 5.71$. The estimates with $J^*$ fixed are all larger than the same estimates with $N_j \to \infty$. This is generally the case, as can be inferred from (2.25).

### F. *The Universe with a Nested Facet*

All the discussion to this point has assumed that the universe of admissible observations follows the $i \times j \times p$ pattern, so that for every $i$ and $j$ a score $X_{pij}$ is defined, and each such score is as likely to be drawn as any other. We shall continue with that model in almost all the rest of our theoretical discussion, but it is necessary to mention other possibilities.

A universe may have the pattern $(j:i) \times p$. That is, in the universe each $j$ goes with one particular $i$ and no other. The commonest example is items nested within tasks (subtests). If $i$ is the digit-span task, there is a universe of items $j$ from which one can draw. The supply of admissible items—02356, 61728, 74731, . . .— is not "indefinitely large," but it is large enough that we need not qualify our statements so as to keep the size limit in mind. One would make a digit-span test by drawing $n_j$ items; for every such item there is an admissible observation $X_{pij_i}$. We have written $j_i$ to indicate that $j$ is drawn from the universe of conditions of $j$ that is dictated by the particular $i$. An item such as 92356 belongs with the digit-span task and no other; it could not possibly be used in connection with a figure-analogies or vocabulary subtest. (To be sure, the series of digits could be used in a paired-associates task, but we need not blur the issue with far-fetched cases.)

Where the universe has the structure $(j:i) \times p$, an observation $X_{pi'j_i}$ is inadmissible: an $i$ item cannot be used with the $i'$ task. This affects our interpretation of components. The grand mean $\mu$ is still an expectation over all $i$ and $j$, but it logically has to be seen as $\underset{p,i}{\mathscr{E}} \, (\underset{j_i}{\mathscr{E}} X_{pij_i})$. The person component is $\underset{i \;\; j_i}{\mathscr{E}\mathscr{E}}(X_{pij_i} - \mu)$; we may continue to speak of $\mu_p$ and $\sigma^2(p)$. The subtest component is analogous, and is again $\mu_i$, with variance $\sigma^2(i)$. Nothing like the $j$ component now exists, because one cannot define $\underset{i}{\mathscr{E}} X_{pij_i}$. There is a component for $j$ within $i$, defined as $\underset{p}{\mathscr{E}} X_{pij_i} - \mu_i$.

Obviously, a universe of the type $(j:i) \times p$ cannot be investigated by a G study in which $i \times j$ or $i:j$. Despite its novel features, this kind of universe presents no great difficulties in analysis or interpretation, as examples will later show (see pp. 203 to 225).

The universe pattern $i \times (j:p)$ or any other "nested in $p$" structure is harder to discuss and is rarely examined. There are conditions relevant to observations on one person and not another. To echo the language of Egon Brunswik, conditions are ecologically tied to the person. When Roger Barker observes a boy as he passes through the behavior settings that make up his environment, it makes little sense to regard the data as a sample of what the boy would do in all possible environments. Sam comes down to breakfast with his widowed mother. He is an only child. His actions cannot be generalized to a universe that includes "breakfast with father and mother and three siblings," which is the situation Herb is exposed to next door. Herb's situation is not and could not be a part of the universe of admissible observations for Sam (unless we wander into science fiction). The universe of admissible observations for Sam differs from that for Herb.

We shall not elaborate on this kind of universe, which is not encountered in later examples. It is an appropriate problem for future work within generalizability theory, since naturalistic observation often does sample from and generalize to an ecological universe. The models developed in this book make sense for such problems only if one assumes a strong null hypothesis. Thus, while Sam and Herb have different acquaintances, one might generalize peer ratings over a universe of all possible peers. But this assumes that the acquaintances Sam makes are randomly selected, without reference to Sam's personality. The factors that in reality cause Sam to make different acquaintances than Herb does are entangled with the ratings Sam receives. Brunswik's protest against designs and models that deny this entanglement was well taken, and points to a limitation in generalizability (and reliability) theory.

## EXERCISES

**E.1.**  In the manner of p. 40, write out the components of score $X_{2Ab}$.

**E.2.**  Prepare a table like Table 2.1 (p. 40) for *one*-facet studies.

**E.3.**  Prepare a diagram for a design $(i \times j):p$ in the manner of Figure 2.2 (p. 36). Give a concrete illustration of a possible study of this sort. (Hint: Consider an oral-examination procedure.)

**E.4.**  Write algebraic expressions for the following parameters of scores $X_{pIJ}$ in terms of $i$ and $j$ rather than $I$ and $J$.

a. $\mu_{IJ} - \mu_I - \mu_J + \mu$

b. $\sigma^2(pIJ)$

c. $\sigma^2(\mu_I - \mu)$

**E.5.** A three-facet study has the design $(k:j:i) \times p$. Locate the 15 definable components of variance within the 7 areas of the Venn diagram below.

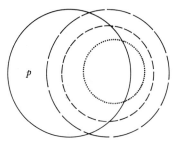

**E.6.** Consider that 60 persons took a 40-item test where each item was scored 1 or 0. The analysis of item scores yielded the following mean squares: for persons, 0.640; for items, 1.920; for residual, 0.120. Calculate estimates of the variance components. What do these estimates describe?

**E.7.** A study in which persons, tasks, and observers were crossed generated the following analysis of variance:

|  | Sum of squares | Degrees of freedom | Mean square |
|---|---|---|---|
| $p$ | 369.593 | 52 | 7.108 |
| Tasks $i$ | 35.970 | 4 | 8.992 |
| Observers $j$ | 18.920 | 1 | 18.920 |
| $pi$ | 412.183 | 208 | 1.982 |
| $pj$ | 69.302 | 52 | 1.333 |
| $ij$ | 11.530 | 4 | 2.882 |
| Residual | 153.180 | 208 | 0.737 |

a. What are the values of $n_p$, $n_i$, and $n_j$?
b. Estimate the variance components assuming $N_i$, $N_j$ very large.

**E.8.** Prepare a figure similar to Figure 2.5 for the $j:(i \times p)$ design whose expected mean square equations are given on p. 49.

**E.9.** Write equations for estimating the variance component for persons and the within-persons component, when a one-way analysis of variance of data has been made of an $i:p$ design.

**E.10.** Develop equations to estimate variance components for Design IV-B.

**E.11.** Treat the information presented in Exercise 7 as if it had come from a study with design $(i:j) \times p$. Estimate components of variance assuming $N_i$, $N_j$ very large.

*TABLE 2.E.1. Ratings of Subjects in an i × j × p Design (after Guilford, 1954, p 282)*

| Subject | Traits by rater | | | | | | | | | | | | | | |
|---|---|---|---|---|---|---|---|---|---|---|---|---|---|---|---|
| | *A* | | | *B* | | | *C* | | | *D* | | | *E* | | |
| | *a* | *b* | *c* | *a* | *b* | *c* | *a* | *b* | *c* | *a* | *b* | *c* | *a* | *b* | *c* |
| 1 | 5 | 6 | 5 | 5 | 5 | 5 | 3 | 4 | 5 | 5 | 6 | 7 | 3 | 3 | 3 |
| 2 | 9 | 8 | 7 | 7 | 7 | 7 | 5 | 5 | 5 | 8 | 7 | 7 | 5 | 2 | 5 |
| 3 | 3 | 4 | 3 | 3 | 5 | 5 | 3 | 3 | 5 | 7 | 6 | 5 | 1 | 6 | 5 |
| 4 | 7 | 5 | 5 | 3 | 6 | 3 | 1 | 4 | 3 | 3 | 5 | 3 | 3 | 5 | 1 |
| 5 | 9 | 2 | 9 | 7 | 4 | 7 | 7 | 3 | 7 | 8 | 2 | 7 | 5 | 3 | 7 |
| 6 | 3 | 4 | 3 | 5 | 4 | 3 | 3 | 6 | 3 | 5 | 4 | 5 | 1 | 2 | 3 |
| 7 | 7 | 3 | 7 | 7 | 3 | 7 | 5 | 5 | 7 | 5 | 5 | 5 | 5 | 4 | 7 |

**E.12.** Guilford reports the data in Table 2.E.1. Seven scientists $p$ in a research organization were rated on five traits $j$ having to do with creative performance. The three raters $i$ were senior scientists in the same group.

Carry out a three-way analysis of variance,[4] estimate the variance components assuming $N$ large, and discuss what their relative size indicates. (It will be illuminating to compare your interpretation with Guilford's presentation based on several two-way analyses.)

**E.13.** Estimate the variance components for the Guilford data under the assumption that the set of traits $J^*$ is fixed and $N_j$ is 5. (Make use of the answers to Exercise 12.)

**E.14.** Determine variance components for the data below, from an $i \times j \times p$ study.

| | Sum of squares | Degrees of freedom | Mean square |
|---|---|---|---|
| *p* | 795.52 | 9 | 87.28 |
| Observers *i* | 200.04 | 2 | 100.02 |
| Occasions *j* | 108.60 | 1 | 108.60 |
| *pi* | 109.34 | 18 | 6.13 |
| *pj* | 406.98 | 9 | 45.22 |
| *ij* | 39.24 | 2 | 19.62 |
| Residual | 111.96 | 18 | 6.22 |

**E.15.** For trait $A$ in Table 2.E.1, apply the jackknife procedure to establish a confidence interval for $\sigma^2(pi,e)$. (Since the computation required is extensive, it is suggested that the reader set up each stage of the problem, check against our formulation, and then use our numerical answer for that stage to formulate the next stage. Only persons with computer facilities should attempt to carry out the many calculations.)

If a computer is not available, take the sum of squares from the answer page.

## Answers

**A.1.** $X_{2Ab} = \mu + (\mu_2 - \mu) + (\mu_A - \mu) + (\mu_b - \mu) + (\mu_{2A} - \mu_2 - \mu_A + \mu)$

$\qquad + (\mu_{2b} - \mu_2 - \mu_b + \mu) + (\mu_{Ab} - \mu_A - \mu_b + \mu)$

$\qquad + (X_{2Ab} - \mu_{2A} - \mu_{2b} - \mu_{Ab} + \mu_2 + \mu_A + \mu_b - \mu)$

**A.2.**

| Design | | | | Compounds |
|--------|----|----|------|-----------|

| Design | Number of observations on | | | Compounds |
|--------|---|---|------|-----------|
|        | $p$ | $i$ | $pi,e$ | |
| $p \times i$ | $n_p$ | $n_i$ | $n_p n_i$ | None |
| $i{:}p$ | $n_p$ | $n_p n_i$ | $n_p n_i$ | 1. within $p$ |

**A.3.** See Figure 2.E.1. Each person has his own examiners and his own set of questions.

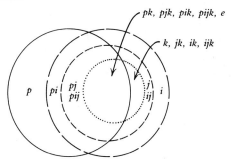

Persons Conditions Conditions $j$

$p$ $\quad$ $i$ $\quad$ $a$ $b$ $c$ $d$

1 $\quad\left\{\begin{array}{l} A \\ B \\ C \end{array}\right.$

2 $\quad\left\{\begin{array}{l} D \\ E \\ F \end{array}\right.$

$(i \times j){:}p$

**FIGURE 2.E.1.** *Answer to Question E.2.3.*

**A.4.** a. $\dfrac{1}{n_i n_j} \Sigma \Sigma \mu_{ij} - \dfrac{1}{n_i} \Sigma \mu_i - \dfrac{1}{n_j} \Sigma \mu_j + \mu$

$\quad$ b. $\dfrac{1}{n_i n_j} \sigma^2(pij)$

$\quad$ c. $\dfrac{1}{n_i} \sigma^2(\mu_i - \mu)$

**A.5.** See Figure 2.E.2.

$pk,\ pjk,\ pik,\ pijk,\ e$

$k,\ jk,\ ik,\ ijk$

$p \quad pi \quad \begin{array}{c} pj \\ pij \end{array} \quad \begin{array}{c} j \\ ij \end{array} \quad i$

**FIGURE 2.E.2.** *Answer to Question E.2.5.*

**A.6.**  $\widehat{\sigma^2}(p) = 0.013$; spread of universe scores. As $\sigma(p)$ is about 0.12, universe scores range over much of the 0-to-1 scale.

$\widehat{\sigma^2}(i) = 0.030$; spread of item means for the population. As $\sigma(i)$ is about 0.18, the items range from extremely difficult to extremely easy. (Maximum $\sigma$, 0.50, occurs if half the items have $\mu_i = 0$ and half have $\mu_i = 1.00$.)

$\widehat{\sigma^2}(pi,e) = 0.120$. This is the combined effect of the person's specific difficulties with particular items and of "chance" variation in his performance. In any single item the effect is fairly large.

[Maximum possible variance from all three components is 0.25. $\sigma^2(X_{pij})$ is 0.163.]

**A.7.**  $n_p = 53$, $n_i = 5$, $n_j = 2$

| Effect | $p$ | $i$ | $j$ | $pi$ | $pj$ | $ij$ | $pij,e$ |
|---|---|---|---|---|---|---|---|
| Estimated $\sigma^2$ | 0.45 | 0.05 | 0.06 | 0.62 | 0.12 | 0.04 | 0.74 |

**A.8.** See Figure 2.E.3.

**A.9.**  $\widehat{\sigma^2}(i,pi,e) = $ MS within persons

$\widehat{\sigma^2}(p) = [(\text{MS } p) - \text{MS within persons}]/n_i$

**A.10.**  EMS $p$ $\quad = \sigma^2(ij,pij,e) + n_i\sigma^2(j,pj) + n_j\sigma^2(i,pi) + n_in_j\sigma^2(p)$
EMS $i, pi$ $\; = \sigma^2(ij,pij,e) \qquad\qquad\qquad\quad + n_j\sigma^2(i,pi)$
EMS $j, pj$ $\; = \sigma^2(ij,pij,e) + n_i\sigma^2(j,pj)$
EMS $ij, pij,e = \sigma^2(ij,pij,e)$

**A.11.**

| | Sum of squares | Degrees of freedom | Mean square | Estimated $\sigma^2$ |
|---|---|---|---|---|
| $p$ | 369.593 | 52 | 7.108 | 0.58 |
| $j$ | 18.920 | 1 | 18.920 | 0.05 |
| $pj$ | 69.302 | 52 | 1.333 | (0) |
| $i, ij$ | 47.500 | 8 | 5.937 | 0.09 |
| $pi, pij,e$ | 565.363 | 416 | 1.359 | 1.36 |

These results are a correct application of the algorithm for the nested design. It will be noted that these results differ substantially from those in Exercise 7, particularly in the vanishing of the $pj$ component.

An investigator sometimes analyzes a study according to a design other than that actually employed in sampling conditions; for example, a design that was actually crossed may be analyzed as if nested, or vice versa. Any such mismatch between the design and the analysis runs grave risk of error. In the present instance, the investigator should be able to verify whether $n_i$ or $n_in_j$ different tasks were used in the G-study observations on any person. The former implies $i \times j$ and the latter implies $i:j$.

**A.12.**

| Effect | $p$ | $i$ | $j$ | $pi$ | $pj$ | $ij$ | $pij,e$ |
|---|---|---|---|---|---|---|---|
| Sum of squares | 94.914 | 9.048 | 46.533 | 98.686 | 51.467 | 12.953 | 56.644 |
| Estimated $\sigma^2$ | 0.94 | 0.07 | 0.09 | 0.01 | 0.35 | 0.29 | 2.77 |

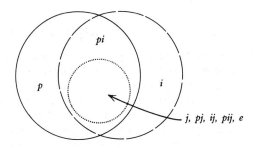

(a)   Components in the design

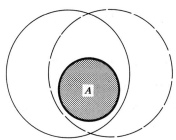

EMS $j,pj,ij,pij,e = A$

(b)   The $p,i,j$ intersection

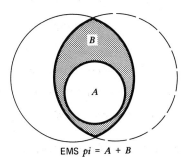

EMS $pi = A + B$

(c)   The $p,i$ intersection

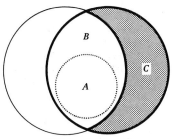

EMS $i - A + B + C$

(d)   The $i$ circle

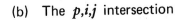

EMS $p = A + B + D$

(e)   The $p$ circle

*FIGURE 2.E.3.   Answer to Question E.2.8.—the $j:(i \times p)$ design.*

**A.13.** Effect $p\,|\,J^*$ $i\,|\,J^*$ $pi,\bar{e}\,|\,J^*$ $pj\!:\!J^*$ $ij\!:\!J^*$ $pij,e\!:\!J^*$
Estimated $\sigma^2$ 1.01 0.13 0.56 0.35 0.29 2.77

**A.14.** Effect $p$ $i$ $j$ $pi$ $pj$ $ij$ $pij,e$
Estimated $\sigma^2$ 7.01 4.02 1.66 $0^5$ 13.00 1.34 6.22

**A.15.** First carry out the anova for the full 7 × 3 data matrix. Then carry out the same analysis for each of the 3 possible 7 × 2 matrices, eliminating one column at a time. Then, carry out the analysis for the 7 possible 6 × 3 matrices, eliminating one row at a time. Finally, carry out the same analysis for each of the 21 possible 6 × 2 matrices. As the exercise calls for information on the residual component only, we are able to tabulate the results of all the foregoing analyses in Table 2.E.2. As is usual in the jackknife formulation, the number in parentheses identifies the row or column deleted from the data. The symbol [ − ] indicates that there was no deletion.

The second stage of the analysis is to form a 7 × 3 matrix of pseudovalues. The formula (2.15) is specialized to:

$$y_{*r.c} = 21 \ln \widehat{\sigma^2}_{(pi,e)[\text{all}]} - 18 \ln \widehat{\sigma^2}_{(pj,e)[\text{row},-]} - 14 \ln \widehat{\sigma^2}_{(pj,e)[,-\text{col}]}$$
$$+ 12 \ln \widehat{\sigma^2}_{(pj,e)[\text{row},\text{col}]}$$

This gives the entry for the residual for row and column. For cell 1,1, one has

$$21 \ln 3.42856 - 18 \ln 3.68888 - 14 \ln 4.64284 + 12 \ln 4.79999$$

or approximately −0.292. Logarithms to any base may be used; natural logarithms

**TABLE 2.E.2.** *Residual Component of Variance in Successive Analyses of Data for Trait A of Table 2.E.1 with Rows and/or Columns Eliminated*

| Row eliminated | Column eliminated | | | |
|---|---|---|---|---|
| | 1 | 2 | 3 | — |
| 1 | 5.73333 | 0.53333 | 4.79999 | 3.68888 |
| 2 | 5.73333 | 0.33333 | 5.53333[a] | 3.86667 |
| 3 | 5.73333 | 0.53333 | 4.79999 | 3.68888 |
| 4 | 6.08332 | 0.33333 | 5.55000 | 3.98888[b] |
| 5 | 2.00000 | 0.53333 | 2.13333 | 1.55555 |
| 6 | 5.73333 | 0.53333 | 4.79999 | 3.68888 |
| 7 | 5.14999 | 0.53333 | 4.88332 | 3.52222 |
| — | 5.16666[c] | 0.47618 | 4.64284 | 3.42856[d] |

[a] Value from anova with person 2 and judge 3 disregarded.
[b] Value from anova with person 4 disregarded.
[c] Value from anova with judge 1 disregarded.
[d] Value from conventional anova, no data disregarded.

[5] Calculated value is −0.045.

were available to us through a convenient computer program. The entire array of pseudovalues appears in Table 2.E.3.

**TABLE 2.E.3.** *Pseudovalues Derived from Tables 2.E.1 and 2.E.2*

| Subject | Rater | | |
|---|---|---|---|
| | 1 | 2 | 3 |
| 1 | 0.34353 | 5.22319 | −0.29206 |
| 2 | −0.50375 | −1.26418 | 0.56677 |
| 3 | 0.34353 | 5.22319 | −0.29206 |
| 4 | −0.35280 | −1.82428 | 0.04276 |
| 5 | 3.24863 | 20.76558 | 5.51967 |
| 6 | 0.34353 | 5.22319 | −0.29206 |
| 7 | −0.11192 | 6.05536 | 0.74664 |

Next, an analysis of variance of the pseudovalues is carried out, with these results:

| Source | Degrees of freedom | Mean square | Component of variance |
|---|---|---|---|
| Row | 6 | 37.083 | 8.3 |
| Column | 2 | 57.756 | 6.5 |
| Residual | 12 | 12.213 | 12.2 |

The equation for the sampling error variance of the residual component, or rather of its logarithm, is similar to (2.16):

$$s^2 = \frac{1}{n_p} \widehat{\sigma^2}(y_{\text{rows}}) + \frac{1}{n_j} \widehat{\sigma^2}(y_{\text{cols}}) + \frac{1}{n_p n_j} \sigma^2(y_{\text{res}})$$

$$= \tfrac{1}{7}(8.29004) + \tfrac{1}{3}(6.50619) + \tfrac{1}{21}(12.21287) = 3.93458$$

$$s = 1.98$$

$$s/n^{1/2} = 1.98/21^{1/2} = 0.43$$

Because the mean of the entries in Table 2.E.3 is 2.31949, the confidence interval is established symmetric about that number. But it will be noted that the distribution of pseudovalues is highly skewed, and this presages difficulty.

For 20 degrees of freedom, the 95% confidence interval is ±2.09 times the standard error of the mean. We arrive at the interval

$$2.319 - 0.899 \leqslant \ln \sigma^2(\text{res}) \leqslant 2.319 + 0.899$$

$$1.420 \leqslant \ln \sigma^2(\text{res}) \leqslant 3.219$$

Hence,

$$4.1 \leqslant \sigma^2(pj,e) \leqslant 25.0$$

This ends the jackknife analysis. It remains to note that the actual value computed from the original matrix is 3.43, which falls outside the confidence interval! All that can be said is that things like this happen. In the table of raw data it can be seen that judge *b* is quite idiosyncratic, notably in his rating of person 5. This necessarily implies that values of components obtained from a G study with three judges will be undependable, and the extremely wide confidence interval calculated is understandable. The next sample might consist entirely of judges like 1 and 3, or of judges like 2; or it might consist of two judges like 2 who produce eccentric values and large residuals, plus one judge like 1 or 3. While there are possible ways of exploring further when so anomalous a result is reached by the jackknife procedure, that is scarcely in point with this example.

# CHAPTER 3
# Inferences from D-Study Data Regarding the Universe Score $\mu_p$

The preceding chapter presented procedures for deriving estimates of components of variance from a G study. We now examine inferences based on these estimates.

One or more experimental designs are under consideration for the D study; we can forecast how well the observed scores obtained under each design will agree with the universe scores of interest. The D study will generate an observed score for each person $p$. This is only one of many scores that could be obtained by applying the same design repeatedly, each time sampling afresh from the universe of conditions of observation. These observed scores depart from the universe score.

It is assumed throughout this chapter that the investigator wishes to generalize over all facets represented in the G study, and that the number of admissible conditions for each facet is large. Discussion of restricted universes of generalization and the associated problem of "hidden" facets is reserved for Chapter 4. Also for the sake of directness, we discuss procedures here without attention to underlying assumptions. The stringent assumptions underlying both the confidence interval technique and the regression technique will be given thorough consideration in Chapter 5. In these applications the classical theory and the established practices in test-score analysis embody much the same assumptions as the theory of generalizability does. Fifty years of experience has shown that much can be accomplished with analytic procedures that employ strong assumptions. Useful though we expect the procedures developed in this chapter to be, it is important to plant a doubt. Chapters 5 and 9 will suggest that some well-known techniques such as the confidence interval for a universe score rest on shaky foundations. The hazard in using these techniques is as great in generalizability theory as in classical theory.

## A. One-Facet D Studies

### The nested design

To begin with a comprehensive example, we discuss the one-facet $i:p$ design. The discussion will be superficial, because to discuss rationale and alternatives would defeat the purpose of this section as an overview. The argument on each point will be developed much more fully in later parts of this chapter and in Chapter 5.

Suppose that a G study has estimated two components of variance: for persons $\sigma^2(p)$ and within persons $\sigma^2(i,pi,e)$. The D study consists of $n_i'$ observations on $n_p'$ persons, different conditions being drawn for each person. It is assumed that the same population and universe are represented as in the G study, but there is no necessity that the number of observations per person be the same.

In accord with statements on p. 28, we define the observed score $X_{pI}$ as the mean over the $n_i'$ observations. (Formulas arising under this definition would be readily modified if the observed score were defined instead as a sum over the $n_i'$ observations.) The expected value of $X_{pI}$ over the universe of conditions is the score $\mu_p$ to which we would like to generalize.

Table 3.1 presents the concepts that conventional reliability analysis would apply to these data and, in the second column, the corresponding concepts from generalizability theory. This design presents the simplest possible case. Conditions are randomly sampled, separately for each person. It follows that if the measuring procedure is carried out twice on the same indefinitely large population of subjects, new conditions being sampled each time: the two population means will be the same, the covariance of the two observed scores will equal $\sigma^2(p)$, and the observed-score variance for the two measurements will be the same in the limit, as more persons and hence more conditions are considered. Thus, even if conditions $i$ and $i'$ lack equivalence in the classical sense, scores arising from the nested design conform to classical assumptions.

Figure 3.1 is a representation, in terms of variance components, of quantities to be discussed in the next several paragraphs. While this figure is very simple, more elaborate diagrams of this kind will be quite helpful with complicated problems. Venn diagram (a) represents all the information in the model. The $p$ circle also represents the score variance in an $i:p$ design with $n_i' = 1$. If $n_i' > 1$, $I$ may be substituted for $i$ in all labels. Where $\mu_p$ is

**TABLE 3.1.** *Comparison of Parameters Considered in Conventional Reliability Theory and Generalizability Theory* ($i:p$ *Design*)

| Term employed in reliability theory and usual symbol | Corresponding concept and symbol in generalizability theory | Remarks |
|---|---|---|
| True-score variance, $\sigma^2(X_\infty)$ | Universe-score variance, $\sigma^2(\mu_p)$ | Directly estimated by $\sigma^2(p)$ from the G study. |
| Observed-score variance, $\sigma^2(X)$ | Observed-score variance, $\sigma^2(X_{pI})$ | To be estimated by combining components of variance for $p$ and $i$, $pi$, $e$. |
| Error variance, $\sigma^2(E)$ | $\sigma^2(\Delta) = \sigma^2(\delta)$ for this design | Consists of the $i$, $pi$, $e$ component of variance. |
| Reliability coefficient,[a] $\rho(X,X') = \rho^2(X,X_\infty)$ | Coefficient of generalizability, the ratio of universe-score and observed-score variances. Denoted by $\rho^2$. | Essentially the same as the Horst (1949) version of the intraclass correlation. In the population, equals squared correlation of observed and universe scores. |
| Confidence interval for true score | Confidence interval for universe score | Determined from $\sigma(\Delta) = \sigma(\delta)$. |
| Regression estimate of true score, equal to $\bar{X}.. + \rho(X,X')(X_{pI} - \bar{X}..)$ | Regression estimate of universe score. Like that at left using $\rho^2(X_{pI},\mu_p)$ as coefficient | Essentially no change. |
| Error of estimate, squared | Error of estimate, squared, $\sigma^2(\varepsilon)$ | Essentially no change. |

[a] In conventional theory, $X$ and $X'$ are "parallel" observations.

the universe score, diagram a may be divided, as shown in diagram b, into true and error components.

The universe-score variance, as Table 3.1 indicates, is directly estimated by the $\widehat{\sigma^2}(p)$ obtained in the G study. The observed score can be regarded as a sum of score components. In the abbreviated code introduced on p. 40, the observed score $X_{pI}$ equals $\mu_p + (\mu_I\sim + \mu_{pI}\sim + e_{pI})$. Because of random sampling, the expression in parentheses is independent of the universe score, hence the observed-score variance equals $\widehat{\sigma^2}(p) + \widehat{\sigma^2}(I,pI,e)$.

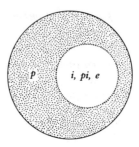

(a)   Total variance of $X_{pi}$ over all admissible observations on all persons

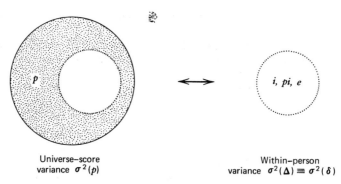

| Universe–score variance $\sigma^2(p)$ | Within–person variance $\sigma^2(\Delta) \equiv \sigma^2(\delta)$ |

(b)   Components of observed score variance

FIGURE 3.1.   *Separation of Kinds of Variance in an i:p Design.*

The error $\Delta_{pI}$ was defined as $X_{pI} - \mu_p$, and hence equals $\mu_I\sim + \mu_{pI}\sim + e_{pI}$. The quantity, $\sigma^2(\Delta)$ for the observed scores in the D study is estimated by $\sigma^2(I,pI,e)$. The error $\delta$ can be defined as $(X_{pI} - X_{PI}) - (\mu_p - \mu)$. (We shall use a slightly different definition on p. 93.) The expression can be rewritten as $X_{pI} - \mu_p - (X_{PI} - \mu)$. In the $i:p$ design each person is observed under different conditions and as $n_i'$ or $n_p'$ increases, the term in parentheses approaches zero. Hence $\delta_{pI}$ approaches $\Delta_{pI}$, and in the population $\sigma^2(\Delta) = \sigma^2(\delta) = \sigma^2(I,pI,e)$.

Under the $i:p$ design the traditional statement adds that the observed-score variance is the sum of universe-score variance and error variance, whether the error is $\Delta$ or $\delta$. This variance includes a variance component for condition means that in strictly classical theory is assumed to be zero. Horst (1949) and Ebel (1951), discussing the reliability of ratings, noted that the conditions means are often unequal in practice, and modified the concept of error variance to take this into account, much as we have. Differences in condition means contribute to the differences among observations on the same person made under various conditions. They also contribute to the differences among persons observed under an $i:p$ design, because one person will draw easier items or more lenient raters than another.

In any application of the $i:p$ design to a large group, the mean value of observed scores will approach the mean of universe scores. As more persons are tested, more conditions enter the mean, and the expected values of the several components of the error score approach zero. This is true even when the conditions are not equivalent. Also, the population variance of observed scores is the same in every application of the design. It is conventional to define the interval estimate of $\mu_{p\bullet}$ as $X_{p\bullet I} \pm \hat{\sigma}(\Delta)$ or $X_{p\bullet I} \pm 1.96\hat{\sigma}(\Delta)$, associated, respectively, with supposed confidence limits of 67 and 95% (see pp. 130–134).

Lengthening the series of observations by increasing $n_i'$ reduces $\sigma^2(\Delta)$. If we write $\sigma^2(\Delta_{pi})$ for the variance with a single condition per person, $\sigma^2(\Delta_{pI}) = \sigma^2(\Delta_{pi})/n_i'$ as is usual for the variance of means of randomly sampled observations.

Because a different set of conditions is drawn randomly in each application of the measuring procedure to person $p$, neither the $i$ nor the $pi$ component covaries with $\mu_p$. Consequently, observed scores have the same population correlation with universe scores in every application of the design, and the population correlations between sets of observed scores are uniform. This, taken with the equivalence of population means and observed variances, implies that in this design, the classical equivalence assumptions apply fully to the $X_{pI}$.

The ratio of universe-score variance to observed-score variance equals $\rho^2(X_{pI}, \mu_p)$. This coefficient of generalizability is estimated by dividing $\widehat{\sigma^2}(p)$ by the estimate of observed-score variance. It is readily shown that the statement made above about the effect of $n_i'$ on $\sigma(\Delta)$ leads to the Spearman–Brown formula, which describes the effect on the coefficient of an increase in $n_i'$. (We shall write the coefficient simply as $\rho^2$ in the remainder of this section.)

The coefficient is an intraclass correlation among observed scores. It differs from the coefficient proposed by Horst for the reliability of ratings only in that Horst followed pre-Fisherian formulas, where we follow Fisher in employing degrees of freedom in our estimation procedure. As Ebel pointed out, the Horst coefficient applies to ratings when there are different raters for each subject (i.e., $i:p$) and not when the same raters rate all subjects (i.e., $i \times p$). These points were clarified in the unpublished work of Buros (1963). Among other contributions, Buros offered a general intraclass correlation formula that takes into account the possibility of varying $n_i'$ and so embodies the Spearman–Brown adjustment. From the one-way analysis of variance one could get

$$(3.1) \qquad r_{\text{intraclass}} = \frac{\text{MS } p - \text{MS within persons}}{\text{MS } p + \dfrac{(n_i - n_i')}{n_i'} \text{MS } wp}$$

This coefficient equals the ratio of estimated universe-score variance to estimated observed-score variance, if the D study has an $i:p$ design. Here, it makes no difference whether we regard the coefficient as a squared correlation between the observed score and the universe score, or as a correlation of two sets of observed scores. The distinction does become pertinent in designs where distributions of observed scores do not satisfy equivalence assumptions.

The regression formula for estimating the universe score, assuming $\rho^2$ and $\mu$ known, is

$$(3.2) \qquad \hat{\mu}_p = \mu + \frac{\sigma(\mu_p)}{\sigma(X_{pI})} \rho (X_{pI} - \mu) = \rho^2 X_{pI} + (1 - \rho^2)\mu$$

When the sample for a D study represents the same population as the sample for the G study, the quantities $\rho^2$ and $\mu$ can be estimated from the G study. Later in this chapter we shall discuss estimation for subpopulations whose parameters are presumed to differ from those of the population represented in the G study.

The variance of errors of estimate is customarily defined by

$$(3.3) \qquad \sigma^2(\varepsilon) = \sigma^2(\hat{\mu}_p - \mu_p) = \sigma^2(p)(1 - \rho^2)$$

Strictly speaking, the third member of the equation equals the second only when population parameters, rather than their estimates, are used in (3.2).

The reader familiar with the Ebel paper will recognize that generalizability theory does not depart from his analysis of this simple design in any important way, though we offer a slightly more general rationale.

It has been assumed that the G study has an $i:p$ design. If the G study were of the type $i \times p$, one could still estimate the quantities discussed. The observed-score variance, for example, is obtained by noting that $\sigma^2(I, pI, e)$ is $1/n_i'$ times the sum of the $i$ and $pi,e$ components of variance. It is possible, then, to use a single G study to obtain information on both the nested and crossed one-facet D studies (Cf. Ebel, 1951, and Rajaratnam, 1960). Multifacet G studies may also estimate the needed components, but to fit facets of the larger study into the one-facet design requires a good deal of judgment. As will become clearer in Chapter 4, a $j$ component may, in the D study, be confounded with $p$ or with $i$; or it may enter the residual. And, if the universe of generalization is defined narrowly as $\mu_{pj\bullet}$, the $j$ component of variance becomes irrelevant.

### The crossed design

Traditional test theory has generally dealt with the crossed $i \times p$ design, because it is usual to apply the same test form to all subjects. A consideration of the $i \times p$ D study draws attention to some major differences between generalizability theory and classical theory.

For the D study, one selects $n'_i$ conditions at random; these are applied to all subjects. The G study has presumably supplied estimates of the components of variance for $p,i$, and $pi,e$. Again, $\widehat{\sigma^2}(p)$ is the desired estimate of universe-score variance. Table 3.2 summarizes the contrasting concepts from reliability theory and generalizability theory. Note the changes from Table 3.1.

If one does not assume equivalence of conditions, then the observed-score variance for the particular conditions employed in the D study can be known only after the D study is carried out. Each set of conditions has its own variance, hence the observed-score variance will differ from one application of the design to another. When evaluating a proposed measuring procedure before carrying out the D study, the only alternative is to estimate the *expected* value of the observed-score variance. That is, one estimates the mean value of the variances that would be found in the course of an indefinitely large number of applications of the design. (A condition used in the D study occasionally is one of those used in the G study, but it is almost never practical to evaluate the generalizability of scores from a specific condition or set of conditions; see p. 101.)

The observed-score variance is, by definition, the variance of the deviation score $X_{pI} - \mu_I$. From the model, this equals $\mu_p\!\sim + \mu_{pI}\!\sim + e_{pI}$. The component $\mu_I\!\sim$ does not enter the deviation score, hence, the observed-score variance has an expected value equal to $\sigma^2(p) + \sigma^2(pI,e)$. In this design $\delta_{pI} = (X_{pI} - \mu_I) - (\mu_p - \mu)$, and $\sigma^2(\delta)$ has the expected value $\sigma^2(pI,e)$. The expected observed-score variance equals the universe-score variance plus *this* expected error variance (Figure 3.2). In Figure 3.2, diagram (*a*), the *left* circle represents the observed-score variance in an $i \times p$ design with $n'_i = 1$; if $n'_i > 1$, $I$ may be substituted for $i$ in all labels. Where $\mu_p$ is the universe score, diagram (*a*) may be divided as in diagram (*b*). The observed-score variance may also be divided as shown in diagram (*c*) of Figure 3.2.

The "error of measurement" $X_{pI} - \mu_p$ or $\Delta_{pI}$ is the sum of the $I$ and $pI,e$ components, as in the nested design. There are several variances of $\Delta_{pI}$ that might be considered:

$$\underset{I}{\mathscr{E}}\Delta^2_{pI} \qquad \text{(over sets of conditions that might be sampled, for } p \text{ fixed)}$$

$$\underset{p\ I}{\mathscr{E}\mathscr{E}}\Delta^2_{pI} \qquad \text{(average, over the population, of the above; variance of } \Delta \text{ for the population and universe)}$$

$$\underset{p}{\mathscr{E}}\left(\Delta_{pI} - \underset{p}{\mathscr{E}}\Delta_{pI}\right)^2 \qquad \text{(over persons, for a particular set of conditions)}$$

**TABLE 3.2.** *Comparison of Parameters Considered in Conventional Reliability Theory and Generalizability Theory ($i \times p$ Design)*

| Term employed in reliability theory and usual symbol | Corresponding concept and symbol in generalizability theory | Remarks |
|---|---|---|
| True-score variance, $\sigma^2(X_\infty)$ | Universe-score variance, $\sigma^2(\mu_p)$ | Directly estimated by $\sigma^2(p)$ from G study. |
| Observed-score variance, $\sigma^2(X)$ | Expected observed-score variance, $\underset{I}{\mathscr{E}}\sigma^2(X_{pI})$ | To be estimated by combining $p$ and $pi$, $e$ components of variance. |
| Error variance, $\sigma^2(E)$ | $\sigma^2(\Delta)$ equals the expected variance within the person | Combines the $i$ and $pi$, $e$ components of variance |
| | $\underset{I}{\mathscr{E}}\sigma^2(\delta) \leqslant \sigma^2(\Delta)$ | Difference between universe-score and expected observed-score variances. (Consists of the $pi$, $e$ component of variance only.) |
| Reliability coefficient, $\rho(X,X') = \rho^2(X,X_\infty)$ | Coefficient of generalizability, the ratio of universe-score and expected observed-score variances. Denoted by $\mathscr{E}\rho^2$. | An intraclass correlation. Interpretable as approximately $\underset{I}{\mathscr{E}}\rho^2(X_{pI},\mu_p)$ if conditions not equivalent. |
| Confidence interval for true score | Confidence interval for universe score | Determined from $\sigma(\Delta)$ in generalizability theory. Classical theory assumes effect of $i$ component absent. |
| Regression estimate of true score, equal to $\bar{X}.. + \rho(X,X')(X_{pI} - \bar{X}..)$ | Regression estimate of universe score, which ideally would be estimated by $\mu + \rho^2(X_{pI},\mu_p)(X_{pI} - \mu_I)$ | Since the regression coefficient for the particular set of $i$ used in the D study cannot be estimated, $\mathscr{E}\rho^2$ must be used. |
| Error of estimate, squared | Error of estimate, squared, $\sigma^2(\varepsilon)$ | Estimated by a formula like the classical one, but this is an underestimate in generalizability theory. |

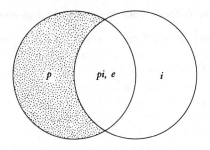

(a)  Total variance of $X_{pi}$ over all admissible
     observations on all persons

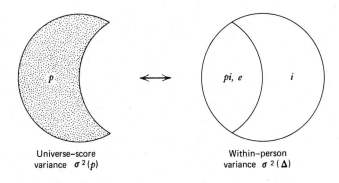

Universe–score                    Within–person
variance  $\sigma^2(p)$           variance  $\sigma^2(\Delta)$

(b)  Division of total variance

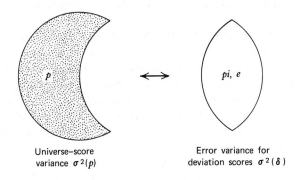

Universe–score                    Error variance for
variance  $\sigma^2(p)$           deviation scores  $\sigma^2(\delta)$

(c)  Division of observed–score variance

FIGURE 3.2.  *Separation of Kinds of Variance in an i × p Design.*

Throughout our work we shall be interested in the second of these and will denote it simply by $\sigma^2(\Delta_{pI})$, or, where no confusion will arise, by $\sigma^2(\Delta)$. Occasionally we shall refer to the first of the three variances and will label it $\sigma^2(\Delta_{pI} \mid p)$ (read "for a fixed $p$"). The third of the definitions, even though analogous to the definition of observed-score variance, can be ignored since it proves to be identical to $\sigma^2(\delta_{pI})$ for condition $I$.

With a crossed D study, $\sigma^2(\Delta)$ is the sum of $\sigma^2(I)$ and $\sigma^2(pI,e)$, and so is identical to $\sigma^2(\Delta)$ in a nested D study with the same $n_i'$. (See Figure 3.2($b$)). How much a person's observed score is likely to differ from his universe score in no way depends on whether the same conditions are used to observe *other* persons. Under the $i \times p$ design, $\sigma^2(\Delta)$ ordinarily exceeds $\mathscr{E}\sigma^2(\delta)$, although the two were equal in the nested design.

Classical theory, assuming uniform condition means ($\mu_I = \mu$), ignores the distinction between $\Delta$ and $\delta$. Lord (1962) pointed out that condition means are unlikely to be equal when tests are not carefully equated. Lord showed that the variance of the within-person error $\Delta$, over nonequivalent tests, differs from the error variance calculated by classical formulas [which is like our $\mathscr{E}\sigma^2(\delta)$]. A confidence interval of the conventional sort has to be defined in terms of $\sigma(\Delta)$, not $\sigma(\delta)$; only if all tests (or other procedures) yield strictly equal means is it appropriate to use $\sigma(\delta)$.[1]

To estimate $\sigma^2(\Delta)$ and/or $\mathscr{E}\sigma^2(\delta)$ one divides $\sigma^2(pi,e)$ and $\sigma^2(i)$ by $n_i'$. This again leads ultimately to the same results as the Spearman–Brown formula. Whereas the Spearman–Brown formula can be brought to bear only in a limited way on designs with more than one facet, the principle of dividing components of variance by the number of conditions generalizes fully.

The coefficient of generalizability has been defined (p. 17) as the ratio of universe-score variance to expected observed-score variance; this is approximately the expected value of the squared correlation of observed score and universe score. The coefficient is denoted by $\mathscr{E}\rho^2$ for convenience. To estimate it, one employs the variance estimates already calculated. This coefficient is an intraclass correlation among observed scores, of the type that discards the condition component of variance from the error term and, hence, from the observed-score variance.

When $n_i' = n_i$ the variance ratio is identical to the coefficient from the Hoyt analysis-of-variance procedure, Kuder–Richardson Formula 20, Cronbach's $\alpha$, and several other well known formulas. The most common procedure has been to compute the variance of scores under each condition— i.e., the several $s^2(X_{pi} \mid i)$—and of total scores on the test, $s^2(X_{pT})$. Then

---

[1] We shall write $\sigma(\delta)$ in place of $[\mathscr{E}\sigma^2(\delta)]^{1/2}$, even though the latter is technically correct.

the universe-score variance for test scores is estimated by $[n/(n-1)][s^2(X_{pT}) - \sum_i s^2(X_{pi}|i)]$.[2] This can be divided by $s^2(X_{pT})$ to estimate the coefficient $\mathscr{E}\rho^2$; the same estimate can be reached from the variance components.

For $n_i' \neq n_i$, our approach yields a coefficient identical to that given by the Spearman–Brown correction of a coefficient from one of the procedures mentioned previously. Ebel recommended application of this type of intra-class formula to G data where the same raters rate all subjects. Buros (1963) offered a version of the formula that takes into account directly the possibility that $n_i'$ differs from $n_i$:

$$(3.4) \qquad r_{\text{intraclass}} = \frac{\text{MS } p - \text{MS } r}{\text{MS } p + \left(\dfrac{n_i - n_i'}{n_i'}\right)\text{MS } r}$$

The more nearly uniform the observed-score variances and the $\rho^2(X_{pI},\mu_p)$ under separate applications of the design, the more closely does the intraclass correlation coincide with $\rho^2(X_{pI},\mu_p)$ for any one application. Monte Carlo studies (Cronbach & Azuma, 1962; Cronbach, Schönemann, & McKie, 1965; Cronbach, Ikeda, & Avner, 1966) show that the discrepancies between the population intraclass correlation and the alternative coefficients listed above are extremely small whenever the number of conditions of a facet is reasonably large in the D study, or the variance within conditions differs little from condition to condition.

The formula for making point estimates of universe scores, expressed in terms of population means, is

$$(3.5) \qquad \hat{\mu}_p = (\widehat{\mathscr{E}\rho^2})(X_{pI} - \mu_I) + \mu$$

In Chapter 5 we shall discuss whether the sample mean from the D study or that from the G study, or a combination, should be used in evaluating the means.

Where conditions are equivalent in the classical sense (equal means, equal variances, and equal intercorrelations) and $n_p$ is indefinitely large, the equation is identical to the regression equation of classical theory. An expected correlation is used in the estimation equation instead of the genuine regression coefficient for the particular condition in the D study. Therefore, if conditions are not equivalent, the estimate is not the best one conceivable.

An equation for estimating the variance of errors of estimate is

$$(3.6) \qquad \widehat{\sigma^2}(\varepsilon) = \widehat{\sigma^2}(p)(1 - \widehat{\mathscr{E}\rho^2})$$

[2] The vertical line is a conditional notation like that used with fixed $J^*$ in Chapter 2. Here, it implies that while $p$ varies, $i$ is held constant.

This is strictly appropriate only where the genuine regression equation applying to the conditions in the D study is used. If conditions are not equivalent, the estimates from (3.5) will have an error greater than (3.6) indicates. Formula (3.6) should be distinguished from $[\widehat{\mathscr{E}\sigma^2(X_{pI})}](1 - \widehat{\mathscr{E}\rho^2})$, which estimates $\mathscr{E}\sigma^2(\delta)$ when applied to this design.

Our numerical results for the $i \times p$ design depart from those obtained under the classical formulas in only one major respect. That is the use of $\hat{\sigma}(\Delta)$ in defining the confidence interval rather than $\hat{\sigma}(\delta)$. When tests are strictly equivalent, and the sample of persons for the G study is large, our theory reduces to the classical theory and our formulas give precisely the same results as traditional ones. [Where $n_p$ is small, there will be some difference between the intraclass correlation and the average of correlations $r(X,X')$, even though conditions are equivalent in the population.] Our theory has more radical implications in multifacet studies.

### B. *The Error* $\Delta$

This section presents in detail the technique for estimating $\sigma(\Delta)$, which is used in establishing confidence intervals.

The decision maker who forms a confidence interval for the person's universe score is presumably interested in its absolute value $\mu_p$ and he takes the observed score as a direct estimate of it. We shall discuss interpretation based on the raw score. Though a similar logic may be applied to standard scores, IQs, etc., this introduces some risk of misinterpretation (see p. 134).

The term $\Delta_{pIJ}$ is written for the discrepancy $X_{pIJ} - \mu_p$. The within-person standard deviation, to which the confidence interval is proportional, is $\sigma(\Delta_{pIJ})$. We can represent the universe score, and consequently $\Delta$, in terms of the score components introduced earlier. Whatever the components of $\Delta$ may be, the variances of these components make up $\sigma^2(\Delta)$. An estimate of the average value of $\sigma^2(\Delta_{pIJ}|p)$ over all persons in the population can be obtained from the G-study estimates of component variances. The square root is taken as an estimate $\hat{\sigma}(\Delta)$. This estimate, subtracted from and added to the observed $X_{p*IJ}$, defines an interval that, according to the model, is likely to contain $\mu_{p*}$. We first review statements made above regarding $\Delta$ in one-facet D studies. This allows us to display a scheme of analysis that will be useful with more complex designs.

### *One-facet studies considered in detail*

In arriving at a confidence interval for $\mu_{p*}$, there is no need to consider whether other persons are observed, or under what conditions they are observed. The within-person design is simply $i:p^*$. Crossing of $p$ with $i$ does not affect $\sigma(\Delta)$, the "within-person" standard deviation. The term $\sigma(\Delta)$ is defined over all sets $I$ that could be drawn from the universe.

In Table 3.3 there are five columns where scores and variances are resolved into components of the one-facet model. It is to be understood that a score or variance indicated at the head of the column is the sum of the components appearing below it. Only the "frequency" column is of a different type.

**TABLE 3.3.**   *Components of Scores in a D Study with the Design $i{:}p$ or $i \times p$ (Generalization to $\mu_\rho$)*

| $X_{pI}$ | $\mu_p$ | $\Delta_{pI}$ | Variance component | Frequency within $p$ | $\sigma^2(\Delta)$ |
|---|---|---|---|---|---|
| $\mu$ | $\mu$ | | | | |
| $\mu_p\sim$ | $\mu_p\sim$ | | $\sigma^2(p)$ | | |
| $\mu_I\sim$ | | $\mu_I\sim$ | $\sigma^2(i)$ | $n_i'$ | $\sigma^2(I)$ |
| $\mu_{pI}\sim, e$ | | $\mu_{pI}\sim, e$ | $\sigma^2(pi,e)$ | $n_i'$ | $\sigma^2(pI,e)$ |

The first column of Table 3.3 contains the basic breakdown of $X_{pI}$ into score components, and the second column contains $\mu_p$ in terms of the components. Since $\Delta_{pI} = X_{pI} - \mu_p$, one subtracts, component by component, to identify the components of $\Delta_{pI}$. For each component of $X_{pI}$ (except the constant $\mu$) there is a variance component that should have been estimated by the G study. The D-study design yields $n_i'$ observations on the $\mu_i\sim$ and $\mu_{pi}\sim,e$ score components. Then, because $\sigma^2(I)$ is the variance of the mean of $n_i'$ values of $\mu_i\sim$, $\sigma^2(I) = \sigma^2(i)/n_i'$. This gives the required entry for the final column. In general, any entry for the $\sigma^2(\Delta)$ column is obtained by dividing the elemental component of variance by the frequency with which the corresponding score component is observed in the D study. Figures 3.1 and 3.2 both indicate that the total variance of all $X_{pi}$ for many persons and conditions decomposes into a person component and a sum of within-person components. The person component is the variance of universe scores, and the within-person variance is the variance of the $\Delta_{pi}$.

In drawing a conclusion about person $p^*$, one would like to know $\sigma(\Delta_{pI}\,|\,p^*)$, the standard deviation of scores for $p^*$ alone. This need not be the same as $\sigma(\Delta\,|\,p)$ for other persons, because $\sigma^2(pi,e\,|\,p)$ may vary from person to person. $[\sigma^2(i)$ is the same for all persons.] In theory, it would be wise to make a direct estimate of $\sigma(\Delta)$ for $p^*$ by observing $p^*$ under a great number of conditions, using those data alone for a G study. Unfortunately, it is almost never practical to do this for a person about whom a decision is being made, because $n_i'$ is almost always small. Rarely, it is practicable to work out a value of $\hat{\sigma}(\Delta)$ for persons having observed scores in a limited range. The usual practice is to estimate the expected value of $\sigma^2(\Delta_{pi}\,|\,p)$ for the population of persons represented in the G study, a population from which $p*$ is presumed to come.

To estimate $\mathscr{E}\sigma^2(\Delta_{pI})$, the entries in the last column of Table 3.3 are simply summed. The square root of that sum is the estimate of $\sigma(\Delta)$. The G study may have used a $i \times p$ design, in which case the required $\sigma^2(i)$ and $\sigma^2(pi,e)$ have been directly estimated. If the G study used an $i:p$ design, only the confounded $\sigma^2(i,pi,e)$ was estimated. But this is simply the variance of $\mu_i\sim + \mu_{pi}\sim + e_{pi}$ and is equal to $\sigma^2(i) + \sigma^2(pi,e)$. Since the divisor for both components is $n_i'$, $\sigma^2(\Delta_{pI})$ can be estimated by the procedure shown in connection with Table 3.3.

The confidence interval is formed by combining a multiple of $\hat{\sigma}(\Delta)$ with the observed score. The usual practice of testers is to employ 1 as the multiplier. Assuming a normal distribution, they conclude that $X_{pI} - \hat{\sigma}(\Delta) \leqslant \mu_p \leqslant X_{pI} + \hat{\sigma}(\Delta)$ with a probability of 0.67. That is, it is presumed that when this technique is applied consistently, one-third of such statements locating $\mu_p$ for individuals will be incorrect. This is a high rate of error, but the developer of psychological and educational tests finds the interval embarrassingly wide if he moves to a more conservative risk level. It is awkward enough to admit that a test of the usual length locates the "true" IQ somewhere in a range from 105 to 115, for instance. To form a confidence interval by raising the multiplier to 1.96 (the value most often used in statistical inference) would generate the embarrassing admission that the IQ is located only within the broad range 100–120. However, the probabilities associated with universe-score confidence intervals are misleading in several ways, as Chapters 4 and 5 will show.

### Two-facet D studies

We move on to the two-facet D study that employs an $i \times j$ design within the person. This design may be any of the three where $i \times j$ appears in the "within-person" column of Figure 2.4 and Table 2.1. There are $n_i'$ conditions of $i$ and $n_j'$ of $j$.

To evaluate $\widehat{\sigma^2}(\Delta)$, Table 3.4 (similar to Table 3.3) is compiled with seven components of variance. (See also Figure 3.3.) If a component such as $\sigma^2(pi)$ has been estimated in the G study, the corresponding $\sigma^2(pI)$ is estimated by dividing $\widehat{\sigma^2}(pi)$ by the entry in the frequency column. Entries for the frequency column can be obtained from Table 2.1, considering a crossed within-person design, setting $n_p = 1$, and, of course, adding primes to show that a D study is under consideration.

The G study introduced in Table 2.2 provides the basis for a numerical example. The analysis in Table 3.5 assumes that the D study will use 10 items and 1 scorer. (Three scorers were used in the G study). We find that

**TABLE 3.4.** Components of Scores in a D Study Where i is Crossed with j within the Person (Generalization to $\mu_p$)

| $X_{pIJ}$ | $\mu_p$ | $\Delta_{pIJ}$ | Variance component | Frequency within $p$ | $\sigma^2(\Delta)$ |
|---|---|---|---|---|---|
| $\mu$ | $\mu$ | | | | |
| $\mu_p\sim$ | $\mu_p\sim$ | | $\sigma^2(p)$ | | |
| $\mu_I\sim$ | | $\mu_I\sim$ | $\sigma^2(i)$ | $n_i'$ | $\sigma^2(I)$ |
| $\mu_J\sim$ | | $\mu_J\sim$ | $\sigma^2(j)$ | $n_j'$ | $\sigma^2(J)$ |
| $\mu_{pI}\sim$ | | $\mu_{pI}\sim$ | $\sigma^2(pi)$ | $n_i'$ | $\sigma^2(pI)$ |
| $\mu_{pJ}\sim$ | | $\mu_{pJ}\sim$ | $\sigma^2(pj)$ | $n_j'$ | $\sigma^2(pJ)$ |
| $\mu_{IJ}\sim$ | | $\mu_{IJ}\sim$ | $\sigma^2(ij)$ | $n_i'n_j'$ | $\sigma^2(IJ)$ |
| $\mu_{pIJ}\sim, e$ | | $\mu_{pIJ}\sim, e$ | $\sigma^2(pij, e)$ | $n_i'n_j'$ | $\sigma^2(pIJ, e)$ |

$\hat{\sigma}(\Delta)$ is 0.93. Compared to the range of 16 points allowed by the grading scale, this implies fairly good agreement. It may or may not be adequate for the intended use of the scale, and if it is inadequate, one would increase $n_i'$ or $n_j'$ or improve the scoring rules.

An investigator might have investigated this design by a G study with a single scorer, having $i$ as the only variable facet. The two-facet information shows the $pj$ component to be a rather large element in $\sigma^2(\Delta)$; this information could not have been obtained from the one-facet G study.

Only a G study with Design VII estimates the variances of all components of $X_{pij}$ separately. If a study has used some other design and estimated

**TABLE 3.5.** Estimation of $\sigma^2(\Delta)$ for a D Study with the Design $i \times j \times p$ ($n_i' = 10$, $n_j' = 1$; Generalization to $\mu_p$)

| Source of variance | Estimate of variance component[a] | Frequency within $p$ | $\sigma^2(\Delta)$ |
|---|---|---|---|
| $p$ | 5.71 | | |
| $i$ | 1.32 | $n_i' = 10$ | 0.13 |
| $j$ | 0.09 | $n_j' = 1$ | 0.09 |
| $pi$ | 2.57 | $n_i' = 10$ | 0.26 |
| $pj$ | 0.21 | $n_j' = 1$ | 0.21 |
| $ij$ | 0.07 | $n_i'n_j' = 10$ | 0.01 |
| $pij, e$ | 1.57 | $n_i'n_j' = 10$ | 0.16 |
| | | | $0.86 = \widehat{\sigma^2}(\Delta)$ |

[a] Calculated on page 44.

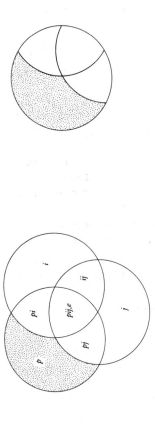

(a) Variance of $X_{pij}$ over all admissible observations on all persons

(b) The observed–score variance may also be divided

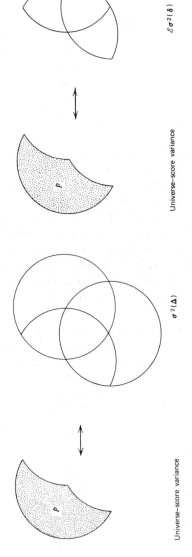

Universe–score variance $\qquad \sigma^2(\Delta)$

Universe–score variance $\qquad \mathscr{E}\sigma^2(\delta)$

(c) Division of total variance

(d) Division of observed–score variance

FIGURE 3.3. *Separation of Kinds of Variance in an* $i \times j \times p$ *D Study.*

certain variance components only in combination, one may still apply the results to the Design VII D study. Because, for example, the frequency within the person for $i$ and $pi$ is the same in the D study, the combined contributions of these effects to $\sigma^2(\Delta)$ can be estimated from any design having $i$ and $pi$ confounded. Similarly, a design with $j$ and $pj$ confounded estimates the combination of those effects. Therefore, a G study with Design IV-B, $(i \times j){:}p$, or Design V-B, $j \times (i{:}p)$ or $i \times (j{:}p)$, provides the estimates needed. (The reader may find it profitable to determine why $j \times (i{:}p)$ is usable but $j{:}(i \times p)$ is not.) If $n'_i$ or $n'_j$ equals 1, estimates from still other designs may be used; for example, if $n'_i = 1$, it is possible to use a G study with design $j{:}(i \times p)$, where $j,pj,ij$, and $pij,e$ are confounded.

If the D study has $j$ nested within $i$, one alters the $j$ and $pj$ entries of the frequency column in accord with Table 3.6 (which is based on Table 2.1). The $j$ and $pj$ components are sampled $n'_i$ times as often as in the crossed D study. Consequently, $\sigma^2(\Delta)$ is smaller. That is, $X_{pIJ}$ is generally closer to $\mu_p$ when a $j{:}i$ within-person design is used for the D study than when a crossed design that collects the same number of observations per person is used.

As might be suspected, weakening the design to $(i,j){:}p$ further reduces $\sigma^2(\Delta)$. Consider a study in which one $j$ is sampled along with each $i$. To obtain the same number of observations per person as before, more conditions of $i$ must be selected. Doing this keeps the components for $J,pJ,IJ$, and $pIJ,e$ the same as before, and reduces the components for $I$ and $pI$. Consequently, where $X_{pIJ}$ is to be taken as an estimate of the value of $\mu_p$, there is an advantage in carrying out the D study by the very weak design that pairs each $i$ with just one $j$.

Measuring procedures most often employ a crossed design. For example, the typical investigator administers the same set of items on two occasions,

**TABLE 3.6.**   *Number of Observations of Each Component of $\Delta_{pIJ}$ as a Function of the within-Person Design of the D Study*

| Score component | Number of observations per person | | |
| --- | --- | --- | --- |
| | $i \times j$ | $j{:}i$ | $i,j$[a] |
| $\mu_i \sim$ | $n'_i$ | $n'_i$ | $n'_i n'_j$ |
| $\mu_j \sim$ | $n'_j$ | $n'_i n'_j$ | $n'_i n'_j$ |
| $\mu_{pi} \sim$ | $n'_i$ | $n'_i$ | $n'_i n'_j$ |
| $\mu_{pj} \sim$ | $n'_j$ | $n'_i n'_j$ | $n'_i n'_j$ |
| $\mu_{ij} \sim$ | $n'_i n'_j$ | $n'_i n'_j$ | $n'_i n'_j$ |
| $\mu_{pij} \sim, e$ | $n'_i n'_j$ | $n'_i n'_j$ | $n'_i n'_j$ |

[a] These entries employ the convention stated in footnote b in Table 2.1. The product $n'_i n'_j$ equals the number of observations per person.

crossing items with occasions. In another study, he collects three protocols of a teacher's classroom remarks, and has each protocol scored by the same four scorers. But he would get a smaller $\sigma(\Delta)$ if he were to break the test into many small parts, each to be given on a different day, or, in the second study, if he collected 12 protocols and had each scored by a different person. Practical considerations limit the use of such designs. It is inconvenient to test on many days; a fragment of a test may be too short to allow for proper warmup; setting up recording equipment in the teacher's classroom a dozen times may be impractical. Our analysis shows a benefit to be gained, however, by weakening the design when one can. These remarks apply, of course, only to the error $\Delta$; weakening the design increases other kinds of error.

If a certain component of $\sigma^2(\Delta)$ is large and an absolute interpretation of $X_{pIJ}$ is intended, it would be wise to sample that effect quite thoroughly in a D study. If an effect is small, on the other hand, an effort to control it in the experimental design, or to sample it extensively probably is not warranted; this type of reasoning will be discussed in Chapter 7.

### C. *Observed-Score Variance and the Error* $\delta$

*Observed-score variance as a function of the D-study design*

Whenever decisions are to be determined by the comparative standing of individuals, one is interested in locating the individual within his group as accurately as possible. The absolute universe score is then unimportant. Research concerning correlations among variables similarly emphasizes comparative rather than absolute standings. To evaluate interpretations of these sorts, one compares the universe-score variance with the expected observed-score variance; the higher the ratio, the more the observed ranks correspond to the ranks of the universe scores. The developments in this section are also pertinent in making point estimates of universe scores.

Estimation of the observed-score variance receives no particular attention in classical theory, because the variance of the scores in the reliability study is directly calculated. For us, however, the G study is a basis for thinking about data that may be collected in future D studies. The observed-score variance that will arise in any one study cannot be directly calculated, but, knowing the proposed design, it is possible to estimate the variance. However, one can do no better than estimate the average of variances for all D studies with such a design in this population and universe. The average is referred to as the expected observed-score variance.

The observed-score variance is defined as the mean square of the deviation score—the person's score minus the mean over persons. The mean is for scores collected according to the design proposed for the D study. As long as the sample variance is an unbiased estimator of the population variance, it makes no difference whether the deviation score is taken from the sample mean or the population mean. Notational problems become quite awkward when we attempt to distinguish the means and variances formed under various designs. Therefore, we shall minimize notational refinements, generally writing $\sigma^2(X)$ for the population variance arising under a single application of whatever design is under discussion, and $\mathcal{E}\sigma^2(X)$ for the expected value over all applications of the design.

Table 3.7 employs a scheme similar to Tables 3.3 and 3.4. This scheme breaks the observed score and the mean for Design VII into components. The score breakdown agrees with Table 3.4. In Design VII all subjects are observed under the same $I$. Consequently, the same $\mu_I$ component enters every score and also the mean, $\mu_{IJ}$. Similarly $\mu_J$ enters the mean. Score components that differ from one person to another become zero in the population mean, as indicated by the blank spaces in the column for components of $\mu_{IJ}$. These are the components that remain in the deviation score formed by subtracting $\mu_{IJ}$ from $X_{pIJ}$. Estimates of the variance components are taken from the G study. The variance for each component of the deviation score is divided by the frequency (within the person) with which that component was observed in arriving at $X_{pIJ}$; this estimates the contribution of the component to the expected variance of observed scores.

**TABLE 3.7.** *Components of Expected Observed-Score Variance in a D Study with the Design $i \times j \times p$*

| $X_{pIJ}$ | Population mean $\mu_{IJ}$ | Deviation score | Variance component | Frequency within deviation score | $\mathcal{E}\sigma^2(X)$ |
|---|---|---|---|---|---|
| $\mu$ | $\mu$ | | | | |
| $\mu_p\sim$ | | $\mu_p\sim$ | $\sigma^2(p)$ | 1 | $\sigma^2(p)$ |
| $\mu_I\sim$ | $\mu_I\sim$ | | $\sigma^2(i)$ | | |
| $\mu_J\sim$ | $\mu_J\sim$ | | $\sigma^2(j)$ | | |
| $\mu_{pI}\sim$ | | $\mu_{pI}\sim$ | $\sigma^2(pi)$ | $n'_i$ | $\sigma^2(pI)$ |
| $\mu_{pJ}\sim$ | | $\mu_{pJ}\sim$ | $\sigma^2(pj)$ | $n'_j$ | $\sigma^2(pJ)$ |
| $\mu_{IJ}\sim$ | $\mu_{IJ}\sim$ | | $\sigma^2(ij)$ | | |
| $\mu_{pIJ}\sim,e$ | | $\mu_{pIJ}\sim,e$ | $\sigma^2(pij,e)$ | $n'_i n'_j$ | $\sigma^2(pIJ,e)$ |

In Figure 3.3, diagram (b) indicates the composition of the observed-score variance. Score components that fall outside the $p$ circle in diagram (a) are the same for all persons under this crossed design. Therefore, they do not contribute to observed-score variance and are omitted from diagram (b).

A similar analysis may be made for any other D-study design. Which components drop out of the deviation score depends on how conditions are

**TABLE 3.8.**  *Number of Observations on Each Component of the Deviation Score as a Function of the Design of the D Study*[a]

| Score component | Design of D study | | | | | | | |
|---|---|---|---|---|---|---|---|---|
|  | VII $i \times j \times p$ | V-A $(j{:}i) \times p$ | V-B $i \times (j{:}p)$ | IV-A $j{:}(i \times p)$ | IV-B $(i \times j){:}p$ | III-A $j{:}i{:}p$ | III-B[b] $(i,j) \times p$ | II[b] $(i,j){:}p$ |
| $\mu_p \sim$ | 1 | 1 | 1 | 1 | 1 | 1 | 1 | 1 |
| $\mu_i \sim$ |  |  |  |  | $n_i$ | $n_i'$ |  | $n_i'n_j'$ |
| $\mu_j \sim$ |  |  | $n_j'$ | $n_i'n_j'$ | $n_j'$ | $n_i'n_j'$ |  | $n_i'n_j'$ |
| $\mu_{pi} \sim$ | $n_i'$ | $n_i'$ | $n_i'$ | $n_i'$ | $n_i'$ | $n_i'$ | $n_i'n_j'$ | $n_i'n_j'$ |
| $\mu_{pj} \sim$ | $n_j'$ | $n_i'n_j'$ | $n_j'$ | $n_i'n_j'$ | $n_j'$ | $n_i'n_j'$ | $n_i'n_j'$ | $n_i'n_j'$ |
| $\mu_{ij} \sim$ |  |  | $n_i'n_j'$ | $n_i'n_j'$ | $n_i'n_j'$ | $n_i'n_j'$ | $n_i'n_j'$ | $n_i'n_j'$ |
| $\mu_{pij} \sim, e$ | $n_i'n_j'$ | $n_i'n_j'$ | $n_i'n_j'$ | $n_i'n_j'$ | $n_i'n_j'$ | $n_i'n_j'$ | $n_i'n_j'$ | $n_i'n_j'$ |

[a] The contribution of each component of variance to $\mathscr{E}\sigma^2(X)$ is inversely proportional to the number shown.
[b] Number of observations fixed at $n_i'n_j'$ pairs per person for comparability to other designs. Normally $n_i' = n_j'$.

crossed with persons. In Designs IV-B, III-A, and II, the population mean of observed scores is $\mu$. Therefore, seven score components from $X_{pIJ}$ carry over to the deviation score and contribute to the variance. Table 3.8 summarizes the way the components of variance enter the observed variance, and presents arrays comparable to the "Frequency" column in Table 3.7. Reciprocals of the frequencies serve as weights for the variance components. If there is no entry beside a component, the component does not contribute to observed variance for the design in question.

The reader is reminded once again of our convention of using average scores rather than totals. For ratings and observations, it is common to state the composite score in the form of an average. All the formulas we have given apply directly to averages over conditions. Test scores, however, are usually totals of item scores. For the sum-of-observations type of composite, the expected observed-score variance given by the procedure just outlined must be multiplied by the square of the number of scores entering the sum in the D study. Occasionally, the observed score may be formed by averaging over one facet and summing over another. Then the observed-score variance obtained by the method outlined must be multiplied by the square of the number of summands, whether these are themselves elementary scores or averages.

The G study generates one particular observed variance. That variance is a single sample from the distribution of variances the design would generate. The $\mathscr{E}\sigma^2(X)$ derived from the G study is unlikely to coincide with the actual observed-score variance in a study made later.

## The error δ

It is convenient to define the error $\delta_{pIJ}$ as the difference between $X_{pIJ} - \mu_{IJ}$ and $\mu_p - \mu$ here. A slightly different value would be obtained if the deviation from the sample mean $X_{PIJ}$ were considered as on p. 76, but this discrepancy is of no consequence. The mean $\mu_{IJ}$ of course is determined by the design for the D study and the sample of conditions selected.

In classical theory, "observed-score variance equals true-score variance plus error variance." In our terminology, the expected observed-score variance in the population equals the universe-score variance plus $\mathscr{E}\sigma^2(\delta)$. This follows from the fact that the observed-score variance is the mean square of the deviation scores; each deviation score may be divided into two components: one the universe score, expressed as a deviation from the population mean, and the other $\delta_{pI}$. The composition of $\mathscr{E}\sigma^2(\delta)$ and its magnitude depend on the experimental design.

In Table 3.8, components of the deviation score for various designs were identified. All the components except $\mu_p \sim$ are components of $\delta$. Therefore, $\mathscr{E}\sigma^2(\delta) = \mathscr{E}\sigma^2(X) - \sigma^2(\mu_p)$. Now it is possible to complete the interpretation of the preceding figures. In Figures 3.1, 3.2, and 3.3, the expected observed-score variance is decomposed into the universe-score variance and $\mathscr{E}\sigma^2(\delta)$. The reader can contrast the components of $\mathscr{E}\sigma^2(\delta)$ with the components of $\sigma^2(\Delta)$.

It is possible to convert the observed deviation score for $p$ into an interval estimate of $\mu_p - \mu$, by writing $(X_{pIJ} - X_{PIJ}) \pm a\hat{\sigma}(\delta)$. For tests that have been rendered equivalent, either by careful construction or by means of a conversion scale, $\delta = \Delta$, and the standard deviation of either may be referred to as the standard error of measurement. Where conditions are not equivalent, the standard error computed by the classical formula resembles $\sigma(\delta)$ rather than $\sigma(\Delta)$. The interval estimate of the deviation score based on $\delta$ appears to have little practical significance. However, $\sigma(\delta)$ is pertinent to the problem taken up in the next section.

## Confidence interval for a difference between persons

Comparative decisions, such as the selection of the best 3 out of 10 applicants for employment, are based on individual differences. The decision that applicant 3 is truly better than applicant 4 can be made confidently if their observed difference is substantially larger than its standard error.

In this type of comparison we have scores $X_{pij}$ and $X_{p'i'j'}$, where $p$ and $p'$ differ, but $i$ and $i'$, or $j$ and $j'$ may be the same. For simplicity, it is assumed that $n_i' = n_j' = 1$. Presumably, one wishes to generalize to the universe-score difference $\mu_p - \mu_{p'}$.

Table 3.9 shows how the error of generalization for the difference can be evaluated. If conditions $i$ and $j$ are nested within persons, there is a different

**TABLE 3.9.** *Components of the Difference between Observed Scores of Two Persons $(n_i' = n_j' = 1$; Generalization to $\mu_p - \mu_{p'})$*

| $X_{pij}$ | $X_{p'i'j'}$ | Error of generalization | Contribution to error variance if: | | |
|---|---|---|---|---|---|
| | | | $i \neq i', j \neq j'$ (Design IV-B, III-A, or II) | $i \equiv i', j \neq j'$ (Design V-B or IV-A) | $i \equiv i', j \equiv j'$ (Design VII, V-A, or III-B) |
| $\mu$ | $\mu$ | | | | |
| $\mu_p$~ | $\mu_{p'}$~ | $\mu_p$~ $- \mu_{p'}$~ | | | |
| $\mu_i$~ | $\mu_{i'}$~ | $\mu_i$~ $- \mu_{i'}$~ | $2\sigma^2(i)$ | | |
| $\mu_j$~ | $\mu_{j'}$~ | $\mu_j$~ $- \mu_{j'}$~ | $2\sigma^2(j)$ | $2\sigma^2(j)$ | |
| $\mu_{pi}$~ | $\mu_{p'i'}$~ | $\mu_{pi}$~ $- \mu_{p'i'}$~ | $2\sigma^2(pi)$ | $2\sigma^2(pi)$ | $2\sigma^2(pi)$ |
| $\mu_{pj}$~ | $\mu_{p'j'}$~ | $\mu_{pj}$~ $- \mu_{p'j'}$~ | $2\sigma^2(pj)$ | $2\sigma^2(pj)$ | $2\sigma^2(pj)$ |
| $\mu_{ij}$~ | $\mu_{i'j'}$~ | $\mu_{ij}$~ $- \mu_{i'j'}$~ | $2\sigma^2(ij)$ | $2\sigma^2(ij)$ | |
| $\mu_{pij}$~,e | $\mu_{p'i'j'}$~,e' | $\mu_{pij}$~,e $- \mu_{p'i'j'}$~,e' | $2\sigma^2(pij,e)$ | $2\sigma^2(pij,e)$ | $2\sigma^2(pij,e)$ |

$i$ and $j$ for each person. Therefore, all score components except $\mu_p\sim$ contribute to the error. The expected variance of $\mu_i\sim - \mu_{i'}\sim$ equals $2\sigma^2(i)$; a similar expected variance is found for each other component.

The next column examines the possibility that $i$ is crossed with persons and $j$ is not. In such a D study the component for $i$ cancels out of the difference score. In the third design, where both $i$ and $j$ are crossed with $p$, the final column indicates that the $j$ and $ij$ components also disappear from the difference score and from the variance.

It is readily seen that the error variance for the difference in each case equals $2\mathscr{E}\sigma^2(\delta_{pij})$ where that variance is properly calculated for the design under discussion. This same general statement holds for designs with larger $n'_i$ and $n'_j$. Therefore, $\hat{\sigma}(\delta)$ gives an indication of the adequacy of the measuring procedure for making *comparative* decisions. Table 3.9 is developed in terms of components of the error (i.e., of the discrepancy between the observed-score difference and the universe-score difference). Only trivial changes in notation would result if deviation scores were used in the development.

If many judgments are to be made regarding the comparative superiority of individuals, a confidence interval could profitably be formed for each person's score, extending, for instance, $\hat{\sigma}(\delta)$ units on either side of the observed score. Then, if the interval for one person is entirely above the interval for another person, the difference in their observed scores is at least 1.41 times the standard error of the difference score. Judgments that the universe score of the higher scoring person is superior to that of the lower scoring person will be correct with probability $\geqslant 0.84$. Criticisms to be made later of such probability statements do not apply here. If the two persons are members of the same group, and therefore the only information on which one can distinguish them is the observed score, the statistical inference is sound.

It is now evident that a design that minimizes $\mathscr{E}\sigma^2(\delta)$ improves the accuracy of conclusions about individual differences in $\mu_p$. All components of $X_{pIJ}$ except that for $p$ can contribute to the error of generalization. Consequently, when individual differences are the concern, a D-study design that eliminates other components from the observed score, or samples them frequently to reduce their contribution to variance, is preferred. The familiar crossed Design VII is good for this purpose, because it brings several potential sources of unwanted variance under experimental control. But III-B—$(i,j) \times p$—proves to be better. Design III-B eliminates the same components as Design VII, and it samples each of the remaining components a greater number of times.

Suppose that each preschool child is to be observed by several child-development students, to provide a score representing differences in aggressive behavior. Generalization over observers and occasions is intended, and it has been decided that a total of 25 observer-hours may be used. Then Design

VII would probably call for sending five girls to visit the school together five times. Design III-B would call for sending 25 girls, each at a different time. In either design all pupils are observed by the observer on each visit. According to Table 3.8, Design III-B would reduce the *pi* and *pj* components of variance markedly. These components are regarded as "errors of observation" by the investigator who is interested in aggressiveness without regard to occasion or observer. Because he wants to minimize the effect of occasion and observer, he prefers Design III-B to VII. Further examples of this line of reasoning will appear in Chapters 6 and 7. (The practical tester has to be concerned with nonmathematical matters also, such as whether 25 adequately qualified observers can be recruited.)

### *Confidence interval for a group mean*

The reader should be familiar with the textbook technique for determining the standard error of a mean. The standard deviation of observed scores is divided by the square root of the number of observations. The resulting standard error can be used to establish a confidence interval for the population mean. It is this technique that the tester has adapted to establish a confidence interval for a universe score.

In experimental research, in educational evaluation, etc., a tester may be primarily interested in the group mean. The textbook teaches him to derive the confidence interval from the standard deviation of observed scores and $n'_p$. This may work well enough for many studies, but it tends systematically to underestimate the confidence interval for the mean the tester should be most interested in. The conventional approach, applied to a score $X_{pIJ}$ and its mean $X_{PIJ}$, establishes a confidence interval for $\mu_{IJ}$, not for $\mu$. The sampling of persons, but not the sampling of conditions, is taken into account.

If the tester wishes to generalize from the sample mean $X_{PIJ}$ to $\mu$, the mean expected over the population of persons and conditions, the error is:

$$(3.7) \quad \Delta_{PIJ} = X_{PIJ} - \mu = \mu_P{\sim} + \mu_I{\sim} + \mu_J{\sim} + \mu_{PI}{\sim} + \mu_{PJ}{\sim} \\ + \mu_{IJ}{\sim} + \mu_{PIJ}{\sim} + e_{PIJ}$$

The experimental design determines how many *i* and *j* enter the means. With the use of the information in Table 2.1, for a D study of Design VII,

$$(3.8) \quad \Delta_{PIJ} = \frac{1}{n'_p}\sum \mu_p{\sim} + \frac{1}{n'_i}\sum \mu_i{\sim} + \frac{1}{n'_j}\sum \mu_j{\sim}$$

$$+ \frac{1}{n'_p n'_i}\sum\sum \mu_{pi}{\sim} + \frac{1}{n'_p n'_j}\sum\sum \mu_{pj}{\sim} + \frac{1}{n'_i n'_j}\sum\sum \mu_{ij}{\sim}$$

$$+ \frac{1}{n'_p n'_i n'_j}\sum\sum\sum (\mu_{pij}{\sim} + e_{pij})$$

Hence, the variance over samples of persons and conditions:

$$(3.9) \qquad \sigma^2(\Delta_{PIJ}) = \frac{1}{n_p'} \sigma^2(p) + \frac{1}{n_i'} \sigma^2(i) + \frac{1}{n_j'} \sigma^2(j)$$

$$+ \frac{1}{n_i' n_j'} \sigma^2(ij) + \frac{1}{n_p'} \mathscr{E}\sigma^2(\delta_{pIJ})$$

The components $\sigma^2(p)$ and $\mathscr{E}\sigma^2(\delta)$ together constitute the expected observed-score variance; the second, third, and fourth terms of (3.9) are omitted from the error variance for the group mean as it is usually calculated in statistics texts.[3]

With other two-facet designs, similar equations apply. Components enter $\mathscr{E}\sigma^2(\delta)$ with different weights in each design, and the multipliers for the $i$ and $j$ components of variance change. If $i$ is nested within $p$, $\sigma^2(i)$ disappears from the equation [but is counted in $\mathscr{E}\sigma^2(\delta)$]. Similarly for $j$ nested, or $i$ and $j$ nested. This nesting reduces the discrepancy between the conventional standard error of the mean and ours. Where the foregoing argument applies, the investigator wishes to generalize beyond the particular conditions $I$, $J$, etc. represented in the experiment or test to a larger universe. Increasing $n_p'$ has only limited ability to increase the power of his experiment; even with the sample of persons indefinitely large, the population-universe mean is not precisely estimated.

### D. *The Coefficient of Generalizability and the Error* ε

Where the universe score is $\mu_p$, $\sigma^2(p)$ alone is the universe-score variance. If multiplication of observed-score variance (e.g., by $n_i'^2$) is required to estimate variance for a total score rather than an average, the estimate for universe-score variance must be similarly multiplied.

### *The intraclass correlation*

The coefficient of generalizability for a certain universe and D-study design is the ratio of the universe-score variance to the expected observed-score variance for that design—an intraclass correlation. It is completely comparable to the traditional reliability coefficient except that full attention has to be given to the universe definition and to the design of the D study. Since

---

[3] Our statements here have much in common with those of Mosteller and Tukey (1968, p. 122 ff.). The flavor of their discussion is given by section headings, to wit: "Hunting out the real uncertainty," and "How $\sigma/\sqrt{n}$ can mislead."

the D-study design need not be the same as that in the G study, one can arrive at a number of different coefficients from a single G study.

The coefficient for a certain D-study design could be obtained directly by applying the design repeatedly to the same large group of persons, a different set of conditions being drawn for each application. Each sample of conditions would produce an array of scores, which could be converted into deviations from the group mean. An intraclass correlation could then be computed for the matrix of deviation scores arising from the several applications of the design. Taking deviations eliminates any effect that is crossed with persons. The coefficient to be estimated is a value for the population and universe; it equals the limit of the intraclass correlation, as the number of persons and the number of applications of the design increase without limit.

One might think of intercorrelating all pairs of columns in the hypothetical matrix of scores from many D studies. The average correlation would be akin to the conventional correlation between two independent measurements for the same persons. The intraclass correlation is quite similar to this in conception, but has the advantage that a single calculation takes into account all pairs of conditions at once. Scores are not standardized within columns as they are in calculating the conventional interclass correlation. Any differences in population variances among the several sets of observed scores will therefore lower the intraclass coefficient to some extent.

Because the relation of observed score to universe score is more fundamental than the relation between independent observed scores, we think in terms of the squared correlation of observed score with universe score. This varies from one application of the design to the next when conditions are not equivalent; the intraclass correlation approximates its expected value. We therefore identify the coefficient as $\mathscr{E}\rho^2(X_{pIJ},\mu_p)$ and use the abbreviated symbol $\mathscr{E}\rho^2$. The expectation is defined over experiments applying the specified design to the population. A fuller notation would indicate the facets and their crossing and nesting in the D-study design, and would give the values of $n_i'$ and $n_j'$. The reader is reminded once again that for any measuring procedure there are many coefficients—one for each design that may be proposed.

The estimate of $\mathscr{E}\rho^2$ is made as in the one-facet study, by dividing the estimate of universe-score variance by the estimate of expected observed-score variance. This is a consistent estimate (Lord & Novick, 1968, p. 202) but not a truly unbiased one. Although a ratio of unbiased estimates of two parameters is not a strictly unbiased estimate of the ratio of the parameters, this is unlikely to introduce appreciable error. With the use of the components estimated on p. 44 from the data of Table 2.2, and assuming a crossed D study with $n_i' = 10$ and $n_j' = 1$, Table 3.10 yields these estimates:

$\widehat{\mathscr{E}\sigma^2}(X)$, 6.34; $\widehat{\sigma^2}(\mu_p)$, 5.71; and $\widehat{\mathscr{E}\rho^2}$, 0.90.

**TABLE 3.10.** *Estimation of $\mathscr{E}\rho^2$, $\sigma^2(\varepsilon)$, and $\mathscr{E}\sigma^2(\delta)$ for a D Study with the Design* $i \times j \times p$ $(n'_i = 10, n'_j = 1;$ *Generalization to* $\mu_p - \mu)$

| Source of variance | Estimate of variance component[a] | Frequency within the deviation score | $\mathscr{E}\sigma^2(X)$ | $\sigma^2(\mu_p)$ |
|---|---|---|---|---|
| $p$ | 5.71 | 1 | 5.71 | 5.71 |
| $i$ | 1.32 | | | |
| $j$ | 0.09 | | | |
| $pi$ | 2.57 | $n'_i = 10$ | 0.26 | |
| $pj$ | 0.21 | $n'_j = 1$ | 0.21 | |
| $ij$ | 0.07 | | | |
| $pij, e$ | 1.57 | $n'_i n'_j = 10$ | 0.16 | |
| | | | 6.34 | 5.71 |

$$\widehat{\mathscr{E}\rho^2} = \frac{5.71}{6.34} = 0.903$$

$$\widehat{\sigma^2(\varepsilon)} = 5.71(1 - 0.903) = 0.55; \ \hat{\sigma}(\varepsilon) = 0.74$$

$$\widehat{\mathscr{E}\sigma^2(\delta)} = 6.34 - 5.71 = 0.63; \ \hat{\sigma}(\delta) = 0.63^{1/2} = 0.79$$

$$\widehat{\sigma^2(\Delta)} = 0.86; \ \hat{\sigma}(\Delta) = 0.93 \ (\text{from Table 3.5})$$

[a] Calculated on page 44.

***Extension to subpopulations.*** Very often $D$ data are collected on a subpopulation rather than on the full range of the population represented in the G study. Ideally, there would be a new G study for the subpopulation, but this is not always practicable. Estimates from the original G study can be modified to fit the subpopulation, as in classical theory. Classical theory assumes that the "error variance" has the same magnitude in the subgroup as in the original population. Having obtained the observed-score variance directly from the subgroup data, the classical approach subtracts the error variance to estimate the true-score variance for the subgroup. The procedure is applied not only to samples that are subgroups of the original population, but also to groups that have a wider range of ability.

A similar correction for "restriction of range" can be made in generalizability theory. It is necessary to assume that all components of $\sigma^2(\Delta)$, or at least all components of $\mathscr{E}\sigma^2(\delta)$ in the D-study subpopulation, equal those in the G-study population. As in classical theory, cases entering the subgroup must be selected without regard to scores on the particular conditions selected for the D study.

It is presumed that the distribution of $\mu_p$ in the subpopulation differs from that in the population. It is necessary to obtain observed scores for the

D-study subpopulation and to calculate the observed-score variance directly. One then subtracts from the observed-score variance the G-study estimate of any component such as $\sigma^2(pI)$ or $\sigma^2(I)$ that contributes to it, to obtain the new $\widehat{\sigma^2}(\mu_p)$. This, divided by the observed-score variance for the D-study sample, gives a value of $\widehat{\mathcal{E}\rho^2}$ for the subpopulation.

The assumption made is one of long standing, and is often plausible. However, if the subpopulation falls near the upper extreme of the scale, ceiling effects may reduce other components of variance as well as $\sigma^2(p)$. At the lower extreme, there may be a floor effect; but, under circumstances where guessing is possible, the residual component may be increased in a low scoring subpopulation.

Testers will undoubtedly continue to use the correction for restricted range as a rule of thumb, because tests are applied in subpopulations where no generalizability study has been made. Empirical studies comparing indirect adjustments of $\sigma^2(p)$ and $\mathcal{E}\rho^2$ with those directly determined from distinctive subgroups within the G-study sample should be made for some typical measures and populations. This will increase our knowledge regarding the extent to which the indirect procedure is misleading. Such studies employing the more complex designs and weaker assumptions of generalizability theory would supplement earlier work under the classical model (Gulliksen, 1950, pp. 197–198).

*Effects of nonequivalence of conditions.* Our model acknowledges the possibility that different conditions will produce scores with nonuniform statistical characteristics. The classical assumptions that tests have equal means, variances, and intercorrelations have been avoided. This limits one's inferences—for example, from a G study with the $i \times p$ design, we estimate the expected value over $I$ of $\sigma^2(X_{pI} \mid I)$. When we do not assume equivalence we can regard this as no better than a rough estimate of the observed-score variance arising under the conditions of $i$ that may be drawn for any one study. Nonequivalence similarly restricts inferences about correlations.

For D-study designs that do not cross $p$ with any other facet, scores obtained under the random-sampling model *completely satisfy the classical equivalence assumptions.* This is true even if there is marked non-equivalence of scores from condition to condition. The scores obtained under every application of the design have the same limiting distribution as $n'_p$ increases. The interclass correlation between scores from any two applications approaches a limiting value equal to the intraclass correlation $\mathcal{E}\rho^2$.

Where there is crossing of $p$ with facet $i$ (or $j$, etc.) observed-score variances may differ from one application of the design to the next, and intercorrelations between pairs of independently obtained observed scores may differ. The intraclass correlation (our coefficient of generalizability) truly equals the mean of $\rho(X,X')$ or the mean of $\rho^2(X,\mu_p)$ only if all observed-score variances are equal. One must be hesitant, then, in taking the coefficient of generalizability as representing the parameter $\rho^2(X,\mu_p)$ for any particular D study with conditions crossed. The hazard is much reduced when many conditions of $i$ and $j$ are sampled for the D study, because this tends to produce equivalence of $I$ and $J$ from different applications of the design.

It would be highly desirable to have numerical experiments with higher-order designs, similar to those that have been made for the one-facet crossed design (see p. 83). These would provide needed information concerning the correspondence of the intraclass correlation to interclass correlations and to the squared correlations of observed score with universe score. Empirical studies with various kinds of data would also be illuminating. One can be sure that no serious discrepancies will arise where the number of conditions of each facet in the D study is large or where the conditions are closely comparable. Ordinarily, except where the facet considered is test items or peer raters, however, an investigator uses only a small number of conditions in his D study.

*Difficulties of evaluating specific conditions.*    The model that we employ recognizes that the conditions employed in the D study have their own specific components $\mu_I - \mu$, $\mu_{pI}\sim$, etc. Therefore, where the design has $I$ crossed with $p$, one might think of evaluating $\sigma^2(\Delta_{pI} \mid I)$, $\sigma^2(X_{pI})$, and $\rho^2(X_{pI},\mu_p)$. Of these, the D data directly estimate only $\sigma^2(X_{pI})$.

In theory one might carry out a G study with $i \times p$ that includes the particular conditions $I^*$ as part of a larger collection of conditions of $i$. One could then infer from this the parameters that indicate how well $X_{pI^*}$ or $X_{pI^* J}$, etc. can be generalized to $\mu_p$. Equations can be developed for this purpose. For example: Write $I'$ for a set of $n_i$ conditions in the G study that has no $i$ in common with $I^*$. Then in a $j:(I \times p)$ study that collects data on both $I^*$ and $I'$, the covariance of $X_{pI^* J}$ with $X_{pI'J}$ estimates $\sigma(X_{pI^*},\mu_p)$; the estimate improves as $n_i$ $(i \in I')$ increases.

The difficulty with such a proposal is that the size of the set $I'$ must be very large to achieve a stable estimate of the specific covariance. It is not worthwhile to attempt this unless the covariance varies from one $I$ to another. Yet the more the parameter varies, the larger the sample of conditions required to make $I'$ representative enough of the universe to obtain stable estimates. A small amount of numerical experimentation leaves us pessimistic about the practical utility of investigations of parameters for specific conditions from conventional crossed designs.

Possibilities that we have not investigated are opened up by the theoretical work of Lord and Novick (1968) who stay within the one-facet model for "nominally" parallel tests. Formally, this is equivalent to our assumption that conditions $i$ are randomly sampled from a large set. Lord and Novick arrive at conclusions generally similar to ours, but they recommend (p. 210) estimating what they call the "generic reliability coefficient"—our $\rho^2(X_{pi*}, \mu_p)$. They suggest that this can be estimated not only by the design discussed above, which is ordinarily of no practical value, but also by an item-sampling design. Essentially, their plan is to divide a large sample of persons at random into three or more subgroups and to observe each subgroup under $i*$ and one other condition $i$ (different for each subgroup). Another tentative proposal is made by Overall (1968). These possibilities have not yet been studied in any detail, and nothing is known about the difficulties that may arise when they are extended to the multifacet universe.

The Lord–Novick argument, which is a modern version of methods explored by Burt (1936, pp. 270–297), becomes particularly appealing when the set of tests is believed to have just one common factor. In this case, intercorrelations among three tests (derived from one large sample or three separate large subsamples) provide a sufficient basis for accurate determination of the correlations of scores for each of the three with the universe score. This model can accept multifacet data, and can recognize the systematic difference in error variance that might be associated with different facets.

### The point estimate of the universe score

A linear function of the observed score gives a better estimate of the universe score than the observed score itself. The function is a weighted average of the person's observed score and the observed mean in some group to which he belongs, as was seen in (3.2) and (3.5). The person's raw score in the D study is weighted by the coefficient of generalizability, and the group mean is weighted by one minus the coefficient. The group mean and the coefficient should be estimated for the same population, universe, and D-study design.

Where there are two or more subpopulations, each has its own equation; the values of $\hat{\mu}$ and $\widehat{\mathscr{E}\rho^2}$ would ideally be determined from separate G studies within the subpopulations. Lacking separate G studies, one is forced to use the correction for range suggested earlier.

In this section we shall employ an estimation equation derived from $\widehat{\mathscr{E}\rho^2}$. *This is not, in general, a genuine "regression" equation* because where nonequivalent conditions are crossed with persons in the D study, the slope and the constant term of the regression equation depend on the particular conditions selected. We proceed for the present without further discussion of the anomalies that may result with nonequivalent conditions, and return to the subject in Chapter 5.

Test theorists have long recognized that when a fifth grader and a fourth grader earn the same score on an achievement test, it is likely that the fifth grader is superior (i.e., has the greater universe score). "Regression toward the mean" is inevitable. The observed score is an indicator of ability, but so is the mere fact of membership in a group with a high (or low) mean, since that fact is a consequence of realities of age and past performance.

One of the most complete discussions of point estimates is that of Truman Kelley (1947, p. 409ff.). In quoting his remarks, we substitute our own notation and equation numbers, and alter some punctuation. He starts with an equation comparable to (3.5), but not recognizing our distinction between G and D data. We write it in terms of our coefficient:

$$(3.5a) \qquad \hat{\mu}_p = (\widehat{\mathscr{E}\rho^2})X_{pI} + (1 - \widehat{\mathscr{E}\rho^2})X_{PI}$$

This [he says] is an interesting equation in that it expresses the estimate of true ability as a weighted sum of two separate estimates—one based upon the individual's observed score $X_{pI}$ and the other based upon the mean of the group to which he belongs $X_{PI}$. If the test is highly reliable, much weight is given to the test score and little to the group mean, and vice versa. Suppose fourth-grade pupil $p$ and fifth-grade pupil $p'$ each score 45 on a test having a reliability of 0.80 in each grade, and that the means and standard deviations for the grades are: $X_{PI(4)} = 40$; $s_{X(4)} = 10$; $X_{PI(5)} = 50$; and $s_{X(5)} = 10$. For pupil $p$ we estimate his true ability thus: $\hat{\mu}_p = 0.80(45) + 0.20(40) = 44$. For pupil $p'$: $\hat{\mu}_{p'} = 0.80(45) + 0.20(50) = 46$. This difference in outcome is certainly sound. We know two things about pupil $p$. The first fact ($X_{pI} = 45$) suggests a true ability of 45, and the second fact (member of group whose mean = 40) suggests a true ability of 40. The best composite of his ability is 44, as given by [3.5a]. Suppose for the single hour when tested pupil $p$ had sat with the fifth grade, would we now use the fifth-grade mean and estimate his true ability as 46? Certainly not, for pupil $p$ is still a fourth-grader. This group membership is not a whim, but a thing as definitely attached to pupil $p$ as is his score 45.

. . .

If the mean and reliability for the group to which the tested person naturally belongs are known, it is always preferable to use the regressed score as the estimate of true ability. Since this best practice is infrequent practice, . . . [most writers use $\hat{\sigma}(\delta)$ as a standard error].

[Further results throw] interesting light upon the classification of individuals by fallible measures. Suppose upon a scholastic test we have fourth, fifth, and sixth grade means of 40, 50, and 60, and that $\sigma(\mu_p)$ for the fifth grade is 10. Assume a normal distribution of ability and a rule which demotes[4] fifth-graders scoring below 40 and promotes fifth-graders scoring above 60. If the reliability of the test is 1.00, we obtain the correct result of 16% below 40 and 16% above 60, and thus, we reclassify 32% of the pupils. If the test has a reliability of 0.50, we find that $\sigma(X) = 14.14$. Employing raw scores, reference to a normal probability table informs us that we would now reclassify 48%, which is an excessive number. If, however, we use regressed scores $\hat{\mu}_p$, we have a distribution whose standard deviation is 7.07 and reclassify 16%, which is a conservative number. It can also be shown, using volumes of a normal bivariate surface, that of the 48% reclassified upon the basis of raw scores, 26/48 did not in truth fall beyond the limits set, and that of the 16% reclassified upon the basis of regressed scores, 6/16 did not in truth fall beyond these limits. In short, the use of a fallible measure at its face value in connection with promotions, classifications, etc., will lead to or create many misplacements,[5] while the use of this same fallible measure properly regressed will create few misplacements. If we will but regress scores and compute standard errors of estimated true scores, we need not hesitate to use an instrument of low reliability.

Substituting a linear function of the observed score for the observed score itself would not alter a decision that simply considers ranks of persons within an undifferentiated group, but if there are subgroups (by sex, education, or other demographic variables) the use of separate regression estimates for each subgroup alters ranks. Application of the appropriate linear function to each individual also alters ranks if some persons have been observed more thoroughly than others. Regression of scores alters decisions that rest on comparisons of individual scores to an absolute standard, or decisions that consider the shape of a score profile.

Estimates of universe scores are rarely made by test interpreters. While there are arguments against regressing scores, the objections are surely no more damaging than the arguments against more commonplace procedures of test interpretation. One is left with the impression that estimation procedures were neglected in the past merely because theorists did not communicate their value to practitioners. The topic is allotted about two pages in texts on test theory, and is left out of the discussion of reliability in texts on psychological statistics or applied testing. Very likely one deterrent to the

---

[4] This reference suggests a more rigid practice in classifying pupils than is to be found in schools currently, but reassignments within today's nongraded schools or in individually prescribed instruction follow a similar logic.

[5] Kelley does not give due recognition to the fact that the number of errors of omission (failures to promote or demote pupils whose true scores are outside the 40–60 range) increase when regressed scores are used. This is a fault of inaccurate measurement, however, rather than of the regression technique itself.

use of estimates is the amount of arithmetical labor it adds to test scoring; the availability of computers should lessen resistance.

It is known that the error ε is smaller on the average than the error Δ or δ. So far as we know, no one has made the more startling point that regressed scores are likely to be more reliable (!) than observed scores. This is the case whenever the linear estimates of universe scores take into account the different means of identifiable subgroups. Then the correlation of $\mu_p$ with $\hat{\mu}_p$ for all cases together exceeds that for $\mu_p$ with $X_{pI}$. (This will not be strictly true if conditions are not equivalent. The following discussion assumes equivalence.)

To develop this point, consider two or more fixed subpopulations, each with a mean universe score denoted by $\mu_P$. For simplicity, assume that all subgroups have the same very large number of cases and the same within-group variance. Call the grand mean $X_{.I}$. The observed score, expressed as a deviation from the grand mean, is $X_{pI} - X_{.I}$; it resolves into two orthogonal components: $(X_{pI} - X_{PI}) + (X_{PI} - X_{.I})$. The first is the within-group (w.g.) deviation score and the second is the between-groups effect. Considering all cases together, the latter is a variable and not a constant. When one determines multiple-regression weights for predicting $\mu_p$ from the two orthogonal variables, each weight is equal to the relevant covariance divided by the variance of the predictor:

$$(3.10) \quad \text{Est.} \, (\mu_p - \mu) = \frac{\sigma^2(\mu_p, \text{w.g.})}{\sigma^2(X, \text{w.g.})} (X_{pI} - X_{PI}) + \frac{\sigma^2(\mu_P)}{\sigma^2(X_{PI})} (X_{PI} - X_{.I})$$

In a large sample the second regression weight approaches 1, $X_{.I}$ approaches $\mu$, and $X_{PI}$ approaches $\mu_P$. The regression estimate of the universe score, then, is an optimally weighted combination of $X_{pI}$ with a good estimate of the score $\mu_P$. The correlation of the estimate with $\mu_p$ (all cases considered) is a multiple correlation, and must be at least slightly greater than the corresponding zero-order correlation of $X_{pI}$ with $\mu_p$.

The increase in the squared correlation has the following form:

$$(3.11) \quad \frac{[\mathscr{E}\sigma^2(\delta)]^2 \times \sigma^2(\mu_P)}{\sigma^2(\mu_p, \text{all cases}) \times \sigma^2(X, \text{all cases}) \times \sigma^2(X, \text{w.g.})}$$

Other things being equal, the greater the separation of group means, the higher the multiple correlation.

Recognize, however, that when $\sigma^2(\mu_p)$ and $\sigma^2(\mu_P)$ are fixed, the advantage of the multiple-regression procedure increases with $\mathscr{E}\sigma^2(\delta)$. When one has an accurate observation procedure, the observed score is an excellent predictor of the universe score and introducing a second variable can add nothing. For further discussion, see p. 151f.

Regressing toward subgroup means is sometimes open to criticism on grounds of social policy. To point to the most obvious example, black and white applicant populations commonly have different means on aptitude or proficiency tests. Regression equations developed from G studies on the racial groups separately are likely to give a white a higher estimated universe score than a black who has the same observed score. While the alteration will, on the average, produce slightly more accurate predictions of relative standing on the criteria, it seems most unlikely that one could convince the black applicant or the FEPC (Fair Employment Practice Commission) examiner that the regression estimate is legitimate and unbiased. For further remarks on this point, see p. 383ff.; see also Novick (1971).

Charges of social injustice are not so likely when tests are used for guidance. It can readily be seen that failure to regress gives the above-average member of a group that has a low average a falsely favorable picture of himself, underestimates the difficulties he will encounter, and so misleads him. The fact remains, however, that when scores are regressed, fewer members of the low-scoring group are given encouragement to set high goals. It is a serious question whether statistically realistic forecasts are to be preferred over optimistic ones, when there is a need to redress social imbalance.

### The error of estimate

Let us return to the one-facet D study with design $I{:}p$. The sample mean estimates $\mu$, the constant term of the regression equation (see p. 141f.). With this nested design, $\rho^2(X_{pI},\mu_p)$ is the same for every application of the design to the same population (or subpopulation) and universe, because each application draws a large random collection of $i$. The regression equation when the parameters are known is:

$$(3.2) \qquad \hat{\mu}_p = \mu + \rho^2(X_{pI} - \mu) = \rho^2 X_{pI} + (1 - \rho^2)\mu$$

The estimate $\hat{\mu}_p$ departs from the actual $\mu_p$ by some amount $\varepsilon_p$ (i.e., this error equals $\hat{\mu}_p - \mu_p$). Assume that excellent estimates of $\rho^2$ and $\mu$ are available. Then:

$$(3.12) \qquad \varepsilon_{pI} = \widehat{\rho^2}(X_{pI} - \mu_p) + (1 - \widehat{\rho^2})(\hat{\mu} - \mu_p)$$

Now $X_{pI} - \mu_p = \Delta_{pI} = \mu_I{\sim} + \mu_{pI}{\sim} + e_{pI}$, and $(\hat{\mu} - \mu_p)$ is the person component. It follows that in the population:

$$(3.13) \qquad \sigma^2(\varepsilon) = (\widehat{\rho^2})^2\sigma^2(I) + (\widehat{\rho^2})^2\sigma^2(pI,e) + (1 - \widehat{\rho^2})^2\sigma^2(p)$$

Recognizing that

$$\widehat{\rho^2} = \widehat{\sigma^2}(p)/[\widehat{\sigma^2}(p) + \widehat{\sigma^2}(I) + \widehat{\sigma^2}(pI,e)]$$

or

$$\widehat{\sigma^2}(p)/\widehat{\sigma^2}(X),$$

we can write:

(3.14)     $$\widehat{\sigma^2}(\varepsilon) = \frac{\widehat{\sigma^2}(p)}{\widehat{\sigma^2}(X)} [\widehat{\rho^2}\widehat{\sigma^2}(X)] + \widehat{\sigma^2}(p) - 2\widehat{\rho^2}\widehat{\sigma^2}(p)$$

and

(3.3)     $$\widehat{\sigma^2}(\varepsilon) = \widehat{\sigma^2}(p)(1 - \widehat{\rho^2})$$

This is the familiar variance for errors of estimate in linear regression.

Both $\hat{\mu}$ and $\widehat{\rho^2}$ come from finite samples, so the regression equation is not exactly known. This enlarges the error of estimate above the value given by (3.3). We encounter an unsolved statistical problem here. There is statistical theory for taking sampling of persons into account in a regression estimate of one observable variable from another. However, this is inadequate for the regression equation (3.5) even under classical assumptions, because the correlation of true score with the predictor (observed score) cannot be directly calculated. It is very possible that statistical theory can be extended to this case, and to a regression equation based on the intraclass correlation. Conceivably it could be extended further to take sampling of conditions into account. Lacking such theory, studies with the jackknife procedure should be made to learn how much the magnitude of errors is underestimated by (3.3) in typical studies.

For the crossed $i \times p$ design, one cannot evaluate the regression equation for the specific condition; the substitutions made will be discussed later (see p. 142). One may calculate:

(3.15)     $$\widehat{\sigma^2}(\varepsilon) \geqslant \widehat{\sigma^2}(p)(1 - \widehat{\mathscr{E}\rho^2})$$

but the equality does not hold even approximately unless $\rho(X,\mu)$ and $\sigma(X)$ are equal for all $I$.

Formulas (3.5) and (3.15) generalize to any number of facets. The observed score in (3.5) becomes $X_{pIJ\cdots}$, and $\widehat{\mathscr{E}\rho^2}$ is calculated as required by the D-study design. It may be worth noting that, for any design,

(3.16)     $$\frac{\widehat{\mathscr{E}\sigma^2}(\delta)}{\widehat{\mathscr{E}\sigma^2}(X)} = \frac{\widehat{\sigma^2}(\varepsilon)}{\widehat{\sigma^2}(p)} = 1 - \widehat{\mathscr{E}\rho^2}$$

(See also numerical example in Table 3.10.)

### E. Reporting and Interpreting the G Study

It may be well to summarize here the various suggestions that have been made regarding the reporting of a G study. Experience will be needed to discover the most useful ways of organizing a report, and it is likely that a sound report will mean little to a reader who does not understand generalizability theory. However, in our opinion, a presentation covering the following information should be far more satisfactory than the sketchy reports of G studies usually offered.

1. Description of data collection, including design, number of conditions of each facet, nature of the conditions sampled (e.g., qualifications and special training of scorers), conditions held constant in all observations, and conditions confounded with a facet deliberately sampled.
2. Number and character of subjects, including pertinent facts about age, sex, educational background, and selective factors.
3. Estimates of all components of variance the G study allows one to evaluate, with a clear indication of the size of unit represented (e.g., whether $i$ stands for a single item or for a 50-item test).

The essential requirement, in the spirit of the *Test Standards*, is to describe the data in such a way that the reader can decide whether the findings apply to the D data he proposes to gather.

Beyond this, the investigator can repackage the data in various ways to show what precision is expected from alternative experimental designs or in generalizing to various universes. There are usually many possibilities, however, and the initial investigator can reasonably be asked to present statistics for only a few likely possibilities. Appropriate summary statistics include $\hat{\mathscr{E}}\rho^2$, $\hat{\sigma}(\Delta)$, $\hat{\sigma}(\delta)$, and $\hat{\sigma}(\varepsilon)$. Whichever of these appear relevant should be given for the likely D-study designs. The person reporting the G study may appropriately go on to advise his reader regarding the designs most likely to improve the precision of information for various purposes, at least cost (see also p. 175).

The three chapters now completed have presented the essential machinery of generalizability analysis and the interpretation of the results. For many readers, an appropriate next step is to skip ahead to Chapters 6 and 7, which deal with numerical examples. Most of the procedures demonstrated there have already been explained in full, and the remainder have been touched upon. Chapter 4 presents some of the more complex reasoning required to take into account fixed facets in the universe of generalization and hidden facets in the G study. This is a logical extension of Chapter 3. Chapter 5 is a further comment on theoretical matters opened up in the present chapter. Specifically, it examines the assumptions and validity of the usual applications of standard errors of measurement and estimating equations.

For another category of reader, it may be appropriate to leap ahead to Chapter 9, which extends Chapters 1–3 into the multivariate theory for dealing with test batteries. This material is considerably more difficult than what precedes it, and for most readers it can best be examined after the examples of univariate analysis are understood.

## EXERCISES

**E.1.** The following passage is from a manual for an aptitude test (slightly modified).

Alternate-forms reliability was determined by giving Form A and Form B in counterbalanced order, within a two-week period, to the same 484 pupils. In Grade 2 the raw-score mean and standard deviation were 98.3 and 23.0 for Form A, 98.4 and 22.8 for Form B. The correlation was 0.89. Data from the two testing sequences were averaged in determining the correlation.

Reorganize the results to estimate the following if possible: $\mathscr{E}\sigma^2(X)$, $\sigma^2(\mu_p)$, $\sigma^2(i)$ for forms, $\sigma^2(pi,e)$, $\sigma^2(\Delta)$, $\mathscr{E}\sigma^2(\delta)$, $\mathscr{E}\rho^2$, $\sigma^2(\varepsilon)$, and the regression equation for estimating $\mu_p$. Consider Table 3.1 or 3.2 as a partial guide. Assume that the forms are carefully equated.

**E.2.** Suppose the pupil is a member of a second-grade class whose mean and s.d. on the form used are 99.0 and 21.0. Considering the information in Exercise 1 and the answers to it, develop an equation for estimating the pupil's universe score.

**E.3.** A G study provides these estimates of components of variance:

| $p$ | $i$ | $j$ | $pi$ | $pj$ | $ij$ | $pij,e$ |
|----|----|----|----|----|----|----|
| 5  | 2  | 1  | 8  | 4  | 1  | 10 |

A D study will have the design $i \times (j{:}p)$; $n_i' = 4$, $n_j' = 1$, $n_p' = 10$. If generalization is to $\mu_p$, calculate estimates of $\sigma^2(\Delta)$, $\mathscr{E}\sigma^2(\delta)$, $\mathscr{E}\sigma^2(X)$, and $\mathscr{E}\rho^2$.

**E.4.** The mother of a preschool child is observed while she and her child follow certain task instructions. One task $i$ asks the mother to tell the child about wild animals pictured on a card, another asks mother and child to converse over toy telephones, etc. Observers $j$ rated tape recordings of the conversations, using several 1-to-9 scales.

In the course of preliminary studies, Leler (1970) investigated generalizability by having 2 observers judge 23 mother–child pairs $p$ on 6 tasks. The design was $i \times j \times p$. Table 3.E.1 gives data for two of the rating scales.

**TABLE 3.E.1.** Information from Generalizability Study of Mother-Child Interactions (after Leler, 1970)

| Scale | Component of variance | | | | | | |
|---|---|---|---|---|---|---|---|
| | $p$ | $i$ | $j$ | $pi$ | $pj$ | $ij$ | $pij, e$ |
| A. Mother's affectionateness | 0.44 | 0.21 | −0.01 | 0.17 | 0.08 | 0.11 | 0.58 |
| M. Child's dependency | 0.00 | 0.08 | 0.04 | 1.10 | 0.27 | 0.04 | 1.91 |

a. Leler intended in her main study to examine the relations of the child's language development (as observed independent of these interaction data) to the variables describing mother–child interaction. In the light of this purpose, which of the following should the generalizability study be most concerned with: $\sigma^2(\Delta)$, $\mathscr{E}\sigma^2(\delta)$, or $\mathscr{E}\rho^2$?

b. Calculate $\mathscr{E}\sigma^2(\delta)$ and $\mathscr{E}\rho^2$ from the data in Table 3.E.1 assuming an $i \times j \times p$ D study, $n'_i = 6$, $n'_j = 2$.

c. What will be the effect on $\mathscr{E}\rho^2$ for scale A of changing $n'_i$ to 4? to 10?

d. What will be the effect on $\mathscr{E}\rho^2$ for scale A of changing $n'_j$ to 1? to 4?

e. Leler discarded scale M in the final study. What explanations can you suggest for its small person component?

**E.5.** In Israel, ability of pupils at the end of secondary-school Hebrew Composition is measured by an essay test that is typically graded by two persons. A generalizability study considered scores assigned by 2 representative examiners to each of 373 papers written by graduates of 11 schools. The analysis done by Pilliner (1965, p. 289) generated estimates of components of variance for schools, examiners, schools × examiners, candidates within schools, and candidates × examiners (within schools).

a. What is the meaning of the component for schools × examiners?

b. Of the candidates-within-schools component?

c. Suppose that all candidates are in competition with each other for a limited number of places in higher education. Suppose further that in the D study many different examiners will take part in the grading, two for each essay. Which score components contribute to error of measurement in this case?

d. Suppose one wants to judge, from the data described in c, which schools produce on the average the best graduates. (Ignore the fact that some draw superior entrants.) Which score components contribute to error of measurement in this case?

e. Pilliner gives these values for the components, in the order stated above: 3.4, 0, 1.3, 13.5, 21.1. What can be learned from an inspection of these, without further calculation?

f. In the G study, the same topics were assigned to all candidates. Does one intend to generalize over topics? How does this modify the interpretation of the G study?

## Answers

**A.1.** As these forms were equated, it seems most reasonable to regard $\sigma^2(i)$ as zero and to consider the intraclass correlation equal to the interclass correlation.

$$\widehat{\mathcal{E}\sigma^2}(X) = \tfrac{1}{2}[(23.0)^2 + (22.8)^2] = \tfrac{1}{2}(529 + 520) = 524.5$$

$$\widehat{\sigma^2}(\mu_p) = 524.5 \times 0.89 = 466.8$$

$$\widehat{\mathcal{E}\sigma^2}(\delta_{pi}) = \widehat{\sigma^2}(pi,e) = 57.7$$

$$\widehat{\sigma^2}(\Delta) = \widehat{\mathcal{E}\sigma^2}(\delta) = 57.7$$

$$\widehat{\mathcal{E}\rho^2}(X,\mu_p) = 0.89$$

$$\widehat{\sigma^2}(\varepsilon) = 467(1 - 0.89) = 51.4$$

Confidence interval (67%) $X_{pi} - 7.6 \leqslant \mu_p \leqslant X_{pi} + 7.6$
Regression estimate $\hat{\mu}_p = 98.35 + 0.89(X_{pi} - 98.35) = 10.8 + 0.89 X_{pi}$

**A.2.** Observed-score variance = 441. Assuming that $\widehat{\sigma^2}(pi,e)$ remains at 57.7, $\widehat{\sigma^2}(p) = 383$, and $\widehat{\mathcal{E}\rho^2} = 0.87$.

$$\hat{\mu}_p = 99.0 + 0.87(X_{pi} - 99.0) = 12.9 + 0.87 X_{pi}$$

**A.3.**

| Component of observed score | Component of population mean? | Component of variance | Frequency within persons | $\sigma^2(\Delta)$ | $\mathcal{E}\sigma^2(\delta)$ |
|---|---|---|---|---|---|
| $p$ | | 5 | 1 | | |
| $i$ | Yes | 2 | 4 | 0.5 | |
| $pi$ | | 8 | 4 | 2.0 | 2.0 |
| $j, pj$ | | 5 | 1 | 5.0 | 5.0 |
| $ij, pij, e$ | | 11 | 4 | 2.75 | 2.75 |
| | | | $\widehat{\sigma^2}(\Delta) =$ | $\sigma^2(\Delta) = 10.25$ | $9.75 = \widehat{\mathcal{E}\sigma^2}(\delta)$ |

$$14.75 = \widehat{\mathcal{E}\sigma^2}(X)$$

$$5/14.75 = 0.34 = \widehat{\mathcal{E}\rho^2}$$

**A.4.**    a. $\mathscr{E}\rho^2$, and perhaps to a lesser degree $\mathscr{E}\sigma^2(\delta)$. The main investigation proposes to compare individual differences in development with the interaction score.

b. Scale A: $\widehat{\sigma^2}(\delta) = 0.12$; $\widehat{\mathscr{E}\rho^2} = 0.79$.

Scale M: $\widehat{\sigma^2}(\delta) = 0.48$; $\widehat{\mathscr{E}\rho^2} = 0.00$.

c. Becomes 0.75; 0.83.

d. Becomes 0.68; 0.85.

e. Perhaps dependency is highly determined by the situation, so that variance in "general dependency" is negligible. It seems unlikely that it is truly zero, as this would imply negative correlations between dependency for many pairs of situations. In Leler's small sample of persons and conditions, the low value of the component of variance could be a vagary of the sample in hand. Perhaps with more cases one would find that the component is comparable to that for other scores. A more extensive study might also find that, while the hypothesis of a trait of general dependency is not powerful, tasks fall into two or three reasonably homogeneous classes, and that dependency does generalize over the class.

**A.5.**    a. School–examiner interaction implies that some characteristic associated with the school (type of pupil? type of instruction?) attracts higher marks from some examiners than from others.

b. The candidate component is a variance over persons, assuming each to have been scored by very many examiners. Examiner variance is ruled out, and so are school differences.

c. Examiners, schools × examiners, candidates (within schools) × examiners. See also Answer $f$ below.

d. Examiners, schools × examiners, candidates within schools, candidates × examiners (within schools). Candidate variance within the school is not error if there is no generalization beyond this year's graduates, all of whom are tested.

e. Examiners seem to be using much the same scale, but disagree markedly in grading the same paper. There is, as expected, large variation among candidates. The variation over schools is remarkably large, considering that the component for the school is an average over the population of its students.

f. Surely the decision maker is interested in performance over topics in general (and also over occasions). Interest is not confined to the topics assigned. But Pilliner's design leaves topics (papers) as a "hidden" fixed facet (see p. 122ff.). The component for candidates determined here includes the candidate–topic interaction. This interaction is actually a source of error in generalizing.

# CHAPTER 4
# Universes
# with
# Fixed
# Facets

In Chapters 2 and 3 attention was directed toward the so-called random model, in which the universe is assumed to include an indefinitely large number of conditions of $i$, of $j$, and of any other facet. The universe score to which the investigator wishes to generalize was consistently taken to be $\mu_p$, the expected value of $p$'s score over all possible conditions. As was mentioned in Chapter 1, the investigator may select a more restricted universe of generalization. He may, for example, propose to infer $\mu_{pJ*}$ from $X_{pIJ*}$, generalizing over only the facet $i$. The interpretations suggested ﹒ɹ Chapter 3 have to be modified when the investigator proposes to generalize over some but not all facets.

Almost the opposite problem arises when the investigator intends to generalize over a facet, but fails to represent it adequately in the G study. He might, for example, wish to generalize over both $i$ and $j$, but collect all G data under a particular condition of $j$. With $j$ held constant in the G study, he learns nothing about the error of generalization that arises from sampling of conditions of $j$. Investigators sometimes fix a facet without realizing it, and if so they reach incorrect conclusions. This chapter will discuss how such a "hidden" or implicit facet confuses the interpretation of a G study.

## A. Generalization Where the Universe Has a Fixed Facet

It is the substantive interest of the investigator that determines how broadly he wishes to generalize. In a study where observers $i$ have rated the creativity shown by pupils on several tasks $j$, for example, there are three pertinent universes. An investigator will generalize to $\mu_p$ if he looks upon creativity as a global variable, transcending particular tasks or raters. The investigator who wants to investigate task-specific creativity will generalize to $\mu_{pj}$, the pupil's universe score on one of the tasks; that is to say, the investigator generalizes over observers, for each task in turn. Still another possibility is generalization over tasks, to a score $\mu_{pi}$. A social psychologist might be interested in what persons a particular observer rates as most creative rather than in treating the observer as a source of random error. There can be no rule about what facets to generalize over. Different degrees and directions of generalization serve different practical and scientific ends.

Until this point, we have used a single asterisk to identify a particular condition or set of conditions. It will now be necessary to consider two distinct sets of conditions in some of our discussions, and hence to make a notational distinction between $J^*$ and $J^{**}$. In this chapter we shall invariably use $J^{**}$ to refer to the set of $n'_j$ conditions used in the D study where these define the universe of generalization. That is, where the generalization is to $\mu_{pJ^{**}}$, from observed score $X_{pIJ^{**}}$. Now a different set of conditions used in the G study should obviously be denoted $J^*$ for the sake of making the required distinction. But when $J^{**}$ itself constituted the set of conditions for the G study ($n_j$ equalling $n'_j$), there is no particular reason to make the distinction. This chapter will employ the symbol $J^{**}$ when the set in the G study is the same as in the D study and universe.

Assume an intent to generalize over facet $i$ and not $j$, to the universe score $\mu_{pJ^{**}}$. Assume that the universe of conditions of $i$ is indefinitely large. There are two basic cases:

1. $J^{**}$ used in G study. That is, the conditions of $j$ used in the G study will also be used as the conditions for the D study. $J^{**}$ is crossed with persons in both studies. A particularly common example is the study of a test score that is a composite of certain fixed subtests. While conditions of this facet are fixed, conditions of some other facets (items within subtests, occasions) are random; this then can be described by a "mixed model." Lord and Novick (1968, Chapter 8) present the theory for this case at some length, under assumptions more restrictive than ours.[1]

2. $J^{**}$ not used in G study. Conditions $j$ for the G study were sampled from an indefinitely large universe of conditions of $j$. (The set sampled

---

[1] Since the "fixed" $J^{**}$ can be thought of as a sample from a much larger set of $j$, the G data could also be interpreted under the random model. The intention to generalize to $\mu_{pJ}^{**}$ is a decision by one particular user of the procedure, and does not preclude broader generalization to $\mu_p$ by someone else. We shall argue later, for example, that even though essentially the same subtests appear in all forms of the Wechsler test, the interpreter may reasonably regard them as samples from the domains of Verbal and Performance tasks.

could be called $J^*$.) The $n'_j$ conditions to be used in the D study may be regarded as a random sample from this same universe, but they have not yet been selected and there is a negligible probability that they will be the ones that appeared in the G study. It is intended to make inferences about $\mu_{pJ**}$ from $X_{pIJ**}$.

From a theoretical standpoint (though not necessarily an operational one) the two cases are distinct. In the former, the two sets of conditions of $j$ are identical, and identical to the universe of conditions of that facet; in the latter, the sets are independently sampled from the universe of conditions. There are a number of intermediate possibilities, and some of these can be discussed briefly after the two basic cases are considered.

### The error Δ

The discrepancy $\Delta_{pIJ**} = X_{pIJ**} - \mu_{pJ**}$ will be considered first. This will be interpreted in terms of the components of $X_{piJ}\,|\,J$ introduced in (2.19) and (2.20).

***D-study designs having $i \times j$.***   We shall consider various D studies in which the $i$ are crossed with the $j$ that compose $J^{**}$. Then we can average over $j$ to obtain $X_{piJ**}$ for each $p,i$ pair. When this is done, however, each of the "within $J$" components defined in (2.20) becomes zero, and the corresponding components of variance vanish.

Consider Design VII; Table 4.1 resolves the observed score $X_{pIJ**}$ and the universe score $\mu_{pJ**}$ into the components defined in (2.19). This table is much like Table 3.3 for a $p \times i$ design with generalization to $\mu_p$. In fact, the present D study can be regarded as having a one-facet $(i \times p)\,|\,J^{**}$ design, with every $X_{pi}$ observed under the same fixed conditions $J^{**}$.

Where the D data use the same set of conditions $J^{**}$ as the G data, there are several ways to estimate the needed variance components: first, a two-way random-model analysis of the $X_{piJ**}$ from the G study, assuming an indefinitely large number of possible $i$; second, a three-way analysis of $X_{pij}$, using mixed-model equations for expected mean squares; third, a recombination of components estimated by the random-model equations applied to a three-way analysis. All three give the same variance estimates for the components listed in Table 4.1 and lead to the same $\sigma^2(\Delta)$.

If $J^{**}$ differs from $J^*$ there is no way to obtain information directly relevant to $X_{piJ**}$ from the G study. However, by assuming that both $J^{**}$ and $J^*$ are randomly sampled sets from an indefinitely large universe (case 2), we can estimate the *expected* variance of observed and universe scores for any set of $n'_j$ conditions. Random-model estimates of variance components

**TABLE 4.1.** *Components of Scores in a D Study Where i is Crossed with j within the Person (Generalization to* $\mu_{pJ**}$)

| $X_{pIJ**}$ | $\mu_{pJ**}$ | $\Delta_{pIJ**}$ | $\sigma^2(\Delta)$ |
|---|---|---|---|
| $\mu_{J**}$ | $\mu_{J**}$ | | |
| $\mu_{pJ**} - \mu_{J**}$ | $\mu_{pJ**} - \mu_{J**}$ | | |
| $\mu_{IJ**} - \mu_{J**}$ | | $\mu_{IJ**} - \mu_{J**}$ | $\dfrac{1}{n_i'}\,\sigma^2(i \mid J^{**})$ |
| $\mu_{pIJ**} - \mu_{pJ**}$ $-\mu_{IJ**} + \mu_{J**}$ $e_{pIJ**}$ | | $\left.\begin{array}{c}\mu_{pIJ**}\sim\\[6pt] e_{pIJ**}\end{array}\right\}$ | $\dfrac{1}{n_i'}\,\sigma^2(pi,\bar{e} \mid J^{**})$ |

are obtained from the G study. In terms of these estimates,

$$(4.1) \qquad \widehat{\sigma^2}(\Delta) = \frac{1}{n_i'}\left[\widehat{\sigma^2}(i) + \frac{1}{n_j'}\,\widehat{\sigma^2}(ij) + \widehat{\sigma^2}(pi) + \frac{1}{n_j'}\,\widehat{\sigma^2}(pij,e)\right]$$

If $n_j' = n_j$, the numerical value from (4.1) is the same as would be obtained under the assumptions of Case 1, even though $J^{**}$ and $J^*$ are different sets of conditions.

A modification of the Venn diagrams previously used may help to give an intuitive picture of the components entering various scores and variances. Consider diagram (a) in Figure 4.1, which refers to the completely crossed design. In earlier presentations the dotted circle representing effects associated with $J$ was shown complete. In Figure 4.1, it is shown incomplete because observed scores do not vary with respect to $J$. The dotted line passing through the region labelled $p \mid J^{**}$ reminds us that that effect includes information on $\mu_{pj}$ as well as $\mu_p$; similarly in the other regions of the diagram. The diagram, with the lower part of the $j$ circle eliminated, includes all components of the observations $X_{pIJ**}$ from which generalization to $\mu_{pJ**}$ is attempted.

If we now ignore the shaded area, which represents the universe score, all the remainder of the diagram represents components of $\Delta_{pIJ**}$. For this design, $\sigma^2(\Delta)$ is made up of the $i \mid J^{**}$ and $pi,\bar{e} \mid J^{**}$ components. (See Figure 4.2 below.) Noting that each of these components is sampled $n_i'$ times in the D study,

$$(4.2) \qquad \sigma^2(\Delta) = \frac{1}{n_i'}\,[\sigma^2(i \mid J^{**}) + \sigma^2(pi,\bar{e} \mid J^{**})]$$

This differs from $\sigma^2(\Delta)$ for generalization to $\mu_p$ because $\sigma^2(pj)$ has moved over to the universe-score variance.

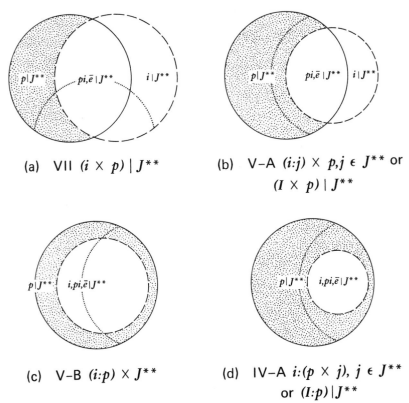

(a)   VII $(i \times p) \mid J^{**}$

(b)   V-A $(i{:}j) \times p, j \in J^{**}$ or
       $(I \times p) \mid J^{**}$

(c)   V-B $(i{:}p) \times J^{**}$

(d)   IV-A $i{:}(p \times j), j \in J^{**}$
       or $(I{:}p) \mid J^{**}$

**FIGURE 4.1.** *Separation of Kinds of Variance in Designs where a Fixed $J^{**}$ is Crossed with Persons.*

Another design that has $i$ crossed with $j$ is V-B, $(i{:}p) \times J^{**}$. As shown in diagram (c) of Figure 4.1, the $j$ component does not contribute to variance, hence is disregarded. The remaining diagram is precisely that for a one-facet $i{:}p$ study, save for a notation recognizing explicitly the use of several fixed $j$. For each $p$ the D study makes $n'$ observations (confounded) of

$$(i, pi, \bar{e} \mid J^{**}).$$

Accordingly, a G study of Design VII using conditions $J^{**}$ gives the estimates required by (4.2). A mixed-model analysis of a Design V-B G study gives $\widehat{\sigma^2}(i, pi, \bar{e} \mid J^{**}) = \widehat{\sigma^2}(\Delta)$. For Case 2, (4.1) is again the suitable general equation.

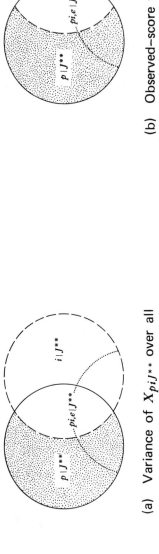

(a)  Variance of $X_{piJ^{**}}$ over all admissible observations on all persons

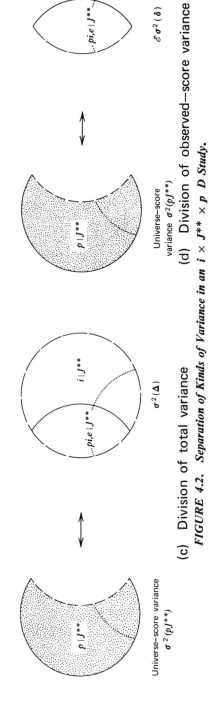

(b)  Observed–score variance

(c)  Division of total variance

(d)  Division of observed–score variance

*FIGURE 4.2.  Separation of Kinds of Variance in an $i \times J^{**} \times p$ D Study.*

**D-study designs having $i:j$.** Again, suppose that $p$ is crossed with $J^{**}$ in the D study, but $i$ is nested. Then we are concerned with Design V-A or IV-A. (Design III-B is simply a version of IV-A.) For scores $X_{pIJ**}$, the design is in effect $(I \times p) \, | \, J^{**}$ and the Venn diagram is almost like that for Design VII. The most general formula, comparable to (4.1), uses values of components estimated by random-model equations:

$$(4.3) \qquad \sigma^2(\Delta) = \frac{1}{n_i' n_j'} \, [\widehat{\sigma^2}(i) + \widehat{\sigma^2}(ij) + \widehat{\sigma^2}(pi) + \widehat{\sigma^2}(pij,e)]$$

$\widehat{\sigma^2}(\Delta)$ is like that for generalization to $\mu_p$, except that the $pj$ and $j$ components are removed.

Design IV-A reduces to $(I:p) \, | \, J^{**}$ and the Venn diagram is almost like that for V-B. Formula (4.3) applies to this design also.

**Observed-score variance, the error $\delta$, and the coefficient of generalizability**

The Venn diagrams, interpreted with the aid of Table 3.8, provide a basis for direct identification of the components of the observed score, the deviation score, and the corresponding variances under any one of the designs. Any component lying outside the circle drawn with a solid line (i.e., any component that does not contain $p$ in its identification) is the same for all persons and is eliminated from the deviation score. Any component within the solid circle is a component of the observed deviation score, and of the observed-score variance.

The components of the deviation score, located within the $p$ circle, divide into two groups: those contributing to the universe score, which correspond to the shaded area, and the remainder, which make up the error $\delta$. The decomposition for Design VII is shown in Figure 4.2. Comparing it to Figure 3.3 (p. 88) we see that the only change is the transfer of $pj$ information to the universe-score component of variance. A similar division could be made for each of the other figures.

Because the number of observations on each component in the D study is the same, whether generalization is to $\mu_p$ or $\mu_{pJ}$, the number of components indicated in Table 3.8 serves as divisor for each component of variance. The expected observed-score variance is the same as that determined in Chapter 3, and the universe-score variance is $\sigma^2(pJ^{**})$, which is estimated by $\widehat{\sigma^2}(p) +$ $[\widehat{\sigma^2}(pj)/n_j']$. The new value of $\mathscr{E}\sigma^2(\delta)$ is the same as that determined in Chapter 3 save for the transfer of $\widehat{\sigma^2}(pj)/n_j'$ to the universe-score variance.

Since the observed-score variance has not been altered, and the universe-score variance contains an additional component, the coefficient of generalizability is larger when $\mu_{pJ**}$ is the universe score instead of $\mu_p$. Counting individual differences specific to conditions $j$ as wanted information produces the increase. With a new value for the coefficient, there will be a new regression equation and a new value for $\sigma^2(\varepsilon)$.

## Intermediate and complex cases

Several problems that have been ignored to this point will now be discussed briefly. We have assumed that the $J$ in the D and G studies are identical, or are randomly sampled from a large number of possible conditions. There is an intermediate case where $J^*$ of the G study defines the universe of admissible observations ($n_j = N_j$), and a subset of these $j$ constitute the $J^{**}$ of the D study. For this case, the components of variance are obtained from a mixed-model analysis of the G data. To use the random-model equations would be incorrect. The expected observed-score variance contains the component:

$$\left(1 - \frac{n'_j}{n_j}\right) \frac{\sigma^2(pj \mid J^*)}{n'_j}$$

(a variance assuming sampling from a finite universe without replacement). Generalization to either $\mu_{pJ*}$ or $\mu_{pJ**}$ could be intended. The universe-score variance would correspondingly be either $\sigma^2(p \mid J^*)$ or

$$\sigma^2(p \mid J^*) + \left(1 - \frac{n'_j}{n_j}\right) \frac{\sigma^2(pj \mid J^*)}{n'_j}$$

The argument is similar when a random set of $n_j$ conditions is sampled for the G study from the $N_j$ conditions, and a set of size $n'_j$ is sampled for the D study. These sets may or may not be identical. The components of variance are correctly estimated by the Cornfield–Tukey equations for sampling from a finite universe (see p. 60f.), but the distinction is unimportant for $N_j \gg n_j$ and $n'_j$.

Mention was made earlier of stratified tests as an example of generalization where there is a fixed facet. Possible test content is divided into strata (e.g., "verbal" and "quantitative" items), and the test constructor is instructed to draw a certain number of items $n_{i_j}$ from stratum $j$, a subuniverse. Thus, *in the universe of items*, $i$ is nested within $j$, and the $j$ are fixed. This model is a much better description of actual test construction than the random-sampling model is. Each test is still seen as a random sample from the universe *of tests* formed by the given sampling rules, applied to the given universe of items. Rabinowitz and Eikeland (1964), Rajaratnam, Cronbach, and Gleser (1965), and Cronbach, Schönemann, and McKie (1965) have developed intraclass-correlation formulas for "stratified-parallel" tests that consider items and persons as sources of random variance.

The formulas as developed originally took various forms, but essentially an analysis of variance was made for the stratum $j$ to get MS res $(j) = \widehat{\sigma}^2(pi,e \mid j)$. Assume that the D study will have $n'_{i_j} = n_{i_j}$, and that all persons will take the same items. Then if the stratum score is a *total* score $n_{i_j}X_{pI_j}$ and the test score $X_{pT}$ is $\sum_j n_{i_j}X_{pI_j}$, for the universe *of tests*,

$$(4.4) \qquad \sigma^2(\mu_p) = \mathscr{E}\sigma^2(X_{pT}) - \sum_j n_j\sigma^2(pi,e \mid j)$$

$$(4.5) \qquad \mathscr{E}\rho^2(X_{pT},\mu_p) = 1 - \frac{\sum_j n_j\sigma^2(pi,e \mid j)}{\mathscr{E}\sigma^2(X_{pT})}$$

$$(4.6) \qquad \mathscr{E}\widehat{\rho^2}(X_{pT},\mu_p) = 1 - \frac{\sum_j n_j \text{ MS res }(j)}{s^2(X_{pT} \mid T)}$$

It was suggested that $\mathscr{E}\sigma^2(X_{pT})$ be estimated directly by the sample variance $s^2(X_{pT})$ if the $n_i$ are fixed. Equation (4.6) is a variant of the long-established Jackson–Ferguson "battery-reliability" formula (1941); using an alpha coefficient (3.4) for the stratum or subtest in their formula produces the same result as (4.6). Where $n'_{i_j}/n_{i_j}$ is the same for all $j$, the value of $\mathscr{E}\rho^2$ changes in accord with the Spearman–Brown formula.

Use can also be made of the mixed-model formulas similar to those treated earlier. However, these formulas have embodied assumptions of uniform $n_i$ over strata (e.g., see p. 221ff.) The stratified test is actually a battery of subtests, and to deal comprehensively with such data requires a much more complicated model. In Chapters 9–10 we shall set forth a multivariate model that applies to the stratified test. It allows for the possibility that one will alter the universe of tests by assigning different lengths or weights to the strata. A single G study may then be applied to many qualitatively different stratified tests.

Throughout this chapter we have assumed that the universe of generalization contains one score $X_{pij}$ for each $pij$ combination, restricted only by the requirement that $j$ belongs to $J^*$ or $J^{**}$. However, one might have the fixed facet nested within person $p$ in the universe. That is to say, for each person there is a fixed set of conditions $J_p$, and generalization of $p$'s score over other conditions of $j$ is not intended; the conditions that enter $J_p$ differ from person to person. When $p$'s spouse fills out a questionnaire, one intends to generalize over the universe of items. Because there is a different spouse for each $p$, spouse $j$ is nested within $p$ or, one might say, $p$ and $j$ are confounded in the universe. One can think of cases where a whole set of $J$ is confounded with $p$; for example, for a college teacher, the courses he teaches may be considered a fixed set of conditions. Generalization over one facet when another facet is fixed within the person was treated briefly by Gleser, *et al.* (1965). We shall not go into this subject here, as no examples of this design occur in the following chapters.

### B. *Implicit Facets in G Studies*

The three-way analysis of scores $X_{piJ\bullet}$ is, as we have said, precisely like the two-way analysis of scores $X_{pi}$. This brings to attention a basic difficulty in interpreting all generalizability (reliability) studies. Unless all facets that might conceivably affect scores are explicitly identified in describing the experimental design, one may easily make too sweeping an interpretation.

Very often, what looks like a one-facet random-model G study actually has unmentioned facets represented in the design in some manner. When several observations are made under various conditions of a facet $i$, some single condition of an additional facet may have been designed into the observing process. For example, when two or more test forms are applied, it is common for them to be administered by the same tester. Testers form a facet; these G data, then, are collected by the design $i \times j \times p$, $n_j = 1$. The situation is essentially the same when several conditions of $j$ are averaged to generate each score, so long as the same set of conditions is used in observing every person. The casual identification of scores with test forms, for example, tends to overlook such facets as tester and occasion.

It is natural to speak of generalizing to a universe score $\mu_p$. This is often a considerable oversimplification, even though most of the literature on reliability in education and psychology speaks of a single "true score" whose properties are those of $\mu_p$. In the most conventional of reliability studies, there are likely to be facets that partly define the universe and yet are not mentioned in describing it.

In *any* study where data are laid out in an $m$-way array (persons, plus $m - 1$ facets), there are further facets along which observations might have been classified. Consider any one such unmentioned facet, $k$. There are three possibilities:

1. $k$ confounded with $p$, or $i$, or some other facet. Each condition of $k$ is associated with a certain person or a certain condition of a specified and systematically varied facet, for instance $i$, so that when a certain $i$ is sampled a particular $k$ is always selected also. It is possible to have $k$ confounded with both $p$ and $i$, etc.

2. A uniform condition $k$ (or a uniform set $K$) is used in every G-study observation.

3. $k$ varies randomly and independently from observation to observation.

Case 3 gives no trouble, because all effects associated with $k$ contribute to the residual component of variance. Any error in generalizing over facet $k$ is properly taken into account.

In Case 1, if the sampling makes $k$ joint with $i$, the interpretation must bring a $k$ into every statement that contains the symbol $i$. Thus, for example, $\mu_{pi}$ is really a $\mu_{pik}$. A variance component such as $\sigma^2(pi)$ is really a component for $\sigma^2(pi,pk,pik)$. Without an experiment that samples $k$ separate from $i$ there is no way of estimating the pure component for $pi$. With such systematic confounding, it is best to express conclusions in terms of the $i,k$ compound. If such raters as the school counselor and the Latin teacher differ because they see pupils in systematically different situations, one can accept the pertinent variance component as reflecting rater differences combined with situation differences, rather than as an error "attributable to raters." Confounding of $k$ with $i$, $j$, or both does not preclude the usual interpretations when the universe score is $\mu_p$. The contribution of variation from $k$ to the various errors is fully recognized, though it is entangled with another facet. Variances for D studies can be estimated only if the same confounding of $k$ with $i$, etc. is retained in the design.

When $k$ is tied to $p$, there is a greater problem. What the analysis reports as a component for $p$ is actually a component combining the $p$, $k$, and $pk$ effects. Likewise, the supposed $pi$ component is augmented by the $ik$ and $pik$ components. In effect, such a study treats $pk$, not $p$, as the subject of inquiry. This is all to the good if $p$ and $k$ are tied together in daily life in the same way they are tied in the investigation—if condition $k$ is, as it were, an invariant aspect of $p$'s environment. If $p$ and $k$ are tied only in the investigation, the study is hard to interpret. An example of the $p$ and $k$ "tied together in daily life" is a study of teachers, in which all observations on a particular teacher are made in the particular classroom where the teacher normally works. Perhaps the teacher would teach differently if he moved to a different physical environment, but the stability of teacher assignments makes it reasonable for scores to reflect the teacher-plus-classroom combination where this is looked on as an independent variable. (Data with this confounding would not be entirely satisfactory as a criterion in a teacher-selection study, however, as one never hopes to predict how well the teacher will do in a particular physical setting he is later assigned to.)

Case 2, where the same conditions of $k$ enter all observations, has perhaps caused the greatest misunderstanding of reliability coefficients. A G study with a single $k$ necessarily estimates components of scores "within $k$." This is also true where several $k$ are present in every score but are not treated as a facet in the analysis. To speak of a retest study over two days and a study applying parallel forms on the same day as both estimating "the" standard error of measurement is clearly misleading. The retest study investigates within-person variation among a universe of observations in which the test form is fixed and occasions vary. The form-to-form study provides information about a universe in which forms vary and the occasion is essentially fixed. Other temporarily fixed conditions such as examiner and scorer also restrict the universe of generalization. In an observational study of pupils, perhaps the subjects' teachers are a fixed, unrecognized, but influential source of variation.

Whenever a condition of a facet is uniform throughout the G study (or is crossed with persons and other conditions but not analyzed as a source of variance) there is no way to estimate what effect the sampling of that facet has on the error of generalization. Therefore, it is most important in reporting a G study to recognize conditions that have been held constant and to qualify any statement about generalizability accordingly.

Results are ordinarily described as simply as possible; therefore it is natural, in a one-facet study, to designate variance components "for persons," and "for test forms." However, one must be careful to recognize shifts in the meaning of common labels and symbols. If a one-facet G study is made that holds constant some condition $k^*$ (Case 2), the estimate designated as $\widehat{\sigma^2}(i)$ is actually $\widehat{\sigma^2}(i \mid k^*)$, for example. The component includes the usual $\sigma^2(i)$ and also the variance component for $\mu_{ik*}$. If $i$ and $k$ vary jointly (Case 1), $\widehat{\sigma^2}(i)$ includes variance from $\sigma^2(i)$, $\sigma^2(k)$, and $\sigma^2(ik)$ components. In a Case 3 study the $ik$ interaction forms part of the residual, along with every other component involving $k$.

## EXERCISES

**E.1.**    Data for rating Mother's affectionateness (Scale A) while mother and child engage in certain tasks were presented in Chapter 3, Exercise 4. Consider here the possibility that the universe of generalization is limited to six tasks, these being the same as the tasks actually used in the D study, but that the investigator intends to generalize over an indefinitely large number of raters, so that the universe score is $\mu_{pI\bullet}$.

a. What is the variance of universe scores?

b. What is $\mathscr{E}\sigma^2(\delta)$ when the same two persons rate all responses collected in the D study?

c. For the same design, what is $\mathscr{E}\widehat{\rho^2}$? Why does this differ from the value obtained in Chapter 3, Exercise 4b.

**E.2.**    An instrument for measuring ego defenses (Gleser & Ihilevich, 1969) describes 10 conflict situations to the subject. After he reads the description, the subject is asked what he would do in such a situation—rather, to select one of five courses of action as most likely and one as least likely. Of the five choices, one represents the defense of projection. He receives a score of 2, 1, or 0, depending on

**TABLE 4.E.1.**   *Analysis of Variance for a Measure of Ego Defenses*[a]

| Source of variance | Degrees of freedom | Mean square |
|---|---|---|
| Person $p$ | 10 | 2.6400 |
| Story $s$ | 9 | 2.5076 |
| Level $\ell$ | 3 | 7.3939 |
| Occasion within person $o, po$ | 11 | 0.3091 |
| $ps$ | 90 | 0.5920 |
| $p\ell$ | 30 | 0.4773 |
| $s\ell$ | 27 | 1.4503 |
| $so, pso$ | 99 | 0.2409 |
| $\ell o, p\ell o$ | 33 | 0.1545 |
| $ps\ell$ | 270 | 0.3300 |
| Residual | 297 | 0.1905 |

[a] Score components are on the scale of $X_{ps\ell o}$.

whether he picks that choice as most likely, or leaves it unmarked, or picks it as least likely. (Scores for other defenses are also obtained, but we ignore them.) The question about actual behavior represents one of four "levels" of response; there are questions regarding three other levels: fantasy, thought, and affect. The entire instrument for measuring tendency to use projection then consists of 10 stories, crossed with 4 levels; the test score is the sum over these 40 responses.

In one study, the test was administered on two occasions approximately a month apart, to eleven persons. Treating the study as having the design $\ell \times s \times (o\!:\!p)$ and analyzing the entire matrix of responses yielded the mean squares in Table 4.E.1.

a. Estimate variance components under the assumption that levels are fixed. Treat occasions and stories as random.

b. Estimate $\sigma^2(\mu_{pL\bullet})$, $\sigma^2(\Delta_{pSL\bullet o})$, and $\mathscr{E}\sigma^2(X_{pSL\bullet o})$ for D data using the total for ten stories, four levels and one occasion. The D-study design is $\ell \times s \times (o\!:\!p)$.

c. Interpret $\widehat{\sigma^2}(p\ell \mid L^*)$.

**E.3.**   The levels of content in the above test can be considered to be a stratification of the response domain into these four levels. Summing over stories for any level then yields a subtest score. Conceivably an investigator might want to use only one subtest score for a particular decision.

a. Estimate the expected subtest observed-score variance.

b. Compute $\widehat{\mathscr{E}\sigma^2}(\delta)$, assuming that the investigator intends to generalize from the subtest score, over the universe of stories and occasions, keeping level fixed.

### Answers

**A.1.**   a.   $\widehat{\sigma^2}(\mu_{pI\bullet}) = 0.44 + \frac{1}{6}(0.17) = 0.47$

b.   $\widehat{\mathscr{E}\sigma^2}(\delta) = 0.04 + 0.048 = 0.088$

c.   $\widehat{\mathscr{E}\rho^2} = 0.84$. This is larger than the value (0.79) computed in Chapter 3, Exercise 4b. Although both coefficients apply to the same D data (i.e., two raters and six tasks), generalization here is over a limited universe and is therefore more accurate

**A.2.** a. All components in Table 4.E.2 except *so, pso, ē* are the same as obtained under the random model.

**TABLE 4.E.2.** *Interpretation of Analysis for Measure of Ego Defenses*[a]

| Source of variance | | Estimate of variance component | Expected variance of observed total score[b] | $\sigma^2(\Delta)$ |
|---|---|---|---|---|
| $p \mid L^*$ | | 0.0248 | 39.68 | |
| $s \mid L^*$ | | 0.0218 | — | 34.9 |
| $ps \mid L^*$ | | 0.0439 | 7.02 | 7.02 |
| $o,po \mid L^*$ | | 0.0017 | 2.72 | 2.72 |
| $so,pso,\bar{e} \mid L^*$ | 1/4 (0.2409) = 0.0602 | 9.63 | 9.63 |
| $\ell \mid L^*$ | | 0.0257 | | |
| $p\ell \mid L^*$ | | 0.0074 | | |
| $s\ell \mid L^*$ | | 0.0509 | | |
| $ps\ell \mid L^*$ | | 0.0698 | | |
| $\ell o,p\ell o \mid L^*$ | | (0) | | |
| $s\ell o,ps\ell o,e \mid L^*$ | | 0.1905 | | |
| | | | 59.05 | 22.86 |

[a] Components are on the scale of $X_{ps\ell o}$.
[b] As the total score is a sum of 40 responses, each component is multiplied by 1600.

b. $\widehat{\sigma^2}(\mu_{pL^*}) = 39.68.$ $\widehat{\mathscr{E}\sigma^2}(X) = 59.05.$ $\widehat{\sigma^2}(\delta) = 22.86.$

c. $\widehat{\sigma^2}(p\ell \mid L^*)$ indicates the extent to which persons vary in their response, depending on the level of response about which inquiry is made. Thus, it has the same meaning as in the completely random model save that $\ell$ is constrained to be a member of the set $L^*$. The estimate is calculated on the assumption that levels are sampled with replacement.

Most often, however, when there is sampling from a finite set, sampling without replacement is assumed. Under this assumption the variance attributable to person-level interaction is decreased by $1/N_\ell$ from the value obtained under the random model.

**A.3.** a. The components entering observed-score variance (for sums over 10 stories) are

| | |
|---|---|
| $p \mid L^*$ | $100(0.0248) = 2.480$ |
| $ps \mid L^*$ | $100(\frac{1}{10})(0.0439) = 0.439$ |
| $o,po \mid L^*$ | $0.170$ |
| $so,pso, \bar{e} \mid L^*$ | $0.602$ |
| $p\ell \mid L^*$ | $(100)(1 - \frac{1}{4})(0.0074) = 0.555$ |
| $ps\ell \mid L^*$ | $100(\frac{1}{10})(1 - \frac{1}{4})(0.0698) = 0.523$ |
| $\ell o,p\ell o \mid L^*$ | $(0)$ |
| $s\ell o,ps\ell o,e \mid L^*$ | $1.429$ |

b. $\widehat{\mathscr{E}\sigma^2}(\delta) = 0.439 + 0.170 + 0.602 + 0.523 + 1.429 = 3.163$

# CHAPTER 5

# Assumptions Underlying Estimates of the Universe Score

To this point, we have devoted minimal attention to mathematical assumptions. Our primary aim has been to show how a multifacet approach, together with the overt distinction between G and D studies deals with the traditional questions of reliability. The only assumption invoked throughout the argument has been the random sampling of conditions and persons. This assumption, in some ways a strong one, is nonetheless weaker than the classical model, which ignores sampling of persons and assumes conditions to be strictly equivalent.

We have treated two kinds of estimate of the universe score: the confidence interval symmetric around the observed score, and the regression estimate. Testers almost invariably use the observed score as if it were an estimate of the universe score (true score). Midway in the history of classical theory, the concept of a confidence interval was added; plotting test scores so as to display a "confidence band" is now a common technique. The regression estimate, though long recognized in test theory, has had little place in testing practice. While developing generalizability theory, we have become very much conscious of the intricate rationales underlying the two approaches. In this chapter we shall trace the logic behind each of them. The criticisms we make apply, for the most part, to interpretations under classical theory as well as to generalizability theory.

The regression and confidence interval estimates are radically different techniques. Once information on the error of measurement is available, it is possible to set up a confidence interval for the universe score when a single person is measured by himself. There is no need to bring in a reference group.

The regression estimate, however, can be made only by bringing estimated parameters for a reference group into the picture. Where the confidence interval relates solely to the distribution of observed scores "within $p*$," the regression estimate relates to the joint distribution of observed and universe scores in the population. This distinction has rarely been pointed out in writings on test theory because the traditional stress on individual differences brings a reference group into many discussions of confidence intervals. Such mixing of concepts is a source of great confusion to beginning students of test theory.

Perhaps matters will be clarified by placing the interpretations in a Bayesian framework. Classical statistics was developed by Fisher, Neyman, and others who advanced beyond the base laid down by Karl Pearson. The classical approach requires the user of statistics to phrase his question formally: "Is it likely that this evidence would be obtained when such-and-such hypothesis is true?" If the answer is "no," the hypothesis is rejected. Fisherian logic does not allow one to consider "the probability that the hypothesis is true." Bayesian logic, which has many advocates today, does make statements about the tenability of the various alternative hypotheses. To do so, it takes advantage of whatever estimate of these probabilities one can make prior to the experiment. In a study where a mean or other measure is wanted, the investigator is asked to state the "prior" probability that the true measure will fall in each interval of the scale. He may arrive at those probabilities by direct tabulation of past experience with similar events or may simply state "beliefs" derived quite indirectly from experience (Mosteller & Tukey, 1968, pp. 160–183). The prior probabilities are weighted into the final solution along with observed values.

The tester seeks to settle upon one hypothesis or a range of reasonable hypotheses about the subject's universe score. The regression estimate of the universe score can easily be seen as a weighted combination of prior information with direct observations. As soon as the tester knows any basic fact about the subject, he knows something about what hypotheses are reasonable.

Knowing that the person is a college student, for example, one can say *a priori* that an IQ in the range of 110–130 points is fairly likely to be his universe score. One would be astonished to find that his universe score is 75. While a universe score of 160 is not out of the question, we would lay odds against such a rare value. These prior probabilities are derived from experience with college students previously tested. Much more definite probabilities are available when we can identify the subject as a senior in College C and the score distribution of that senior class is known. The distribution may be still more sharply defined if, for instance, he is an honor student, and we can obtain the score distribution for that group.

The scores in such a distribution do not give literally "prior" information, if all the persons' scores become available at once. From a Bayesian standpoint, however, they are priors because looking at the distribution permits the tester to make some judgment about the subject before he turns to the observations made directly on him. Test interpreters of course do make judgments from priors very often; this is most clearly seen in the statement that a certain test score is suspect "because it occurs so rarely among such persons as this one." The regression estimate is one technique—less subtle than most of those used in Bayesian statistics—of using prior information along with test evidence to reach a more accurate conclusion about the universe score.

Though the confidence interval arose in the classical framework, it can be given a Bayesian interpretation. The tester marks off a multiple of $\hat{\sigma}(\Delta)$ on each side of the observed score and takes the resulting interval as his conclusion about the universe score; thus, no weight at all is given to prior information. The Bayesian theorist considers this reasonable only when the distribution of priors is nearly flat: "gentle," to use the Mosteller–Tukey term. Sometimes the tester has little or no prior basis for judging what universe score to expect, except as the scoring rules set a ceiling and a floor. If that is the case, the tester cannot "weigh in" the prior information.

It is precisely this situation in physical science that makes it appropriate to report an observation as, for example, $13.005 \pm 0.002$ grams. The scientist has some prior knowledge; merely from looking at the specimen he can be sure that its weight is closer to 10 grams than to 500 grams. However, he has little basis on which to say that, in the range 12.50–13.50, one value is appreciably more likely than another. His region of uncertainty is wide relative to the errors of measurement, and his prior probabilities within that region are equal (or nearly so). Taking priors into account in the Bayesian fashion would make no difference; to the right of the decimal place, the balance reading is all the evidence there is.

The psychological tester has taken over the "plus-or-minus" technique from physical measurement, without realizing that his situation is fundamentally different. *He* almost always knows something about the score distribution to be expected for a population from which his subject comes, and *his* error of measurement is fairly large. When a tester anticipates that his subject's IQ (universe score) is in the range of 120–130 points but can make no finer prior judgment, the range over which his priors are equal is about $2\sigma(\Delta)$. In contrast, the physical scientist's range of flat priors was 500 times the standard error.

The preceding paragraph implies that the regression technique is advantageous, but there is perhaps a case to be made for the confidence-interval method. If it requires fewer (or more acceptable) assumptions than the regression estimate, it will be preferred in many applications. Therefore, a detailed look at assumptions is required.

### A. The Logic of Confidence Intervals

The complications of generalizability theory will be sidestepped in this section. Emphasis on a familiar model will make the crucial points a good deal more evident. Assume strictly parallel measures $X_{p1}$, $X_{p2}$, etc. Assume a population mean of 0 and a variance (over persons) of 1.00. Suppose that the coefficient of generalizability is 0.75, hence the standard error of measurement is $0.25^{1/2}$ or 0.50. A confidence interval is stated in the form $X_{pi} - 0.50 \leqslant \mu_p \leqslant X_{pi} + 0.50$. The standard error will be referred to as $\sigma(\Delta)$, not $\sigma(\delta)$, though under the assumption of equal means $\sigma(\Delta)$ does equal $\sigma(\delta)$. This section does not discuss confidence intervals based on $\sigma(\varepsilon)$.

### The rationale in general statistics

Establishing a confidence interval for a universe score rests on the statistical rationale for an interval estimate of a population mean. Because that rationale is presented in many statistics texts (Hays, 1964, p. 287 ff.; McNemar, 1969, pp. 99–105), we need only summarize it here. It is usually assumed that an indefinitely large number of samples of size $n$ could be drawn from the population and that the sample means so collected would form a normal distribution. The standard deviation of sample means $\sigma(\bar{X})$ is equal to the standard deviation of the individual scores in the population, divided by $n^{1/2}$.

The assumption of normality implies that the distance of a sample mean from the population mean will rarely be two or three times as large as $\sigma(\bar{X})$. In 33% of the samples, the sample mean is outside the range defined by marking off $\sigma(\bar{X})$ on each side of the population mean. If we have an estimate $\hat{\sigma}(\bar{X})$, the observed mean $\pm\hat{\sigma}(\bar{X})$ is an interval estimate of the population mean. Because in 67% of all such analyses the interval will contain the population mean, it is called a "67% confidence" interval. Other multiples of $\hat{\sigma}(\bar{X})$ correspond to other confidence levels. If the interval $\bar{X} \pm 1.96\sigma(\bar{X})$ is used, the normal distribution assumption implies that the corresponding confidence level is 95%.

The argument is modified if the $t$ distribution is used to recognize the fact that $\sigma(\bar{X})$ is estimated rather than known. In the usual statistical analysis there is a single sample of size $n$ from which $\sigma(\bar{X})$ is estimated. If $n$ is as large as 5, the interval $\bar{X} \pm \hat{\sigma}(\bar{X})$ encloses very nearly 67% of all sample means, and to derive the precise confidence level from the $t$ distribution instead of the normal would have only a trivial effect on the interpretation. When the sample is smaller than 5, the interval given by $\pm1.96\sigma(\bar{X})$ includes appreciably less than 95% of the sample means, according to the $t$ distribution (Mosteller & Tukey, 1968, pp. 84–85).

To adapt the statistical rationale to measurement problems, the test theorist notes that the act of measurement produces samples from a distribution of admissible scores for the person, a distribution that has the universe score as its mean. The observed score is a sample mean, whether it is based on a single event and the sample size is 1, or is a composite of a whole series of observations, for instance, of $n_i$ item scores. (The observed score may be a total over $n$ observations instead of a mean; if so, this would require small changes in wording of many statements below.)

The distribution of scores from an indefinitely large number of observations on the same person is a within-person distribution of scores. This distribution, like that for single observations, has the universe score as its mean. An inference about the within-person distribution has nothing to do with differences between persons or with the universe-score variance. The standard deviation of possible sample means (observed scores) for a single person is analogous to the standard error of sample means drawn from a single population.

There would be no essential difficulty in applying the usual rationale if a sizeable number of observations on $p*$ were available from which to estimate $\sigma(X_{p*I}) = [\mathcal{E}_I(X_{p*I} - \mu_{p*})^2]^{1/2}$. It would have to be assumed that each observation is randomly sampled from the within-$p*$ distribution, and that the distribution of $X_{p*I}$ is normal. These assumptions do not seem unreasonable for test scores each of which is based on a large number of items. However, one may have only a single $X_{p*i}$ for the person observed in the D study, and if $n_i'$ is moderate in size, one must employ the $t$ distribution. This would produce a wider interval than the tester ordinarily establishes.

### The tester's assumption of uniform within-person distributions

The familiar way around the difficulty of obtaining $\sigma^2(\Delta_{pI})$ separately for each person is to add an assumption. Assuming parallel tests, $\sigma^2(\Delta_{pI})$ for all $p$ and $I$ together is the same as the variance $\sigma^2(\Delta_{pI} \mid I)$ for any one test. Now the assumption is added that $\sigma^2(\Delta_{pI}) = \sigma^2(\Delta_{pI} \mid p)$ for any $p$ in the population. This, or some slightly weaker statement about approximate equality, is the basis on which confidence intervals are justified in most extant test theory.

The assumption of a uniform $\sigma(\Delta \mid p)$ for all $p$ may or may not be justified by the facts. If the assumption were valid, the standard deviation across persons would be the same at each level of $\mu_p$. Scatter diagrams for test against retest frequently show that this assumption is untenable. Most commonly, it appears that the within-person standard deviation is large for persons who find the test difficult, and small for persons with fairly high universe scores. But just the reverse is reported at times; the Stanford–Binet evidently has a much smaller $\sigma(\Delta)$ for a child whose score is below average than for a high scorer.

Suppose intervals are established on the basis of $\hat{\sigma}(\Delta)$ estimated for the population, but the $\sigma(\Delta \mid p)$ actually vary from person to person. Then, by the ordinary statistical reasoning, the confidence level must be larger than 0.67 for persons whose $\sigma(\Delta \mid p)$ is relatively small, and *vice versa*. Concretely, suppose that $\sigma(\Delta)$ is 10, and the interval is made 20 points wide ($\pm 10$ points). Then for the subset of persons whose $\sigma(\Delta \mid p)$ is 15, the interval spans only $\pm 0.67\sigma$ and the confidence level is 0.50. For the subset whose $\sigma(\Delta \mid p)$ is 5, the interval spans $\pm 2\sigma$ and the confidence level is 0.95.

The foregoing discussion suggests that the tester has had little basis for stating what confidence level can be associated with the interval estimate for a person.

### Problems of interpretation

***Does the interval for p\* contain*** $\mu_{p*}$***?*** Lord and Novick have shown (1968, pp. 511–512) that confidence intervals are likely to be misleading in the interpretation of scores even when the assumptions the tester usually makes are fully satisfied. For the moment, assume that there is a population of persons to which person $p*$ belongs, that universe scores in that population are normally distributed, and that the within-person distribution of observed scores is normal and has the same standard deviation for every person. Assume also, for simplicity, that the parameters $\sigma^2(\mu_p)$, $\mu$, and $\sigma^2(\Delta)$ are known. Then the joint distribution of observed and universe scores will be bivariate normal. If confidence intervals of the form $X_{p*i} \pm \sigma(\Delta)$ are now set up, the intervals will contain $\mu_p$ for two-thirds of the persons, which is the risk chosen in setting up an interval of this width. What is often overlooked is that the risk that $\mu_{p*}$ will fall outside the interval is greater than one-third for individuals whose observed scores are far from the mean. This is offset in the overall odds by the fact that for persons with observed scores near the mean the universe score falls within the interval with probability greater than two-thirds.

To be specific, consider further the test with $\sigma(X) = 1$, $\mu = 0$, $\rho^2 = 0.75$, hence $\sigma(\Delta) = 0.50$. Suppose $p*$ has an observed score of $+4$; an extreme value deliberately chosen. The interpreter following customary methods is likely to say to $p*$: "The chances are two out of three that your score is in the range 3.5–4.5." Are these the right odds?

The odds stated by the interpreter apply to the whole population of persons. But we already know that $p*$ belongs to the subclass of persons whose observed score is 4, and the odds in that subclass are not two out of three.

Consider the slice of the bivariate distribution where $X = 4$. This is a distribution of universe scores; it is normal, with mean 3 and standard deviation 0.43, in accordance with equations (3.2) and (3.3). We do not know which score in the distribution belongs to $p^*$. But each score belongs to a person for whom the same interval estimate of the universe score would be given; for what fraction of these persons does the universe score actually fall in the range 3.5–4.5? Practically none have universe scores above 4.5, as that is more than 3 standard deviations above the mean of the distribution; but 3.5, the lower edge of the interval, is 1.15 standard deviations above the mean and therefore about 87% (!) of the persons who are told that their universe scores fall between 3.5 and 4.5 actually have universe scores below 3.5.

The reader, knowing that scores as high as $+4$ standard deviations are unlikely, may not be impressed by this example. While we chose an extreme value to produce a dramatic result, there will be similar but lesser contradictions of expectation when the observed score is much less extreme. Moreover, if one's measure has a low coefficient of generalizability, dramatic contradictions will be obtained for scores that occur more frequently. Suppose that the coefficient is 0.40. Then 81% of persons with an observed score of $+2$ will have universe scores below the lower edge of the confidence interval. The interval will contain $\mu_p$ for 63% of the persons having an observed score of 1, and this might seem at last to be a case where a 67% interval is working out about as it should. Surely, however, the subject who is told that the chances are two in three that his true score is in the range 0.23–1.77 is left with the impression that his chance is one in six of falling above the range and one in six of falling below. Actually, the cases with $\mu_p$ outside this interval are almost all on the low side of it.

In the examples so far, the universe scores have tended to fall outside the interval. Just the opposite occurs also, as is necessary to return the overall error rate to the theoretical value. On a test with a coefficient of 0.40, a person with an observed score of 0 is said to have $\mu_p$ between $-0.77$ and $+0.77$. This will be true for 88% of such persons, not 67%.

In statistics, an interval ranging over $\pm 1$ standard deviation is said to "include the mean with probability 0.67." That is, when such intervals are established for a large number of means, the mean will lie within the interval for about 67% of the intervals so formed. There is no basis for asserting that the probability is 0.67 for *any one* designated distribution. The tester is in a position, granted the several assumptions, to say that for two-thirds of the persons for whom confidence intervals are formed, the interval will contain $\mu_p$. This is not logically the same as saying that for persons like $p^*$, or for $p^*$ himself, the odds are two out of three. Lord and Novick discuss the point at some length (1968, p. 512), stressing that "no confidence statement can be made about a particular, nonrandomly chosen examinee in whom we happen to be interested. Nor can any confidence statement be made about those examinees who have some specified observed score."

The paradoxes considered are not very paradoxical. Where a population can be differentiated into subpopulations whose members are distinguishable on any nonchance basis, one expects probabilities for the subclasses to be different. The demonstration of the shift in odds as a function of $X$ rests basically on regression toward the mean, a phenomenon with which all users of correlation are familiar. So long as the mean of $\mu_p$ is less than $X_{pi}$ for persons with $X_{pi} > \mu$, and greater than $X_{pi}$ for persons with $X_{pi} < \mu$, there will be effects such as we have illustrated. The departures from the population odds will be modified by various changes in the assumptions (e.g., by nonuniform error distributions), but departures there will be— unless the $\mu_p$-on-$X$ regression coincides with that for $X$ on $\mu_p$.

Obviously, if $\rho^2$ is near 1.00, the two regressions will nearly coincide and subclass probabilities will not shift appreciably from one level of $X$ to another until the most distant tails of the $X$ distribution are considered. A second possibility is that the universe-score distribution is rectangular.[1] Where that is the case, and errors are normally distributed around $\mu_p$, the regression line is a much-flattened ogive, running close to the $X$-on-$\mu_p$ regression line until it hooks away as the end of the $\mu_p$ range is approached. The ends of the $\mu_p$ range are asymptotes. Only for persons with $X$ near or beyond the extremes of the $\mu_p$ range does regression toward the mean distort probabilities. These persons will be a small fraction of the group if $\sigma^2(\Delta)$ is small relative to $\sigma^2(\mu_p)$. In general, the interval estimate of a universe score can defensibly be interpreted in terms of the specified risk level if prior probabilities are fairly flat, and not otherwise.

*What scale does the estimate refer to?* Several commercial tests arrange for the tester to display a pupil's score as a "band" rather than a point. There may be a table in the test manual that converts a raw score of 66 into a band of 63–69, for example, to take into account the error of measurement. The band for the test score is plotted onto a record sheet that usually embodies a percentile (or standard-score) conversion. Alongside the scale-point for 63 one may read that the percentile equivalent is 82; beside 69, that the percentile equivalent is 91. So the pupil is told that his universe score probably lies between percentiles 82 and 91. Percentiles of what?

The scaling procedure is such that the percentiles refer to an observed-score distribution. The pupil, then, is told that his *universe* score is unlikely to

---

[1] We have also given thought to severely skewed and bimodal distributions. In such cases one is likely to have regression toward the mean in some arrays and regression away in others. The odds will again vary from subgroup to subgroup.

fall below the 82nd percentile for *observed* scores. This is scarcely a useful statement. Universe scores have a smaller standard deviation than observed scores. While 63 may fall at the 82nd percentile for observed scores, it falls higher (88th percentile, perhaps?) among universe scores. It is not meaningful to place the universe score against the observed-score distribution.

It seems neither feasible nor desirable to convert the confidence interval into a norm-referenced statement. One cannot ordinarily get good data on the distribution of universe scores. To apply universe-score percentiles to the confidence interval, moreover, is to bastardize the argument. Once one admits the presence of a reference group for whom the prior distribution is known and non-flat, there is no justification for reporting a result consistent with flat priors.

*Comparing the score to an absolute standard.* The difficulties examined above are less troublesome in certain kinds of decisions where an absolute standard is invoked. One can, for instance, defend the common practice of using bands to make an inference that the difference between two universe scores departs from zero. That procedure will be treated in Chapter 10, since it is a multivariate problem. As will be seen, the common procedure (in effect) sets up an interval estimate for the observed difference score symmetric about the hypothesized (null) universe-score difference. This is relatively free from the faults of stating that $\mu_p$ for an ordinary difference falls within an interval. Nevertheless, in Chapter 10 we shall argue against displaying the bands on a profile sheet.

Intervals may be used in judging a single score against an absolute standard. It may be agreed, for example, that algebra students who cannot solve 75% of the equations in a specified universe should receive further training on such problems before moving to advanced topics. The decision could be based on whether $X_{pI} \geqslant 75$ (assuming that scores are expressed as percentages), but this ignores errors of observation. A safer procedure is to set up a band symmetric around the standard, for instance, from $75 - 1.96\sigma(\Delta)$ to $75 + 1.96\sigma(\Delta)$. Then one would hold back students whose scores fall below the band and assign advanced work to students whose scores fall above the band; that is, whose universe scores are very probably in the lower or upper region, respectively. It would be sensible to test further those whose scores fall within the band. After each stage of testing, the band narrows, because larger $n_i'$ implies a smaller error. Elementary cases of this kind of sequential testing were discussed by Cronbach and Gleser (1965, pp. 69–85, 91–96), and a recent practical application of such techniques by Ferguson (1970) makes their utility in instructional tests quite evident. We may also refer to the work by Mathur & Kumar (1969), who derive from the classical model a sequential procedure relating the confidence interval to the amount of additional testing the investigator is willing to undertake. They indicate the extent to which the procedure reduces erroneous decisions, assuming that average testing cost (on a per-subject basis) is fixed.

To revert to our example: Where the correct value of $\sigma(\Delta \mid p)$ is used and the within-person distribution is normal, 95% of the persons whose (unknown) universe scores precisely equal the standard will be held for further testing. The probability is 0.025 that such a person will be held back, and 0.025 that he will be sent to advanced work without further testing. The probability is less than 0.025 that a person for whom $\mu_p > 75$ will be held back, or that one for whom $\mu_p < 75$ will erroneously be sent to advanced work. The procedure does not guarantee that 97½ persons out of 100 sent into advanced work will be up to the standard. If there are very many weak students, and very few above the standard, then the number misclassified into the advanced group may be large; indeed, it may be larger than the number who are properly assigned there.

Bayesian methods are able to take base rates into account. One can also take into account the relative seriousness of the two types of misclassification. Such considerations would lead to a decision rule that does not use cutting scores symmetric about the standard.

*Considerations added by generalizability theory*

In developing the theory of Chapters 1–3 the equivalence assumptions of classical theory were avoided. The model and the equations that lead to estimates of variance components depend solely on the assumption that conditions are randomly and independently sampled from the universe. The first consequence of this weakening of the model is that in many D studies components for $i$, $pi$, and the like enter the error $\Delta$; this makes necessary our distinction between $\Delta$ and $\delta$.

The likelihood of unequal condition means was obvious in early studies of the reliability of ratings, as early as the 1920's. Much more recently, Lord (1955a, 1962) noted that when tests are formed by random sampling of dichotomous items, the variation in item difficulty introduces a component into the error of measurement that classical theory ignores. Generalizability theory provides a systematic way of identifying and estimating such components of error. It goes further, and considers components arising from the sampling of more than one facet. This complicates reasoning about confidence intervals.

Either classical theory or generalizability theory allows for the possibility that a G study will be carried out on the particular person $p^*$ about whom conclusions will be drawn. An adequate multifacet study of this kind is harder to carry out than a study that observes $p^*$ under conditions of a single facet. Therefore, it is likely that the $\sigma^2(\Delta)$ derived from a sizeable sample of persons will be taken as $\sigma^2(\Delta \mid p^*)$ for any $p^*$. Employing a single value of $\sigma^2(\Delta)$ for all persons is perhaps more risky under a multifacet model than under the classical model, because some of the added components probably vary from person to person.

Fixing accurate confidence levels for an interval based on $\hat{\sigma}(\Delta)$ is difficult. It is convenient to assume that the $\Delta_{pIJ}$ are normally distributed, but this is a stronger assumption than it was in the classical case. While the $pi$, $pj$, and $pij,e$ components may reasonably be thought of as normally distributed, assuming normality for $i$ and $j$ components is more questionable. Furthermore, values of $n_i$ and $n_j$ in G studies are often small, which means that the $i$ and $j$ components are not well estimated. If normality of score components is assumed, the $t$ distribution provides a precise small-sample theory for arriving at confidence levels.

Forming an interval estimate of the mean $\mu$ for a population on the basis of sample data is a legitimate application of confidence-interval theory. The textbook approach forms such an interval directly from $X_{PI*}$ and $s(X_{pI*})$, for some sample from the population.[2] Only sampling of persons is considered. This is sound enough as a way of establishing a confidence interval for the condition mean $\mu_{I*}$, but it does not allow for sampling of conditions. A procedure that recognizes all sources of variance in $X_{PI}$ is required when both $p$ and $i$ are sampled. This procedure (p. 96) is open to few of the objections that apply to the interpretation of confidence intervals for single persons. The priors are likely to be flat. The assumptions that conditions for G and D studies are randomly drawn from the same universe, and persons for the studies drawn from the same population may or may not be acceptable. The most severe difficulty is that confidence levels are not soundly determined from the normal distribution when the number of degrees of freedom for any sizeable component of variance is small in the G study. In the face of this difficulty a jackknife procedure is probably advisable.

## B. *The Logic of Regression Estimates and Similar Equations*

At various points in Chapter 3 equations were presented for making a point estimate of a person's universe score. Variants of these equations are listed in Table 5.1. These will now be helpful in elaborating on points barely touched upon in Chapter 3. Each one of the equations shown has a simpler counterpart that estimates the deviation score $\mu_p - \mu$ rather than the absolute value $\mu_p$; to obtain the right-hand side of the counterpart equation, simply ignore the first term of the equation in the table. For example, in the

---

[2] In this chapter we use a single asterisk, dropping the double-asterisk convention of Chapter 4.

**TABLE 5.1.**    *Alternative Equations for Estimating a Person's Universe Score*

| Formula[a] | Remarks[b] |
|---|---|
| Equation for a nested one-facet D study: | |
| (5.1)    $\hat{\mu}_p = X_{PI} + [\widehat{\rho^2(X_{pI},\mu_p)}](X_{pI} - X_{PI})$ | Cf. Table 3.1 and (3.2) Cf. Table 5.2: 1, 2a, 3a, 4a |
| Equations for D studies with $I^* \times p$: | |
| (5.2)    $\hat{\mu}_p = X_{PI} + [\widehat{\rho^2(X_{pI*},\mu_p)}](X_{pI*} - X_{PI*})$ | Cf. Table 3.2 and Table 5.2: 2b, 2c |
| (5.3)    $\hat{\mu}_p = X_{PI} + \text{Est}\left[\dfrac{\sigma(\mu_p)}{\sigma(X_{pI*})}\rho(X_{pI*},\mu_p)\right]$ $\times (X_{pI*} - X_{PI*})$ | Cf. Table 5.2: 4b, 4c |
| (5.4)    $\hat{\mu}_p = X_{PI} + \dfrac{\widehat{\sigma^2(\mu_p)}}{\widehat{\sigma^2(X_{pI*})}}(X_{pI*} - X_{PI*})$ | Cf. Table 5.2: 3b, 3c |
| (5.5)    $\hat{\mu}_p = X_{PI} + [\widehat{\mathscr{E}\rho^2(X_{pI},\mu_p \mid I)}](X_{pI*} - X_{PI*})$ $= X_{PI} + \dfrac{\widehat{\sigma^2(\mu_p)}}{\widehat{\mathscr{E}\sigma^2(X_{pI})}}(X_{pI*} - X_{PI*})$ | Cf. (3.5a). This "estimation equation" applies to all crossed designs of Table 5.2. |

[a] $X_{pI*}$ is the D-study mean. The value of $X_{pI}$ may be determined from the G study or the D study or both together (see text).
[b] The cross-references to Table 5.2 indicate the conditions that call for application of each equation.

series of equations in Table 5.1 (5.1–5.5), (5.2) changes to $\text{Est}(\mu_p - \mu) = \widehat{\rho^2}(X_{pI*} - X_{PI*})$. A number of cross-references are made between Tables 5.1 and 5.2; these are used to trace the assumptions and operations behind each formula. As we proceed, we shall clarify how various quantities entering the equations are to be estimated.

Equations such as these take into account the fact that the subject belongs to a population (or subpopulation[3]), a fact the procedures discussed in

[3] Throughout the next several pages we shall assume that equations for the population sampled in the G study are under discussion. We return later to equations for distinguishable subgroups.

**TABLE 5.2.** *Possibility of Estimating Parameters Required in the Regression Equation and the Estimation Equation, under Various Assumptions*

| Assumptions regarding the universe and the designs for G and D study | Regression equation[a] No. | D-study mean | Universe mean | Slope | Estimation equation No. | D-study mean | Universe mean | Slope |
|---|---|---|---|---|---|---|---|---|
| 1. Conditions equivalent in means, $X$-on-$\mu_D$ slopes, and error variances $\sigma^2(\varepsilon)$. No restrictions on design of D study. | (5.1) | + | + | + | (5.1)[b] | | | |
| 2. Conditions equivalent in $X$-on-$\mu_D$ slopes, and error variances, but not in means. | | | | | | | | |
|   a. D study has nested design. | (5.1) | + | + | + | (5.1)[b] | | | |
|   b. D study has $I^* \times p$; and $I^*$ is present in G study along with other conditions. | (5.2) | + | + | + | (5.2)[b] | | | |
|   c. D study has $I^* \times p$; $I^*$ is only set of conditions in G study, or $I^*$ is absent from G study. | (5.2) | + | +[c] | + | (5.5) | + | + | + |
| 3. Conditions equivalent in $X$-on-$\mu_D$ slopes only. | | | | | | | | |
|   a. D study has nested design. | (5.1) | + | + | + | (5.1)[b] | | | |
|   b. D study has $I^* \times p$; and $I^*$ is present in G study along with other conditions. | (5.4) | + | + | + | (5.4)[b] | | | |
|   c. D study has $I^* \times p$; $I^*$ is only set of conditions in G study, or $I^*$ is absent from G study. | (5.4) | + | +[c] | + | (5.5) | + | + | + |
| 4. No equivalence assumption. | | | | | | | | |
|   a. D study has nested design. | (5.1) | + | + | + | (5.1)[b] | | | |
|   b. D study has $I^* \times p$; and $I^*$ is present in G study along with other conditions. | (5.3) | + | + | No[d] | (5.5) | + | + | + |
|   c. D study has $I^* \times p$; $I^*$ is only set of conditions in G study, or $I^*$ is absent from G study. | (5.3) | + | +[c] | No | (5.5) | + | + | + |

[a] + indicates that parameter can be estimated; *No* indicates that it cannot.

[b] Regression equation is used; no special estimation equation required.

[c] If D study uses same conditions as G study, estimate of $\mu$ is poor when conditions are not equivalent (see text). If $I^*$ is not in G study, there is no way to check for possible differences between G- and D-study populations.

[d] While no feasible method is now available, in principle such an estimate can be made.

Section 5.A ignore. In classical theory, it is assumed that tests are fully equivalent and that the parameters of the joint distribution of one observed score with another are known; from these, the population value of the regression coefficient $\rho_{XX'}$ is obtained. Under these assumptions, the regression of true score on observed score is the same for every test. A reliability study is used to obtain values of $\widehat{\rho^2}$ and $X_{PI}$ for entry in (5.1). In classical theory, no distinction is made between G and D studies, and there is no need to ask whether the two values of $X_{PI}$ should be taken from the reliability study, or whether one or both should be taken from the D study. Generalizability theory greatly complicates the estimation problem when it recognizes that conditions may not be equivalent and considers any set of conditions to be a sample from a universe.

There are no particular difficulties in making a regression estimate when the D study uses a nested design. As there will be a new condition for each person, there is no thought of adjusting his score for the "constant errors" associated with that particular condition; these are manifestly impossible to separate from the person's own characteristics. In the joint distribution of universe scores with observed scores from a nested design with $n_i$ uniform over persons, the regression of $\mu_p$ on $X_{pI}$ has the slope $\sigma^2(p)/\sigma^2(X_{pI})$. This is a function of all $p$ and $i$, and can be written as $\rho^2(X_{pI}, \mu_p)$. Equation (5.1) applies. While (5.1) is written in terms of the nested one-facet study, the equation would apply equally well if subscripts for $j$ and other facets were added. The weaker assumptions of generalizability theory, then, create no problems in making regression estimates for a nested D study that are not inherent in the classical model also. Generalizability theory does offer greater flexibility in determining regression equations for a variety of nested D-study designs from a single G study.

### Difficulties introduced by weak assumptions

Table 5.2 lists a variety of ways in which the classical assumptions may be weakened, and summarizes points to be developed in the following section. As assumptions become progressively weaker, it becomes more and more difficult to obtain a satisfactory regression formula.

Let us start with the most troublesome case, presented in line 4c of the table. There is a crossed $I^* \times p$ D study, and conditions $I^*$ are not represented in the G study. The regression equation, (5.3), calls for parameters of the joint distribution of $\mu_p$ and $X_{pI^*}$, but very little is known about that specific distribution. As indicated in Chapter 3, the most practical solution is to substitute $\widehat{\mathscr{E}\rho^2}$ as the slope, shifting to (5.5). This use of the average slope makes (5.5) an estimation equation, not a genuine regression equation. It is only an approximation to the desired regression equation unless a strong equivalence assumption is made. Line 4c of Table 5.2 is read as follows: The regression equation, (5.3), calls for two terms that are difficult or impossible to estimate. The three terms of the estimation equation, (5.5), can all be estimated. An examination of line 1 also will help us to explain the format of the table. This line refers to the strict classical case. For (5.1), the genuine regression equation, all needed parameters can be estimated when the assumptions hold. There is no need to discuss a substitute "estimation" equation. The same is true of nested D studies (lines 2a, 3a, 4a), no matter what the assumptions about equivalence of conditions.

***The constant term.***   Each equation adds a fraction of the person's deviation score to the estimated universe-score mean, to get the estimate of his universe score. The constant term of the equation is written in two parts. One constant is the first member of the equation (our best estimate of the universe-score mean $\mu$). The other, used in forming the deviation score, is a product of the regression slope and the mean in the D study.

If conditions have the same mean (Case 1), (5.1) applies; the mean in the G study or the mean in the D study may be taken as an estimate of either or both constants. When one is satisfied that the persons in both studies are random samples from the same population, the mean based on the greater number of observations might reasonably be employed. A combination of the two means makes fuller use of the data. Each mean has its own standard error and is to be weighted in inverse proportion to that error.

In general, for G studies, the variance of sample means over successive applications of the design:

$$(5.6) \qquad \sigma^2(X_{PI} - \mu) = \frac{1}{n_p}\,\sigma^2(p) + \frac{1}{n_i}\,\sigma^2(i) + \frac{1}{n_p n_i}\,\sigma^2(pi,e)$$

Assuming equivalent conditions, $\sigma^2(i)$ becomes zero. There is an analogous equation in $n_p'$ and $n_i'$ for the D study. Write $\omega_{(G)}$ and $\omega_{(D)}$ for the two standard errors, and write $X_{PI(G)}$ and $X_{PI(D)}$ for the two sample means. Then the best weighting is

$$(5.7) \qquad \hat{\mu} = \frac{\omega_{(G)}\omega_{(D)}}{\omega_{(G)} + \omega_{(D)}}\left[\frac{X_{PI(G)}}{\omega_{(G)}} + \frac{X_{PI(D)}}{\omega_{(D)}}\right]$$

Even though we rely on the random-sampling assumption in all our theoretical development, perhaps the investigator will wish to lessen his dependence on it. If the D-study mean departs from the G-study mean by a modest amount [relative to $\hat{\sigma}(p)$], employing the D-study mean alone in forming deviation scores guards against a possible systematic difference between the G and D samples. It is hard to advise whether to take as the first $X_{PI}$ of the equation the D-study value or the weighted value from (5.7).

The decision rests in part on the size of each study, and in part on the investigator's commitment to the various assumptions. If the means of the G and D studies differ substantially, a reexamination of assumptions is critically needed.

For a nested design the argument is similar. The weighting formula has the form of (5.7), but the standard error of the mean in a nested design is:

$$(5.8) \qquad \omega^2_{(G)} = \frac{1}{n_p} \widehat{\sigma^2}(p) + \frac{1}{n_p n_i} \widehat{\sigma^2}(i) + \frac{1}{n_p n_i} \widehat{\sigma^2}(pi,e)$$

In a D study, one has the same equation with $n'_i$ and $n'_p$.

With crossed designs and nonequivalence, the mean from the D study is used in forming deviation scores. The weighted average from (5.7) is presumably the best estimate of the universe-score mean. However, when the G study and D study use the same conditions, one obtains no information on $\mu_{I*} - \mu$. In effect this is taken to be zero.

*The slope.* Estimating the slope of the regression equation is a straightforward matter if one can assume strict equivalence, or uniform $X$-on-$\mu_p$ regression slopes along with uniform error distributions for various conditions. It is also a simple matter if the D study is nested. The ratio of estimated universe-score variance to estimated observed-score variance serves quite adequately as a slope for (5.1) or (5.2).

However, when weaker assumptions are made, and the D study employs a particular set of conditions $I*$, serious problems arise. The ordinary G study is unable to give a usefully precise estimate of the specific $\rho^2(X_{pI*},\mu_p)$. The conditions $I*$ would have to be used in the G study along with a considerable number of other conditions. We noted earlier the suggestion of Lord and Novick that one may be able to estimate the specific coefficient by an item-sampling design which distributes conditions other than $I*$ over subgroups of subjects. This possibility deserves development, though it obviously cannot be applied routinely to deal with all the different sets $I$ that might be used in D studies.

In (5.4), dividing the estimated universe-score variance by the actual observed-score variance for condition $I*$ is suggested. This does indeed give the regression slope if the $X$-on-$\mu_p$ slope (designated $b_i$ in the development below) is the same for all conditions. Then nonequivalence of $\mu_p$-on-$X$ slopes arises solely from differences in the variances $\sigma^2(\mu_{pi}\sim,e \mid i)$. We are skeptical that this intermediate equivalence assumption can be justified, and hence point to (5.4) for its interest rather than its practical value.

If the investigator who avoids an equivalence assumption must fall back on the estimation equation when he has a crossed design, what are likely to be the consequences of this substitution? More bluntly, how bad are his estimates of $\mu_p$? To examine this, we modify the basic model of (1.1) in which score $X_{pi}$ was divided into components $\mu_p\sim$, $\mu_i\sim$, and $\mu_{pi}\sim,e$. Write $b_i$ for the $X$-on-$\mu_p$ regression slope,[4] and define $v_{pi}$ by the following equation:

$$(5.9) \qquad \mu_{pi}\sim = (b_i - 1)\mu_p\sim + v_{pi}$$

This divides the interaction into two portions, one that is correlated with $\mu_p$ and one that is independent. The equation could be rearranged: $v_{pi} = \mu_{pi} - b_i\mu_p - \mu_i + b_i\mu$. Using these symbols, the model (1.1) becomes

$$(5.10) \qquad X_{pi} = \mu + b_i(\mu_p - \mu) + (\mu_i - \mu) + v_{pi} + e_{pi}$$

This is not a change in the model. According to Chapter 1 the expected value of $\mu_{pi}\sim$ is zero, and hence, the expected value of the covariance over persons $\sigma(\mu_p\sim,\mu_{pi}\sim)$, considering all $i$, equals zero. This does not imply that each separate covariance is zero.

Unless one is dealing with tests carefully constructed to be parallel, it is entirely likely that the interaction for some conditions will covary positively with the universe score, hence $b_i > 1$, and that others will have a negative covariance and $b_i < 1$. It is well known that some judges, for example, extend their ratings over the full range of a scale, while others tend to crowd their ratings into a narrow range. That is to say, some judges are emphatic and some are conservative in reporting information. The result is an interaction component $\mu_{pi}\sim$ that is correlated with $\mu_p$; indeed, $\mu_{pi} - v_{pi}$ is perfectly correlated with $\mu_p$. Classical theory assumes $b_i$ to be 1 for all $i$. We weaken this to $\mathscr{E}b_i = 1$ (which must be the case if $\mathscr{E}X_{pi}$ is to equal $\mu_p$).

Using an estimation equation rather than a regression equation is made more hazardous by an increase in $\sigma^2(b_i)$, and less hazardous by an increase in $n_i'$. [Because $b_{I\bullet} = (1/n_i') \sum_{i\in I\bullet} b_i$, the variance of $b_I$ equals $(1/n_i')\sigma^2(b_i)$.] How much does the result from (5.5) depart from the result (5.3) would give if it could be used? To explore this numerically, we set up a hypothetical problem:

For person $p^*$, $\mu_{p*} = 1$. Conditions in the universe fall into three equally frequent classes, defined by allowing $b_i$ to equal 1.2, 1.0, or 0.8. All conditions have the same error variance $\sigma^2(v,e)$. All conditions have the same mean, equal to zero. $\sigma^2(\mu_p) = 1$.

The variation assumed for the $b_i$ is appreciable, since the largest value is 1.5 times the smallest.

Figure 5.1 presents detailed results for each of nine different combinations of the three $b_i$ with three values of $\sigma^2(v,e)$. We shall explain later how the

---

[4] A coefficient $\beta$, similar to $b$, appears in the Lord–Novick formulation for nominally parallel tests (1968, p. 209).

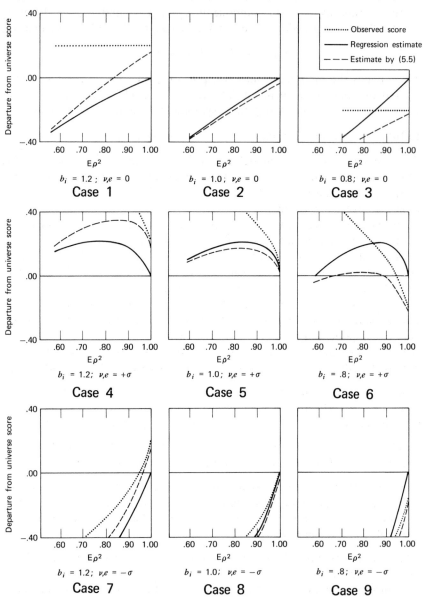

**FIGURE 5.1.** *Error in Estimating Universe Score from the Regression Equation and from the Estimation Equation (5.5), under Certain Assumptions. The Universe-Score Scale has Mean 0 and Standard Deviation 1. The Variable Error Takes One of Three Values: 0, $+\sigma(\nu,e)$, $-\sigma(\nu,e)$.*

calculations were made. Only the higher values of $\mathscr{E}\rho^2$ are represented in the figure. The left portions of the curves are readily constructed, because in any panel both of them extend steeply downward to the point $(-1,0)$. The reader may find much of interest in the details of the relationships, but the main interpretation has to focus on how each approach works out on the average.

The departures of the estimate from the universe score, without regard to sign, were averaged over all nine cases. The average discrepancies for both the estimation equation and the regression equation have a trend about like this:

| $\mathscr{E}\rho^2$ | 0 | $\cdots$ | 0.60 | 0.70 | 0.80 | 0.90 | 0.95 |
|---|---|---|---|---|---|---|---|
| Mean absolute error | 1.00 | $\cdots$ | 0.46 | 0.40 | 0.32 | 0.23 | 0.16 |

This may be compared with the discrepancy between the observed score and the universe score, for which the trend is:

| $\mathscr{E}\rho^2$ | 0 | $\cdots$ | 0.60 | 0.70 | 0.80 | 0.90 | 0.95 |
|---|---|---|---|---|---|---|---|
| Mean absolute error | $\infty$ | $\cdots$ | 0.59 | 0.47 | 0.36 | 0.25 | 0.19 |

The estimate from (5.5) has, at all levels of $\mathscr{E}\rho^2$, a greater mean departure from the universe score than the regression estimate; but the difference is negligible unless $\mathscr{E}\rho^2$ is large:

| $\mathscr{E}\rho^2$ | 0–0.85 | 0.90 | 0.95 | 1.00 |
|---|---|---|---|---|
| Difference in mean absolute error | $\ll 0.01$ | 0.02 | 0.06 | 0.14 |

The correspondence of the two estimates would have been less close if we had employed a wider spread of $b_i$. Also, use of a different $\mu_{p*}$ would change the results somewhat. Nonetheless, we conclude that the use of (5.5) in place of (5.3) will generally give satisfactory results.

Let us now present enough detail to enable the reader to derive these results, and also to work out results for other hypothetical problems. For any condition,

$$(5.11) \qquad \sigma^2(X) = b_i^2\sigma^2(\mu_p) + \sigma^2(v,e)$$

Because $\sigma^2(\mu_p) = 1$ and $\sigma^2(v,e)$ is uniform,

$$(5.12) \qquad \mathscr{E}\sigma^2(X) = \mathscr{E}b_i^2 + \sigma^2(v,e)$$

$\mathscr{E}b_i^2 = (1/3)(1.44 + 1.00 + 0.64) = 1.03$. For purposes of this illustration take $\sigma^2(v,e)$ to be 0.09. For this universe, with $n_i' = 1$ we obtain $\mathscr{E}\sigma^2(X) = 1.12$. For $\mathscr{E}\rho^2$ we evaluate the usual formula: $\sigma^2(\mu_p)/\mathscr{E}\sigma^2(X)$ is $1/1.12 = 0.89$,

If the condition drawn has $b_i = 1.2$, (5.11) indicates that $\sigma^2(X) = 1.44 + 0.09 = 1.53$. The regression slope $b_i/\sigma^2(X)$ is $1.2/1.53 = 0.78$. This is to be compared with the 0.89 of the estimation equation. Likewise, if $b_i = 0.8$, $\sigma^2(X) = 0.73$ and the regression slope is 1.09. The actual $\mathscr{E}\rho^2$ for a given $\sigma^2(v,e)$ is found by calculating $b_i^2/\sigma^2(X)$ to get $\rho_i^2$, and averaging over the three values of $b_i$. For $\sigma^2(v,e) = 0.09$ we get $\rho^2 = 0.94$, 0.92, and 0.87. which averages 0.91.

Consider Case 4 of the figure, where $b_i = 1.2$ and the regression slope is 0.78. We have $\mu_{p*} = 1$. When $\sigma^2(v,e) = 0.09$, $X_{p*i} = 1.2 + \sigma(v,e) = 1.50$. When we multiply this by the regression slope we obtain 1.18, which is 0.18 greater than $\mu_{p*}$. Multiplying by the estimation slope gives 1.34, and a discrepancy of 0.34. A point at (0.91, 0.18) helps locate the solid line of the figure; a point at (0.91, 0.34) lies on the broken line. Other values of $\sigma^2(v,e)$ are used to get additional points.

Estimating universe scores resembles the forecasting of criteria. But if a faulty prediction equation is put into use, it is automatically called into question at a later time when actual outcomes do not accord reasonably well with the predictions. There will never be observations of the universe score, and if the estimation equation is faulty, that fact may not be discovered. It is possible, for example, that the equation is highly accurate when initially established. However, changes over time in the population mean, universe-score variance, or correlation among conditions may make it inapplicable. The possibility of change should be recognized in any attempt to use results from a G study in subsequent years, but the estimation equation may be more sensitive to such changes than are other applications.

### Interpreting the estimate

The estimate may be interpreted on the basis of norms, or may be used to forecast criterion performance, or may be used to describe how well a universe of content has been mastered. We shall discuss the difficulties and puzzlements involved in such interpretations. It is to be emphasized that the problems are present whether one assumes the strong classical model or our weaker one. To simplify the discussion we shall write this section in terms of strictly equivalent conditions.

*Norm-referenced interpretation.* The score $\hat{\mu}_p$ locates the person on the scale common to both observed scores and universe scores. But the norms for observed scores do not apply. The percentile rank of $\hat{\mu}_{p*}$ in the distribution of *estimated* universe scores $\hat{\mu}_p$ is the same as the percentile rank of the observed score $X_{p*}$ in the $X$ distribution. Surely, a norm-referenced interpretation should be based on the location of the estimate within the distribution of *actual* universe scores $\mu_p$—not the distribution of $\hat{\mu}_p$ or of $X_{pi}$.

The distribution of $\mu_p$ is unknown. The most plausible approximation ordinarily available is a normal distribution with mean $\hat{\mu}$ and variance $\sigma^2(p)$. (Information about higher moments of the universe-score distribution, and therefore a check on its normality, may in principle be obtainable when the G data are very extensive. See Lord & Novick, 1968, pp. 232–233, 245–248; Lord, 1969).

If deviation scores $\mu_p - \mu$ are estimated, one can if he desires introduce a standard-score scale for universe scores. Suppose this to have mean 50 and standard deviation 10. Then one could express $\sigma(p)$ in raw-score units and assign the value 60 to an estimated universe score that falls $\sigma(p)$ units above the mean, and so on. A simple table could be prepared for any reference population in which a G study has been conducted. If one has reasonable confidence in the estimate of $\mu$, then the table could convert an absolute estimate $\hat{\mu}_p$ to a standard-score scale for which $\sigma(\mu_p) = 10$.

The tester has long been trained to read a standard score of 60 as implying a percentile rank of 83, and to make similar interpretations for other scores. *If* actual universe scores were known, that interpretation would apply to normally distributed universe scores. When universe scores have a standard deviation of 10, the estimated scores have a standard deviation of $10(\mathscr{E}\rho^2)^{1/2}$. This implies that the tester will have fewer extremely high and extremely low standard scores to interpret than he has when working with ordinary standard-score transformations of observed scores. That is good, because with a fallible test one has no warrant for making extreme statements. It is strange to think that one might test a representative group of job applicants and conclude that no one ranks above the 90th percentile. It is less strange when the statement is recast as follows: "Given the fallibility of the test, our best judgment locates this person (the highest scorer) as falling somewhere around the 90th percentile of the universe-score distribution."

Pursuing this line of reasoning leads to fresh questions about matters long taken for granted. Should the deviation IQ be defined, as in the past, so that observed IQs have a standard deviation of 16? This amounts to defining a unit of measurement in terms of a fallible operation. An accurate mental test and an inaccurate one, both scaled to yield IQs with a standard deviation of 16, will have different standard deviations for $\mu_p$. If two forms of any test so scaled were averaged, the standard deviation of observed IQs for this more accurate procedure would be less than 16. Paradoxes like this make it obvious that a unit one intends to use in describing individuals and in stating functional relations should be defined in terms of universe scores and not in terms of a fallible operation. This argues strongly for changing norming practice, see also p. 257.

While any one measure belongs to various universes of generalization, it would seem to be feasible to select the universe that most interpreters are likely to have in mind and to select the reference population most of them are interested in, and to calculate $\sigma(\mu_p)$ for that universe and population. This would then be used in place of $\sigma(X)$ to define the unit for the standard-score scale. If that is done, percentiles should similarly be inferred for the universe-score distribution, though this requires an assumption of normality unless very extensive G-study data are available.

*Criterion-referenced interpretation.*    The usual form of criterion-referenced interpretation draws on past experience to report what criterion performance is typical of persons having a given observed score. This may take the form, for example, of a regression equation for predicting grades from an aptitude measure, or of an expectancy table that gives the grade distribution corresponding to each score level. Provided that all persons are members of the same population, and that a test of the same length will be used both in the follow-up study from which data on expectancies come and also in measuring individuals in the D study, there is no value in making any adjustment for error of measurement. One might estimate universe scores for persons in the follow-up study, and express criterion expectations as a function of estimated universe scores. But this tabulation, entered with the estimated universe score for a person $p^*$, will report precisely the same expectation for him as the observed-score table did. The same would be true of an expectancy table relating actual universe scores to criterion scores.

The situation is different when some persons are measured more thoroughly than others. Suppose that a population of college freshmen has a mean of 600, that a certain person has the observed score 800, and that the regression line indicates an expected grade average of 3.50 for him. Consider now another person who is tested twice, earning scores of 750 and 850. Perhaps the regression line indicates expected outcomes of 3.20 and 3.80 for these scores individually, or 3.50 for the average. This interpretation is unsound, as a person who earns a high score twice probably has higher aptitude than a person who earns that score once. If universe scores were estimated, the single test score of 800 would be regressed to, say, 770 and the score based on two testings to 784. It appears, then, that when some persons will be measured more thoroughly than others it is appropriate to alter the expectancy table or criterion-on-test regression function to take this into account.

It is possible to get the criterion-on-universe-score regression function. The empirical follow-up data give the covariance of criterion score with observed score, which estimates the expected covariance of criterion score with universe score. Dividing this by the universe-score variance calculated from a G study gives the slope relating criterion score to universe score. Such a correction requires that the follow-up study and the G study employ the same sample, or, if separate samples are used, that they be reasonably large and unquestionably from the same population.

If the follow-up data show a curvilinear regression of criterion on observed score, one can easily take this into account when the observed score is interpreted directly. We have no way to translate the curvilinear function into the comparable function relating criterion score to universe score (though perhaps methods for this purpose could be developed). Similarly, while one might reconstruct an expectancy table relating criterion to universe score by making strong assumptions of bivariate normality, one has no way to convert the typical heteroscedastic expectancy table into the comparable table for universe score.

In establishing a regression function relating criterion score to universe score, estimating universe scores for the individual, and interpreting them with the aid of the function, one must invoke a large number of assumptions. The hazards of the procedure are therefore considerable. Practically, it may be sufficient to warn test interpreters that any expectancy table or criterion-on-observed-score regression is accurate only for test data in which $n_i'$ is the same as in the original follow-up study. For a limited number of individuals a different $n_i'$ will have been used; a sophisticated interpreter will be able to take this into account without actually going through the rather treacherous two-stage regression. Fortunately, the predictor variables that enter into criterion-referenced interpretations are usually highly reliable. Therefore, an effort to take error of measurement into account formally would produce little difference in the final interpretation.

*Content-referenced interpretation.* Interpretations that consider the observed score as representative of the universe of content from which a test is sampled are becoming increasingly prominent. This is seen in many forms of instruction where the lessons a student is given next week will be determined by his score on this week's test. Where the computer is used to regulate instruction, the tests employed may be designed on a sequential basis that administers far more items to some students than others. The decision rule may be, for example, that students who can solve 75% of a domain of quadratic equations are ready to proceed to a new topic. If the test is a true sample from that domain, the estimated universe score, expressed in terms of percentage correct, is a better basis for the decision than the observed score.

The estimation equations offered by generalizability theory can be used to estimate the universe score. In the instructional situation, however, one may feel that such a complication is unnecessary because wrong decisions resulting from the direct interpretation of the observed score may not be costly. If estimation equations are to be used, it may be well to consider a multivariate estimation procedure of the sort introduced in Chapter 10.

Where measuring procedures are sequential, it is necessary to recognize that generalizability is greater for those pupils who are measured most thoroughly. If the sequential procedure simply draws additional items at random in measuring certain persons, then our equations apply; one forms the $\widehat{\mathscr{E}\rho^2}$ of (5.5) for various values of $n_i'$ and applies to each person the equation that corresponds to the $n_i'$ used in observing him. The typical sequential procedure in educational measurement, however, selects easier or harder items for the second, third, etc. stages of testing, depending on the person's success at previous stages. To interpret the resulting scores calls for a Bayesian analysis; this kind of analysis is still in the process of development. (See Novick, Jackson, & Thayer, 1971.)

### Interpretation of $\sigma(\varepsilon)$

The error of estimate $\varepsilon$ indicates in a gross sort of way the adequacy with which the observed score forecasts the universe score. Even when the population parameters $b_i$ and $\sigma^2(\delta)$ are the same for all conditions, however, the inferences made from the G study are only approximate. The regression coefficient is estimated from a finite body of data, and therefore is affected to some extent by sampling error. Therefore, regression equations derived from successive, independent G studies would vary in both slope and intercept, marking out a pattern of criss-crossing lines with a more-or-less hyperbolic envelope. When conditions are not equivalent, the estimation equations (5.5) will show a similar pattern, but the variation from one sample to another will be greater, unless $n_i'$ is large.

For persons whose observed scores are near the population mean, the prediction of the universe score is relatively accurate. However for persons far from the mean, the variability of possible estimation or regression functions is substantial, and $\sigma(\varepsilon)$ (which ignores sampling of persons and conditions in the G study) gives much too conservative an account of the error of estimate. It follows that a simple confidence interval of the form $\hat{\mu}_p - \sigma(\varepsilon) \leqslant \mu_p \leqslant \hat{\mu}_p + \sigma(\varepsilon)$ is not useful.

Our reservations about the direct interpretability of $\sigma(\varepsilon)$ do not apply in the D study where conditions are nested within persons. Under those circumstances the relevant parameters are likely to have been well estimated and the slopes are uniform over D studies. A confidence interval symmetric around $\hat{\mu}_p$ [using the value of $\sigma(\varepsilon)$ appropriate for a nested study] may then be interpreted with some confidence, except where ceiling and floor effects call the assumption of linearity into question.

### Estimation on the basis of subpopulation means

Kelley's original introduction (1923) of the regression estimate of true scores placed considerable emphasis on the fact that the estimate takes into account the mean of whichever group the person belongs to. Therefore, it draws different conclusions about, for example, a fourth grader and a fifth grader having the same observed score. We have pointed out that the regression equation can be interpreted as a multiple-regression equation that assigns appropriate weights to the individual information and the group information, greater weight being placed on the former as the coefficient of generalizability becomes greater.

Whether subpopulation parameters should be employed in place of population parameters in the equations of Table 5.1 involves three distinct questions. One is the political question referred to on p. 106: Will persons tested and institutional decision makers regard a procedure as fair if it leads to different decisions for persons who earn the same test score? We postpone discussion of this to Chapter 11. Another is a straightforward numerical question: Does the use of subgroup information alter scores by an appreciable amount? The third question is complex: To what extent can one have confidence in the assumptions underlying the use of subgroup regresssion equations?

*Practical significance.* The use of subgroup information produces a $\hat{\mu}_p$ for which the correlation $\mathscr{E}\rho^2(\hat{\mu}_p,\mu_p)$ ought to be larger than the correlation $\mathscr{E}\rho^2(X_{pi},\mu_p)$, which equals the correlation $\mathscr{E}\rho^2(\hat{\mu}_p,\mu_p)$ for the $\hat{\mu}_p$ obtained from the population estimation equations. Equation (3.11) showed that this increase depends on the separation of the subgroup means and on the value of $\mathscr{E}\sigma^2(\delta)$. It is useful now to consider some specific numerical values.

Suppose that $\sigma^2(\mu_p$, within groups) is 200 and $\sigma^2(\mu_P$, between groups) is 50. This is a moderately large separation of groups; the two means differ by about one within-group standard deviation. Then for various values of $\mathscr{E}\sigma^2(\delta)$ we have:

| $\mathscr{E}\sigma^2(\delta)$ | $\widehat{\mathscr{E}\rho^2}(X_{pi},\mu_p)$ | $\widehat{\mathscr{E}\rho^2}(\hat{\mu}_p,\mu_p)$ | Increment resulting from use of subgroup information | Increment as proportion of $1 - \widehat{\mathscr{E}\rho^2}(X_{pi},\mu_p)$ |
|---|---|---|---|---|
| 0 | $250/250 = 1.000$ | 1.000 | 0.000 | — |
| 20 | $250/270 = 0.925$ | 0.926 | 0.001 | 0.01 |
| 50 | $250/300 = 0.833$ | 0.839 | 0.006 | 0.04 |
| 100 | $250/350 = 0.714$ | 0.734 | 0.020 | 0.07 |
| $\infty$ | | 0.000 | 0.200 | $0.200^5$ | 0.20 |

We used (3.11) to calculate the increments, and added these to the second column to get $\mathscr{E}\rho^2(\hat{\mu}_p,\mu_p)$. The last column indicates the fraction by which one would have to increase $n_i'$ to get the same increment in $\mathscr{E}\rho^2(X_{pi},\mu_p)$.

[5] This value equals $\sigma^2(\mu_P)/\sigma^2(\mu_p$, all cases).

None of these increments makes for appreciable improvement in the usefulness of the score. That is, with a group separation of $\sigma(\mu_p)$ or less, one gains nothing from the controversial procedure of regressing toward the subgroup mean.

One further set of examples will be useful. Keep the within-group variance at 200 and raise the between-group variance to 200. Then the difference in means is twice the within-group standard deviation—a large value indeed. These results follow:

| $\mathscr{E}\sigma^2(\delta)$ | $\widehat{\mathscr{E}\rho^2}(X_{pi},\mu_p)$ | $\widehat{\mathscr{E}\rho^2}(\hat{\mu}_p,\mu_p)$ | Increment | Increment as proportion |
|---|---|---|---|---|
| 0 | 400/400 = 1.000 | 1.000 | 0.000 | — |
| 20 | 400/420 = 0.952 | 0.954 | 0.002 | 0.04 |
| 50 | 400/450 = 0.889 | 0.900 | 0.011 | 0.10 |
| 100 | 400/500 = 0.800 | 0.833 | 0.033 | 0.17 |
| 200 | 400/600 = 0.667 | 0.750 | 0.083 | 0.25 |
| ∞ | | 0.000 | 0.500 | 0.500 | 0.50 |

Even with this quite large separation of groups, the gains in accuracy are modest unless the original coefficient of generalizability is low.

In general, it does not appear that regressing toward subgroup means will improve estimates of individual universe scores by a great deal. However, the above calculations consider the average improvement over all cases. It should be remembered that regressing toward the group mean will make an appreciable difference in the ranking or the absolute value of $\hat{\mu}_p$ for those persons belonging to the lower-scoring subpopulation who have exceptionally large observed scores, and also for low-scoring persons within the higher-ranking population.

Regression toward subgroup means has been suggested by Lord (1960) and by Porter (1967) as a means of improving the interpretation of experiments where persons are not assigned to treatments at random. It is suggested that using $\hat{\mu}_p$ (from 3.10) as a covariate will do more to counteract the bias introduced into the experiment by nonrandom assignment than will use of $X_{pi}$ as a covariate. While this is not invariably the case, the procedure does seem to be advantageous. The numerical analysis above does not imply that regressing toward the subgroup mean has little effects in *this* application. The slope of the regression of dependent variable on covariate universe score is greater than that for the regression on covariate observed score, the two slopes having the ratio $1/\widehat{\mathscr{E}\rho^2}$. This will not produce a large numerical shift in the adjusted between-groups difference in outcome. But if the covariate is only moderately reliable, the change will be enough in many experiments to change a significant $F$-ratio to a nonsignificant one or vice versa.

Use of subpopulation equations introduces no special problem in norm-referenced interpretation or content-referenced interpretation beyond those discussed earlier. In criterion-referenced interpretation, it is necessary to recognize the possibility that there is a different criterion-on-universe-score regression equation for each subpopulation. This, of course, will not be known unless a follow-up study is carried out in a sample from each subpopulation.

*Dependability of interpretation.* Our earlier discussion of the pitfalls in interpreting estimation equations was restricted to the population estimation equation. Certain additional hazards are introduced by the use of equations for subpopulations. Although the following statements are brief, they raise questions of considerable importance.

On page 141 we discussed the difficulty of estimating the constant term of equations in Table 5.1. Estimating the universe score mean (or that of $\mu_{I^*}$ — $\mu$) for a subpopulation is even more difficult, unless one has carried out a G study within the subpopulation. This has no serious consequences when one is primarily interested in comparing two subpopulations, as in the Lord–Porter technique. In that application, any fault in the estimate of $\mu_{I^*} - \mu$ affects both groups in the same way.

In determining the slope for the regression equation (5.1) the classical theory has to make a correction for "restriction of range" (see page 99) which assumes that the error variance is the same in all subpopulations. With non-equivalent conditions, we can make a similar correction only by making the much stronger assumption that all components of variance except that for persons are the same in the two subpopulations.

### The Bayesian approach to estimation of universe scores

It was mentioned in Chapter 2 that Bayesian statistical methods are beginning to be applied to the analyses this monograph calls for. It will be useful here to introduce some notes on the Bayesian replacement of the regression equation offered by Novick and his colleagues. To do this, we quote from a summary statement recently prepared by Novick and Jackson (1970, pp. 473–475). We have modified the notation to conform to that used elsewhere in this book. A number of comments made following the quotation will assist the reader to understand the key differences between this approach and ours.

Unfortunately, there is a difficulty in attempting to apply the Kelley formulation [see p. 103] in most practical applications because the population mean is typically not known before measurements are taken and hence the regression formula cannot be used in its given form. In effect, what is needed is a regression estimate based not on the person's observed score and a *known* mean observed score, but rather one based on the person's observed score and the average observed score of a random sample of people from the population. Results of this type are available in the framework of Bayesian methods with normality assumptions. The first of these was given by Lindley (see the discussion in Stein, 1962); later a fuller development was given by Box and Tiao (1968) and by Lindley (see Novick, 1969). These estimates are of the form, $wX_{pI} + (1 - w)X_{PI}$, i.e., a weighted average depending on weights $w$, the person's observed score $X_{pI}$ and the mean observed score $X_{PI}$ in the sample. In this formulation the quantity $w$ is an estimate of the reliability of the observed score. It tends to unity as the number of observations on the person increases without limit and to zero as the number of observations on the person tends to zero. For intermediate cases its value depends on the relative number of observations on the particular person, the number of observations on all persons and also on the number of persons on whom observations are available.

For our purposes it will be useful to consider Bayesian estimates obtained by Lindley since this method easily generalizes to the case of unequal replications. Under moderate conditions and using the specific prior distribution suggested by Novick (1969) to characterize a situation in which we have no prior information, Lindley shows that the mode of the conditional distribution (the posterior Bayes distribution) of the true scores $\mu_{p*}$ after obtaining all observed scores can be calculated as the solution of the $n_p$ equations

$$(5.13) \qquad n_p n_i \frac{\mu_{p*} - X_{p*I}}{\sum s_p^2 + \sum (X_{pI} - \mu_p)^2} + \frac{(n_p - 1)(\mu_p - \mu_P)}{\sum (\mu_p - \mu_P)^2} = 0$$

where $X_{pi}$ is the $i$th observation on the $p$th person, $n_p$ is the number of persons, $n_i$ is the number of replications on each person, $s_p^2 = \sum (X_{pi} - X_{pI})^2/n_i$, $X_{pI} = \sum_i X_{pi}/n_i$ and $\mu_P = \sum_p \mu_p/n_p$ and where it is assumed that $\mu_p$ are not all equal. These are the simplest of the *Lindley equations*. Because the quantity $\mu_{p*}$ is a part of the mean value $\mu_P$, these equations cannot be solved directly. An approximate solution to these equations for large $n_p$ and $n_i$ having the general form described above is

$$(5.14) \qquad \hat{\mu}_{p*} = \frac{\widehat{\sigma^2(\mu_p)}}{\widehat{\sigma^2(\mu_p)} + \widehat{\sigma^2(E)}/n_i} X_{p*I} + \frac{\widehat{\sigma^2(E)}/n_i}{\widehat{\sigma^2(\mu_p)} + \widehat{\sigma^2(E)}/n_i} X_{PI},$$

$$p^* = 1, 2, \ldots, n_p$$

where $\hat{\sigma}$'s are the usual ANOVA estimates and $X_{PI} = \left(\sum_p X_{pI}\right)/n_p$.

For small samples, equations (5.13) and (5.14) do not give the same results. Further details of a method for obtaining the exact solutions to the Lindley equations by iteration, the modest conditions under which they are valid and reasons for preferring the Lindley method are given by Novick, Jackson and Thayer (1971). A generalization of these equations to include the case of unequal replication numbers and including technical improvement to guarantee convergence was provided by Lindley (1969).

The true score estimates given in (5.13) were obtained from a *Bayesian structural model* which assumes that the observed scores for each individual are normally distributed with mean equal to that person's true score and with homogeneous error variance $\sigma^2(E)$ and that the true scores are normally distributed with mean $\mu$ and variance $\sigma^2(\mu_p)$. It was further assumed that there was no information available [prior to the G study] about the true or error score variances or the mean true score. Formally this was accomplished by using the indifference prior distributions for $\mu$, $\sigma^2(\mu_p)$ and $\sigma^2(E)$ suggested by Novick (1969) as developed from the work of Novick and Hall (1965). These indifference priors consist of independent uniform distribution on $\mu$, log $\sigma(E)$ and log $\sigma(\mu_p)$. However, if some prior information is available either about the distribution of true or error scores, this information can be incorporated into the prior distribution using the procedure suggested by Novick (1969) as developed from the work of Novick and Grizzle (1965). Often it is useful and sometimes it may be essential to do this. However, it seems to be true that when the number of persons being tested is large, prior information can be *largely* disregarded (Novick, Jackson, & Thayer, 1971).

The choice of the prior distribution for this analysis reflects prior information and beliefs (or lack of them) concerning the mean true score in the population, the spread of true score values and the average variability within persons. These, in total, imply a prior distribution on the individual true scores $\mu_p$. After obtaining observations on persons we have a new Bayes distribution for the $\mu_p$ and we also have a new Bayes distribution for the mean true score, the variance of the true scores, and the variance of the error scores, and all of this information is available to guide any decision that must be made at any stage of testing. Lindley's methods and the very similar ones of Box and Tiao provide improved techniques for estimating true and error score variances and reliability. The details are given in a paper by Novick, Jackson, & Thayer (1971).

The point to be emphasized here is that at any point in the data gathering the Bayes distribution for any particular $\mu_p$ reflects more than just the observations on person $p$. Rather it reflects the combined information relevant to all of the $\mu_p$. Thus after one obtains information on some $\mu_p$, he is no longer completely uninformed about a new $\mu_{p*}$; rather the prior distribution for this new $\mu_{p*}$ would effectively be the estimated distribution of $\mu_p$ values in the population of people. As has been seen, the effect of this is to regress estimates of true score towards a common mean. This regression provides the Bayesian solution to a number of statistical problems. Thus for this rather complex Bayesian structural model the actual use of a vague prior for data analysis seems appropriate when the number of persons is large, but for less complex models objections can be raised (e.g., Novick & Grizzle, 1965). This is so because the buildup of information is much more rapid with the structural model than with simpler models.

This statement embodies an essentially classical assumption, strengthened by the assumptions of normal distributions. The remarks would apply to a one-facet nested design or to a crossed design with complete equivalence. While the distinction between the G and D studies is not mentioned, such a concept is actually embodied in the Bayesian approach. Where we have discussed the Bayesian point of view earlier in this chapter, we pointed out that one could regard the distribution of scores for other persons as "prior information." In this statement by Novick, the information on other persons is taken as essentially simultaneous with the observations on $p*$. All references to prior information in the passage quoted have to do with information collected in the past. The statement is worded to suggest that the investigator starts his G study with a blank slate, and that $p*$ is included as one case within the G study. If one were to make our separation between G study and D study, he could treat the G-study information as prior to that in the D study whether the D study is based on one person or more. In that event, the prior distribution should not be disregarded. The variance estimates in (5.14) are of the kind we use in this monograph, following the methods of Chapter 2. Elsewhere in the paper, Novick and Jackson emphasize that Bayesian analysis of variance may give more satisfactory estimates when the data contradict the strong assumptions embodied in the solution above. It remains for future work to determine how much practical difference the application of Bayesian methods for this basic case will make, and to determine whether the methods can practically be extended to some of the complex designs with which this monograph deals.

*Summary*

This chapter has traced the intricate arguments required in making inferences about the universe score. We have seen impressive evidence that interpretation even under the simple model of classical theory brings in assumptions or logical leaps that are difficult to justify. Generalizability theory, by bringing many of these logical difficulties to the surface, performs a considerable service. On the whole, however, the more elaborate model of generalizability theory complicates the problem of making inferences.

The practice of forming confidence intervals symmetric around the observed score is open to severe criticism, even when the G data are extensive and the interval is properly based on the error $\Delta$ rather than $\delta$. The determination of $\hat{\sigma}(\Delta)$ does provide a superficial but useful statement regarding the extent to which observed scores are likely to depart from universe scores. Going on to form an interval estimate for the individual is a dubious practice because any confidence level attached to the estimate is likely to be misleading.

The confidence-interval technique can be justified as a way of studying a few particular questions. It is an appropriate method of summarizing information on a group mean. The confidence level that the multifacet model calls for will be known only roughly, however, unless one can assume normality of score-component distributions. Second, $\hat{\sigma}(\Delta)$ can be used to advantage when one attempts to test a hypothesis about the universe score instead of attempting to estimate it. This is exemplified in the inspection of score differences to learn which ones depart reliably from zero, and in the use of sequential methods for selection or classification.

The linear equation for estimating the universe score with the aid of $\widehat{\mathscr{E}\rho^2}$ is a regression estimate only under strong assumptions, essentially like those of the classical theory. Although marked nonequivalence makes such estimation equations misleading, we are inclined to recommend cautious use of such estimates of the universe score. For most subjects the estimate will be closer to the truth than the observed score. The benefit from regressing scores toward subpopulation means, however, rarely will be great enough to justify the extensive assumptions required.

## EXERCISES

**E.1.** In 1960, Mahalanobis suggested that nationwide examinations to select among college applicants in India be formed by sampling from a universe of multiple-choice items stratified on content and difficulty. He would select a set of items independently (with replacement) for each applicant. This would exclude any possibility of copying another student's responses. The same large pool of questions could be used for several years, permitting maintenance of uniform standards.

a. Is it reasonable to suppose that all students would receive examinations of comparable difficulty?

b. Das (1967) develops the proposal, showing how $\widehat{\sigma^2}(\Delta)$ would be computed and used to form an interval for the applicant's universe score, symmetric around his observed test score, "The student's true intelligence or scholastic attainment is expected to lie, with 95% or 99% confidence, within these limits." This chapter has criticized the formation of confidence intervals of this kind. If the test is used to select the best 100 applicants out of 1000, is the interval helpful in decision making?

c. Das recommends that each student be given three stratified-parallel tests. (Each of these might be formed by drawing just one item from each content-difficulty stratum.) From these scores, Das would estimate $\sigma(\Delta \mid p)$ for each student separately. Assuming that this general type of information is wanted, what are the arguments for and against calculating $\hat{\sigma}(\Delta \mid p)$ rather than $\hat{\sigma}(\Delta)$?

**E.2.** The *National Longitudinal Study* of *Mathematical Abilities* applied many short tests to large groups of students who had studied from one or another textbook. Each test was reported on a standard-score scale, where the mean for the entire group of students (all textbooks together) was set at 50 and the standard deviation of observed scores was set at 10. A typical comparison chart shows the means for several textbook groups on each of the measures. For example, on the test *Rational Numbers* the range of these means is 46–56; and on *Multiplication of Fractions*, the range is 32–67. One is tempted to conclude that textbook differences produce large differences in learning on fractions, and rather small differences on items dealing with rational numbers.

Do the standard-score scales permit this kind of interpretation? Is information on generalizability of the scales in any way pertinent? (Assume that each mean is based on a large sample and that the textbook groups were comparable in background.)

**E.3.** John Doe (p. 18) is a resident of California, an electrician, a person with a $15,000 income, and a Republican. In a study of political radicalism, he (along with several hundred other adults) is given an attitude scale. If one wished to adopt the suggestion that universe scores be estimated by subpopulation regression equations, what subpopulation should be considered in the case of Doe?

**E.4.** In certain types of individually paced instruction, a student is advanced to a new unit of study only after he demonstrates a specified level of mastery on the preceding unit. The standard may be, for example, "can perform two-digit multiplications with 95% accuracy." Logically, this standard would seem to refer to the person's universe score, and hence, one faces the problem of designing a test with adequate precision.

Sequential testing is suggested by Cronbach and Gleser (1965, pp. 91–96) for decisions of this kind. The proposal, in its simplest form, is as follows: Set up a band of width $2a\hat{\sigma}(\Delta)$ symmetric around the standard. If the observed score, expressed in per cent, lies above the band, advance the student. If it lies below the band, return him to further instruction on the multiplication unit, and let him "come up for promotion" after another week of training. If the score is within the band, administer a second random-parallel test. After this second test, average the two observed scores. Calculate $\hat{\sigma}(\Delta)$ for the double-length test, and set up around the standard a band proportional to *this* $\hat{\sigma}(\Delta)$. This band is narrower than the one for the single test. Apply the same decision rule to the averaged score. Some fraction of the students will be sent on to a third test, and the same process is applied to those scores. Ultimately all students will have been classified as above or below the standard, and assigned to the advanced or basic instruction. Is this procedure (recommended some years ago) sound, in view of the present Chapter 5?

## Answers

**A.1.   a.** This can be assured if the items have been pretested for difficulty and the difficulty scale is finely stratified, or if a substantial number of items is drawn from each stratum. It would be possible to determine, from $\sigma^2(i)$ within a stratum and the number of strata, the number of items required to reduce the systematic error $\mu_I - \mu$ to any desired level.
**b.** The transformation of the observed score into the interval estimate has no direct bearing on the selection decisions. These decisions are based on the rank order of subjects; one would select the same 100 subjects whether he examined the observed score, or the lower ends of the confidence intervals, or the scores regressed toward the mean of all applicants. Reporting the interval estimate to the selectors might warn them of the uncertainty associated with judgments of the last candidates to make up the quota, and so motivate them to seek additional information on those cases. Therefore, the interval might be helpful even though the preselected confidence level does not apply. (Though one may attempt to set 99% confidence intervals, he cannot say that for individuals at a particular observed-score level the probability is 99 out of 100 that their universe scores lie within the interval.)
**c.** In theory, $\sigma(\Delta)$ will vary from person to person. The person with the universe score of 100% will have $\sigma(\Delta) = 0$ and, in general, the more extreme universe score will be associated with a smaller $\sigma(\Delta)$.

To make a G study with three tests for every subject in the D sample will be difficult when each test must be long, on account of the total number of content × difficulty strata. Moreover, with only three or four tests per person, one is not likely to obtain an accurate estimate of the idiosyncratic value of each person's $\sigma(\Delta \mid p)$.

It appears far more practical to conduct a single G study on one sample of students, and to carry that information forward to future cases. The nature of the test construction makes that extrapolation to a new sample quite legitimate. It would be well, however, to calculate $\hat{\sigma}(\Delta)$ separately for subsamples in different ranges of observed score, to check on the possibility that the value changes appreciably. If it does, one would apply to person $p$ the $\hat{\sigma}(\Delta)$ corresponding to his observed score.

**A.2.**   In the first place, the scales do not have a meaningful common metric. There is no reason to think that a difference of one standard deviation on the test of fractions is "equal to" one standard deviation on the *Rational Numbers* test. (The standard deviation of IQ is 16 points. The standard deviation of height of adult men is about 3 inches. Is 3 inches equal to 16 IQ points? Is such a question meaningful?)

Even so, one might be interested in examining the overlap of distributions for different textbooks on the same scale. The apparent finding that the means are crowded together on the *Rational Numbers* test is an artifact of the standardization.

Suppose there are just two textbooks, for which the universe-score means are 20 and 30 (respectively) on both variable 1 and variable 2. Suppose that the universe-score standard deviation for each variable is 10. Now suppose that the test of variable 1 is highly fallible, so that the observed-score distribution has a standard deviation of 100, most of which arises from error. The standardizing operation transforms the raw-score scale by dividing scores by 100; the difference between the textbook means on $v_2$ will shrink from 10 points to 0.1 point. Suppose that $v_2$ is perfectly reliable; then the observed-score standard deviation will be 10, and the standardization will not alter the difference between means on $v_2$. In general (other things being equal), a less reliable scale will tend to report smaller differences in standardized observed score between individuals or groups. It appears that the only way to make the intended examination of overlap is to estimate the mean $\mu_P$ and $\sigma(\mu_p)$ separately within each textbook group. The raw score is appropriate for use in that calculation.

**A.3.**  There will not be a sufficient sample of California Republican electricians with a \$15,000 income to serve as reference group, even though logically it is the correct one. There is a reasonable chance that the national sample will include a few dozen persons with an income near \$15,000, or a few dozen in the skilled crafts. And it will have a hundred or more Republicans. Perhaps a group of 50 Republicans with income \$8000 to \$20,000 can be found; that would be a fairly suitable reference group for Doe.

It should be possible to form a regression equation that would assign appropriate weights to the means of several reference groups [e.g., $0.7X_{pi} + 0.2$ (mean for Republicans) $-0.4$ (mean for skilled craftsmen) $+0.1$ (mean for \$15,000 income)]. The theory for such estimates has not been explored, however.

**A.4.**  Yes; the sequential procedure employs sound statistical inference. It asks whether the evidence (observed score) would be likely to arise, given the hypothesis that the person's universe score is at or below the standard. The probability can be denoted by $P(E \mid H)$. When one advances the person, the evidence $E$ has allowed him to reject the hypothesis $H$.

Constructing a symmetric band around the standard is not open to the objections made to a confidence interval symmetric about the observed score. However, the number of erroneous decisions matches the risk the investigator intends only if, among persons whose universe scores equal the standard, the observed scores are normally distributed with a standard deviation equal to $\hat{\sigma}(\Delta)$.

The procedure is weaker than a Bayesian procedure because it makes no use of information on the scores of other subjects. A Bayesian procedure asks about the distribution of universe scores among persons who have the same observed score (or series of observed scores). To paraphrase the question asked in the first paragraph above: Is the hypothesis that the universe score equals a certain value likely to be true, given the evidence of the observed score? (I.e., what is $P(H \mid E?)$ Because Bayesian analysis takes additional information into account, Bayesian analysis of sequential data can classify subjects just as accurately as the non-Bayesian procedure can, giving fewer tests, on the average. That is to say, Bayesian analysis will assign more persons after the first stage of testing than will the basic sequential procedure, and at each later stage will have fewer persons awaiting further testing.

# CHAPTER 6
# Illustrative Analyses of Crossed Designs[1]

## A. A Test for Aphasic Patients

### Description of instrument and basic data

The *Porch Index of Communicative Ability* (PICA; Porch, 1966, 1970) is an individual test designed for use by speech pathologists. It is intended for initial diagnosis of patients with aphasic symptoms and for measuring the change or lack of change during treatment. There are three sections, calling respectively for oral, gestural, and graphic responses; these sections are to be regarded as fixed modes of response.

Each section of PICA is divided into subtests (Figure 6.1). and the mean score over 18 subtests is taken as an "overall" score. There are four oral, eight gestural, and six graphic subtests, hence the weighting of sections in the overall score is not uniform. Each subtest consists of 10 items. In any item an object (e.g., a comb) is presented and the subject is directed to respond in some manner. Directions for a few illustrative subtests are approximately as follows:

a.   I.   Oral. Tell me what you do with this object.

b.   II.   Gestural. [Points at object] Pick it up and show me how you can use it.

c.   III.   Gestural. [Hands object to patient] Show me how you can use it.

.

.

.

---

[1] Partially nested designs appear at two points in this chapter (pp. 173 and 174).

**161**

**Figure 6.1.** *Structure of the Porch test* (*PICA*)

| Section | Subtest[a] | Object | | | | Subtest score |
|---|---|---|---|---|---|---|
| | | 1 | 2 | ⋯ | 10 | |
| Oral[b] | I Tells what to do with object | ×[c] | × | ⋯ | × | ×[d] ⎫ Score on |
| | IV Tells name of object | × | × | ⋯ | × | × ⎪ Oral |
| | IX Sentence completion | × | × | ⋯ | × | × ⎬ section[e] |
| | XII E says name; S repeats | × | × | ⋯ | × | × ⎭ |
| Gestural | II Shows how to use object | × | × | ⋯ | × | × ⎫ |
| | III Shows how to use object | × | × | ⋯ | × | × ⎪ |
| | V Reads name and matches to object | × | × | ⋯ | × | × ⎪ |
| | VI Points out object whose use E states | × | × | ⋯ | × | × ⎬ Score on Gestural |
| | VII Matches printed name to object without reading | × | × | ⋯ | × | × ⎪ section |
| | VIII Matches printed picture to object | × | × | ⋯ | × | × ⎪ |
| | X Points to object E names | × | × | ⋯ | × | × ⎪ |
| | XI Matches duplicate object with object | × | × | ⋯ | × | × ⎭ |
| Graphic | A. Writes use of object | × | × | ⋯ | × | × ⎫ |
| | B. Writes name of object | × | × | ⋯ | × | × ⎪ |
| | C. E says name, S writes | × | × | ⋯ | × | × ⎬ Score on Graphic |
| | D. E spells name, S writes | × | × | ⋯ | × | × ⎪ section |
| | E. Copies name from model script | × | × | ⋯ | × | × ⎪ |
| | F. Copies drawing of object | × | × | ⋯ | × | × ⎭ |
| | | | | | × | Overall score[f] |

[a] Order of administration is I to XII, then A to F.
[b] Called Verbal by Porch.
[c] Each score is a numeral on a 1–16 scale.
[d] Average over 10 objects.
[e] Each section score is an average over subtests.
[f] Overall score is an average over subtests.

d. IX. Oral. Finish the sentence [read by the examiner]: "You lock a door with a · · ·."

e. A. Graphic. Write down what you use that for.

The same 10 objects (or their names or pictures) are used in every subtest; hence modes of response are crossed with stimuli. The objects are to be regarded as sampled from a universe of objects.

Every response is scored on a 16-point scale. A score of 6, for example, goes to a response that is "intelligible but incorrect," and 11 to a response that is "accurate but delayed and incomplete." Testers score responses as they are given. The subtest score is the average of item scores, that is, of performance on the 10 objects under the subtest directions. The subtests are administered in a fixed order, progressing from those that offer a minimum of cues to those that provide lavish cuing. Some subtests are essentially repetitions of earlier subtests with added prompting. Each subject is put through the entire sequence of tests, though the early tests are quite difficult for the severely impaired person, and the later tests are quite easy for the person whose speech involvement is slight. It is probably most reasonable to regard the subtests as fixed, because the range of possible communicative tasks is limited. Nevertheless, for the sake of illustrating techniques, in some analyses we shall regard subtest tasks as randomly sampled from a universe of tasks.

Porch collected data for his studies from a series of clinical testings in a large hospital. Certain atypical cases were eliminated, but the remainder may be regarded as run-of-the-clinic sample. Two subsamples are used here. The first was a series of 30 cases tested by one clinician while two additional clinicians observed through a mirror and independently scored the performance (sample 1). The other subsample (sample 2) consisted of 40 cases tested twice. Some of these retests were requested for clinical purposes, but others were obtained specially to get data on stability. The interval between tests was less than two weeks. Tests were generally but not invariably given by the same tester.

Porch studied the accuracy of generalization over scorers, over objects, and over occasions. The facets were treated in separate one-facet studies in the original report (Porch, 1966). We have been able to reanalyze the data by the more illuminating multifacet method and also to examine some questions that fell outside Porch's area of interest. Porch's one-facet analyses are reviewed for the sake of comparison.

### Analysis 1: Patients × scorers

A one-facet analysis of sample 1 extracted mean squares for scorers $j$, patients $p$, and residual for each subtest in turn, for each section, and for the overall score.

**TABLE 6.1.** *Selected Results of One-Facet Analysis with Scorers Treated as Randomly Sampled*

| Subtest | Mean square (degrees of freedom in parenthesis) | | | Estimate of variance component | | | $\sigma^2(\Delta)$ | Estimate for D study, $j \times p$; $n'_j = 1$ | | | |
| | Patients $p$ (29) | Scorers $j$ (2) | Residual (58) | $p\|I$ | $j\|I$ | $pj,\bar{e}\|I$ | | Expected observed-score variance | | Coefficient of generalizability | |
| | | | | | | | | Uniform $j$ | $j$ joint with $p$ | Uniform $j$ | $j$ joint with $p$ |
|---|---|---|---|---|---|---|---|---|---|---|---|
| I | 45.37961 | 0.33100 | 0.17651 | 15.07 | 0.005 | 0.176 | 0.18 | 15.24 | 15.25 | 0.99 | 0.99 |
| II | 16.62950 | 0.24044 | 0.25584 | 5.46 | —[a] | 0.256 | 0.26 | 5.72 | 5.72 | 0.96 | 0.96 |
| III | 18.27670 | 3.26744 | 0.36963 | 5.97 | 0.097 | 0.370 | 0.47 | 6.34 | 6.44 | 0.94 | 0.93 |
| Oral section | 49.67344 | 0.09902 | 0.09020 | 16.53 | 0.000 | 0.090 | 0.09 | 16.62 | 16.62 | 0.99 | 0.99 |
| Graphic section | 17.17864 | 0.93297 | 0.18979 | 5.66 | 0.025 | 0.190 | 0.22 | 5.85 | 5.88 | 0.97 | 0.96 |
| Overall score | 16.67547 | 0.11494 | 0.03103 | 5.55 | 0.003 | 0.030 | 0.03 | 5.58 | 5.58 | 0.99 | 0.99 |

[a] Negative value, treated as zero.

This G study employed a $j \times p$ design. The mean squares given in Table 6.1 were calculated by us from data supplied by Porch. Because it is necessary to subtract one mean square from another, we find it advisable to carry calculations for mean squares to an unusually large number of decimal places, rounding off when estimates for components of variance are reached. The variance components indicate the magnitude of person effects, scorer effects, and a residual. Because objects are held constant in the scores $X_{pIj}$ analyzed in this study, the estimates are for components within the set of objects. For this reason, components have been labeled $p \mid I$, $j \mid I$, and $pj,\bar{e} \mid I$. Occasions vary from person to person; therefore, the person component is confounded with any occasion effect or person $\times$ occasion effect.

For Subtest I, $\hat{\sigma}(p \mid I)$ is almost 4, which implies that, on this difficult subtest, universe scores range over the whole 16-point response scale. The scorer effect is negligible, and the residual rather small. The value of $\hat{\sigma}(\Delta)$ being only 0.4, one can be satisfied with the accuracy of scoring; 67% of the observed scores fall within 0.4 of the corresponding universe score. To determine the possible value of adding scorers, one divides $\sigma^2(\Delta)$ by various $n'_j$. With three scorers, for example, $\widehat{\sigma^2(\Delta)} = 0.06$ and $\hat{\sigma}(\Delta) = 0.25$.

The error components for Subtests II and III are somewhat greater. For Subtest III, $\hat{\sigma}(\Delta)$ is 0.7, which suggests that the scoring rules for that subtest could profitably be revised. Porch carried out a further analysis on each item separately, as a way of learning more about the locus of scoring errors. The person components of variance for Subtests II and III are much smaller than for Subtest I, indicating that patients are more nearly uniform in ability to perform these easier tasks.

Section scores are based on a large number of responses and the error components are correspondingly smaller. Scoring of graphic responses is evidently harder for judges to agree upon than oral ones in the present stage of development of the method. Oral scores have a much greater $p \mid I$ component of variance than graphic scores.

This is a clinical instrument, and interpretations are almost invariably to be made about one person at a time. The profile is treated as a description of the person on an absolute scale; because the raw-score profile is to be interpreted, $\sigma(\Delta)$ is a suitable index of the seriousness of scorer error.

The variance over persons and the coefficient of generalizability are of minor interest. Because the constant errors of scorers are negligible, the observed-score variance is nearly the same when scorers change from person to person as when the scorer is held fixed. Individual differences in PICA scores are excellently generalizable over scorers, where these scorers are trained by Porch's methods. The very high coefficients result from the wide range of pathology in a clinic sample, as well as from the excellent scoring rules.

While PICA is a wide-range instrument, its chief purpose is to make precise differentiations that will aid the therapist in planning a treatment for the current patient, somewhat different from that for the next patient who has the same gross pathology. Therefore, the degree to which PICA differentiates within the total population of patients is of little relevance save as an upper limit to its ability to make useful clinical differentiations. The coefficient estimated here is $\mathscr{E}\rho^2(X_{pIj},\mu_{pI})$, either with the same scorer for all persons or with varying scorers. The fact that the $j$ component is small warrants mixing together, in any study of a group, records obtained from different scorers.

### Analysis 2: Patients × objects

The second one-facet analysis of sample 1 treated objects as randomly sampled. The average score $X_{piJ}$ assigned by the three scorers to each response was analyzed; here, the set $J$ is implicitly fixed. The occasion is again confounded with the person. The results are tabulated in Table 6.2. The mean squares are very much larger than those in Table 6.1 simply because the data come from a 30 × 10 score matrix whereas those used in Table 6.2 come from a 30 × 3 matrix.

While the component for persons is rather close in size to that in Table 6.1, we must stress that the two tables define these components differently. In two-facet notation, the so-called person component of Table 6.1 is $\sigma^2(p,pI)$ and that of Table 6.2 is $\sigma^2(p,pJ)$.

The object component is rather large for Subtests II and III. This argues for presenting the same test objects to all subjects when person-to-person comparisons will be made or when norms will be used in interpreting. Interestingly, the object component is negligible for Subtest I; this may suggest something about the manner in which severe aphasic impairment affects the retrieval mechanism for speech responses. To pursue this lead, one would try to explain why difficulty varies markedly from object to object in II and III and not in I.

The error of generalization from one object to the universe of objects is ordinarily larger than the error of generalization from any one scorer to the universe of scorers. This is a reflection both of the unpredictability of an aphasic's response at any moment and of his specific difficulty with particular objects. The efficiency of the patient varies from moment to moment and this produces object-to-object variability, but it does not cause scorings of the same response to differ. Since there are 10 objects, the $i \mid J$ and $pi,\bar{e} \mid J$ components of error shown in the table are reduced by a factor of 10 in the subtest score. The quantity $\sigma(\Delta)$ is around 0.5 to 0.7, which implies that the observed score is not often more than one unit distant from the score that would be obtained by testing with an extensive set of objects.

TABLE 6.2. Selected Results of One-Facet Analysis with Objects Treated as Randomly Sampled

| Subtest | Mean square (degrees of freedom in parenthesis) | | | Estimate of variance component | | | Estimate for D study, $I \times p$; $n'_i = 10$ | | |
|---|---|---|---|---|---|---|---|---|---|
| | Patients $p$ (29) | Objects $i$ (9) | Residual (261) | $p \mid J$ | $i \mid J$ | $pi, \bar{e} \mid J$ | $\sigma^2(\Delta)$ | Expected observed-score variance | Coefficient of generalizability |
| I | 150.94825 | 2.92300 | 2.80677 | 14.81 | 0.00 | 2.81 | 0.28 | 15.09 | 0.98 |
| II | 55.43166 | 35.45926 | 3.60484 | 5.18 | 1.06 | 3.60 | 0.47 | 5.54 | 0.94 |
| III | 60.92235 | 43.28317 | 3.09210 | 5.78 | 1.34 | 3.09 | 0.44 | 6.09 | 0.95 |
| Oral section | 164.72227 | 1.02585 | 0.91491 | 16.38 | 0.00 | 0.91 | 0.09 | 16.47 | 0.99 |
| Graphic section | 57.26209 | 5.33083 | 0.84778 | 5.64 | 0.15 | 0.85 | 0.10 | 5.73 | 0.99 |
| Overall score | 55.30614 | 1.69871 | 0.19800 | 5.51 | 0.05 | 0.20 | 0.02 | 5.53 | 0.99 |

As Porch proposes to use an $i \times p$ design in practice, we calculate an observed-score variance and a coefficient only for that design. The data would also permit calculations for the $i:p$ design and for values of $n_i'$ other than 10. While these relatively conventional one-facet studies give an adequate answer to the chief questions about errors arising from sampling of scorers and test objects, a multifacet analysis teases out interactions, separates the $p$ component from $pI$ and $pJ$, and is more compact than two separate analyses.

It is possible to analyze one subtest at a time or several subtests together. Since our purpose is to illustrate procedure, two additional analyses are carried out:

Analysis within a subtest. Subtest III is chosen because its large scorer effect is of interest; scorers and objects are facets, crossed in the G data with each other and with patients.

Analysis within a section. The Graphic section is chosen. Subtests $k$, objects $i$, and scorers $j$ are facets, crossed with each other and with patients. (Most instruments have items nested within subtests, but the common objects of PICA create a crossed design.) A composite estimate of the various components over all subtests is needed. An analysis is carried out in which subtests are considered to be fixed and another in which subtests are considered to be samples of an indefinitely large number of response modes.

It may be worth mentioning that the analysis demonstrated within a section (Analysis 4) could also be carried out for the entire test, simply by ignoring the division of the test into sections. This procedure is not as satisfactory as the multivariate procedures to be developed in Chapters 9 and 10.

### Analysis 3: Patients × scorers × objects

A multifacet analysis was made of the scores generated by 30 patients, 3 scorers, and 10 objects under the directions for Subtest III. The mean squares for this study were presented in Table 2.2. The mean squares for $p$, $i$, and $j$ in the two-facet analysis agree with those from Tables 6.1 and 6.2, after allowance is made for the fact that the one-facet mean squares are scaled down by factors of 10 and 3, respectively. For the two-facet study components of variance are given in Table 3.5.

The components have been redefined in going to the multifacet analysis. Consider the "component for persons." In Table 6.1 $p$ was confounded with $pI$; in Table 6.2 with $pJ$. In the two-facet study the $p$ component is separated out. The person component of variance in each table is a universe-score variance, but the universe changes as follows:

Table 6.1: Universe of scorers, 10 objects fixed.
Table 6.2: Universe of objects, 3 scorers fixed.
Table 2.2: Universe of objects and scorers.

The $pi$ interaction thus contributes to the universe-score variance in Table 6.1, but not in Table 6.2 nor in the two-facet study. This interaction emerges explicitly in the multifacet analysis, and the "person" component of variance is correspondingly reduced. The scorer component is numerically much the same in Table 6.1 and the two-facet study, though in the former it includes the object–scorer interaction. A comparable statement can be made about the component for objects. The two-facet study, like the one-facet study, has hidden facets, notably occasions and testers.

In the two-facet study, the large effects are those for patients, for objects, for patients × objects, and for residual. Scorers evidently are not a source of appreciable variance. The person–object effect implies that some items (an item being defined by the object together with the subtest directions) are much more troublesome for a given patient than their general difficulty and his general impairment would suggest. To be sure that there is a genuine person–object effect, however, it would be necessary to present the same object more than once. In retrospect, we see that Porch would have learned somewhat more about components of variance, with an equal expenditure of effort, if he had presented half of his objects twice, with a rest period between trials. Occasion effects could then be appraised. (It is unusual to have a residual smaller than an interaction, but it is understandable in this study. The residual represents discrepancies between scorer ratings of the same performance, after correcting for scorer main effects and simple interactions. Where the subject performs inconsistently, the scorers agree in their reports of that inconsistency; hence, the variation is assigned to the patient–object interaction.)

The large components of variance in $X_{pij}$ are reduced by using 10 objects. With 10 objects and a single scorer, the estimated $\sigma^2(\Delta)$ (using values from p. 87 and dividing by the proper $n'$) is the sum of $\widehat{\sigma^2}(I)$, 0.13; $\widehat{\sigma^2}(J)$, 0.09; $\widehat{\sigma^2}(pI)$, 0.26; $\widehat{\sigma^2}(pJ)$, 0.21; $\widehat{\sigma^2}(IJ)$, 0.01; and $\widehat{\sigma^2}(pIJ,e)$, 0.16. Hence, $\widehat{\sigma^2}(\Delta)$ equals 0.86 and $\hat{\sigma}(\Delta)$ is 0.93 points on the 16-point scale. This statement assumes that the observed score will be taken as an estimate of $\mu_p$. If it were taken as an estimate of $\mu_{pI}$, the $pI$ contribution of 0.26 and the $I$ contribution of 0.13 would not enter the error variance. In that case, $\hat{\sigma}(\Delta)$ would drop to 0.69, which is in agreement with the $\widehat{\sigma^2}(\Delta)$ of 0.47 in Table 6.1.

By what reasoning can one judge whether $\mu_p$ or $\mu_{pI}$ is the more appropriate universe score?[2] If the tester is attempting to describe the person's ability to perform the gestural task of Subtest III, no matter what the stimulus object is, $\mu_p$ is clearly the target score. But if there is to be only one form of the Porch test, meaning may come to surround a "Porch score" of, perhaps, 11 that to some extent reflects the stimulus characteristics of these specific objects. If the Porch score is considered as describing response to these particular objects, $\mu_{pI}$ is the universe score, and the correspondingly smaller $\sigma(\Delta)$ is pertinent.

The expected observed-score variance depends on the design of the D study. Not all designs make practical sense; it seems most unlikely that one would nest scorers within objects, for example. The designs $I \times J \times p$, $I \times (J:p)$, and $(I \times J):p$ might plausibly be used. Table 6.3 presents estimates of observed-score variance and the coefficient for Subtest III, assuming that 1 scorer and a 10-item test produce each patient's score.

**TABLE 6.3.**  *Expected Observed-Score Variances and Coefficients of Generalizability for Subtest III for Various Experimental Designs*

| Design of D study | Components entering observed-score variance | Expected observed-score variance ($n'_i = 10$, $n'_j = 1$) | Coefficient, if universe score[a] is | |
|---|---|---|---|---|
| | | | $\mu_p$ | $\mu_{pI}$ |
| Objects × scorers × patients | $p, pi, pj, pij, e$ | 6.34 | 0.90 | 0.94 |
| Objects × (scorers: patients) | all but $i$ | 6.44 | 0.89 | 0.93 |
| (Objects × scorers) :patients | all | 6.57 | 0.87 | —[b] |

[a] Variance of $\mu_p$ estimated as 5.71. Variance of $\mu_{pI}$ estimated as 5.97.
[b] It is highly unlikely that one would generalize to $\mu_{pI_p}$, where objects defining the universe differ from person to person. In such a case, the universe-score variance includes $p$, $pI$, and $I$ components.

The first point to note in Table 6.3 is that the coefficient for the $I \times j \times p$ design with generalization to $\mu_p$ is 0.90. This value is smaller by a practically significant amount than the values of 0.94 and 0.95 from Tables 6.1 and 6.2. Taking two sources of error into account simultaneously has given a less

[2] It should be noted that there is a hidden fixed facet, $k^*$, since one is not generalizing beyond Subtest III here. Therefore, to be strict in designations, these universe scores should be identified as $\mu_{pk*}$ and $\mu_{pIk*}$.

flattering picture of the test than the conventional one-facet study did. In effect, the two-facet study estimates the correlation expected if responses to one set of objects judged by one scorer are compared with responses to another set of objects judged by another scorer. This is surely of greater interest than either of the one-facet coefficients.

If one is interested in measuring individual differences within a run-of-the-clinic sample, it makes little difference whether the $i \times j \times p$ design or the $i \times (j{:}p)$ design is used. The third design, with objects different for each patient, seems at first glance not be to much worse than the first two. The drop in the coefficient from 0.90 to 0.87 implies, however, that the third design with 10 objects is no more effective in estimating $\mu_p$ than the first design would be with only 7 objects.

The Spearman–Brown formula was capable of evaluating the effect of altering $n_i'$ in Analysis 2 or $n_j'$ in Analysis 1. It cannot be used, however, when the design has more than one random facet, because different divisors apply to the several components of variance. "Doubling the length of the test" (i.e., doubling $n_i'$) reduces the components for $I$, $pI$, $IJ$, and $pIJ,e$—but not those for $J$ and $pJ$. If the test is extended to infinite length, the coefficient of generalizability with $\mu_p$ as the universe score does not go to 1.00 as the Spearman–Brown rule implies. With increasing $n_i'$, it is only the coefficient for generalization to $\mu_{pJ}$ that has 1.00 as its limit.

### Analysis 4: Patients × scorers × objects × subtests

A four-way variance analysis of scores on the Graphic section, in which 10 objects are crossed with 6 subtests (modes of response), generates the mean squares given in Table 6.4. The components of variance have been estimated under assumptions of subtests random and subtests fixed. There is little numerical difference between the components for this test under the two models.

The two components for $p$ deserve special consideration. In generalizing over subtests the universe score is $\mu_p$. With generalization to a fixed set of subtests the universe score is $\mu_{pK*}$. The person component of variance with fixed subtests is $\sigma^2(p \mid K^*)$; because it includes the $pK$ interaction of the random model, this variance is larger than $\sigma^2(p)$. In this example the $pK$ interaction is the average of interaction components for six subtests, hence its variance (0.29) is one-sixth of the variance for the $pk$ component (1.71).

The estimation formulas for the mixed model differ from the random-model formulas, for each component identified with an $a$ in the last column of Table 6.4. Where the random model has:

$$(6.1) \qquad \widehat{\sigma}^2(pij) = \frac{\text{MS } pij - \text{MS } pijk, e}{n_k}$$

**TABLE 6.4.**  *Estimates of Variance Components for Graphic Section*

| Source of variance | Degrees of freedom | Mean square | Estimate of variance component if subtests are | |
|---|---|---|---|---|
| | | | Randomly selected from large collection | Fixed |
| Patients $p$ | 29 | 1030.71678 | 5.30 | 5.59[a] |
| Objects $i$ | 9 | 95.95480 | 0.13 | 0.15[a] |
| Scorers $j$ | 2 | 55.97745 | 0.02 | 0.02[a] |
| Subtests $k$ | 5 | 2458.43769 | 2.65 | 2.65 |
| $pi$ | 261 | 15.26016 | 0.41 | 0.74[a] |
| $pj$ | 58 | 11.38810 | 0.15 | 0.16[a] |
| $pk$ | 145 | 59.17398 | 1.71 | 1.71 |
| $ij$ | 18 | 3.12108 | 0.00 | 0.01[a] |
| $ik$ | 45 | 18.24349 | 0.12 | 0.12 |
| $jk$ | 10 | 5.05145 | 0.01 | 0.01 |
| $pij$ | 522 | 1.88159 | 0.15 | 0.31[a] |
| $pik$ | 1305 | 7.00967 | 2.01 | 2.01 |
| $pjk$ | 290 | 1.71136 | 0.07 | 0.07 |
| $ijk$ | 90 | 1.38559 | 0.01 | 0.01 |
| Residual | 2610 | 0.99167 | 0.99 | 0.99 |

[a] These are to be interpreted as components with $K^*$ fixed [i.e., as $\sigma^2(p \mid K^*), \ldots \sigma^2(pij,\bar{e} \mid K^*)$]. The remaining components in this column are "within $K^*$."

the mixed model has

$$(6.2) \qquad \widehat{\sigma^2}(pij,\bar{e} \mid K^*) = \frac{\text{MS } pij}{n_k}$$

In estimating the component for $pi \mid K^*$, MS $pij$ is subtracted from MS $pi$; similarly with $pj$ and $ij$. The mixed model analysis assumes that the six subtests comprise the universe of subtests. Therefore it provides direct estimates of the variance of $\mu_{pK^*}$, $\mu_{piK^*}$, etc. Table 6.5 shows how to compute $\sigma^2(\Delta)$ assuming fixed subtests. With subtests fixed, generalization is to $\mu_{pK^*}$.

While we have treated various PICA scores in accord with the univariate theory developed in preceding chapters, the instrument is multivariate, and methods to be developed in Chapters 9 and 10 could appropriately be applied. These probably would not improve generalization over objects and scorers appreciably, because the coefficients of generalizability obtained by univariate analysis are high. In view of the considerably larger error variance

**TABLE 6.5.** *Composition of Error Variance for Graphic Section if the within-Person Design is* $i \times j \times k$ ($n_i' = 10$, $n_j' = 1$, $n_k' = 6$; *Generalization to* $\mu_{pK*}$)

| $X_{pIJK*}$[a] | $\mu_{pK*}$ | $\Delta_{pIJK*}$ | Estimate of variance component[b] | Frequency within $p$ | $\sigma^2(\Delta)$ |
|---|---|---|---|---|---|
| $\mu_{K*}$ | $\mu_{K*}$ | | | | |
| $\mu_{pK*}\sim$ | $\mu_{pK*}\sim$ | | | | |
| $\mu_{IK*}\sim$ | | $\mu_{IK*}\sim$ | 0.15 | 10 | 0.02[c] |
| $\mu_{JK*}\sim$ | | $\mu_{JK*}\sim$ | 0.02 | 1 | 0.02[c] |
| $\mu_{pIK*}\sim$ | | $\mu_{pIK*}\sim$ | 0.74 | 10 | 0.07 |
| $\mu_{pJK*}\sim$ | | $\mu_{pJK*}\sim$ | 0.16 | 1 | 0.16 |
| $\mu_{IJK*}\sim$ | | $\mu_{IJK*}\sim$ | 0.01 | 10 | 0.00[c] |
| $\mu_{pIJK*}\sim, \bar{e}$ | | $\mu_{pIJK*}\sim, \bar{e}$ | 0.31 | 10 | 0.03 |

$$0.30 = \widehat{\sigma^2}(\Delta)$$

$$0.26 = \widehat{\mathscr{E}\sigma^2}(\delta)$$

[a] Components written by analogy to (2.19).
[b] Values calculated in Table 6.4.
[c] Components not entering $\sigma^2(\delta)$.

associated with occasions (to be discussed shortly), multivariate estimation of the subtest universe score might well offset some of the error arising from sampling of occasions. The retest data available to us are not sufficient to warrant that kind of analysis for PICA.

In generalization to $\mu_p$ (over subtests), the $K$ and $pK$ components would make large contributions to $\sigma(\Delta)$. If generalization over subtests were intended, it would be desirable to use a much greater number of subtests, possibly with fewer items per subtest. For example, consider three possible values of $n_i'$ and the possibility of using different objects for each subtest ($i:k$) while holding the total sample of behavior to $n_i'n_k' = 60$. Table 6.6 gives values of $\widehat{\sigma^2}(\Delta)$ for the Graphic section of PICA under the random model, for various designs. Generalization to $\mu_p$ is clearly improved when the number of subtests is increased. Nothing important appears to be gained by varying objects from subtest to subtest, even though this would increase the sampling of $pi$ and other components. To increase the number of subtests while holding the overall testing effort constant would make the several subtest scores less useful, because the error in generalizing to the several $\mu_{pk}$ increases as $n_i'$ drops.

Table 6.6 has substantive implications. The large subtest component is expected, because some responses are especially hard for the aphasic patient

**TABLE 6.6.** *Values of $\sigma^2(\Delta)$ for the Graphic Section as a Function of D-Study Design ($n_i'n_k' = 60$; Generalization to $\mu_p$)*

| Source of variance | Estimate of variance component[a] | Design: $i \times j \times k \times p$ | | | | Design: $(i{:}k) \times j \times p$ | | | |
|---|---|---|---|---|---|---|---|---|---|
| | | Frequency ($n_j' = 1$) | $n_i'=10$ $n_k'=6$ | $n_i'=6$ $n_k'=10$ | $n_i'=3$ $n_k'=20$ | Frequency ($n_j' = 1$) | $n_i'=10$ $n_k'=6$ | $n_i'=6$ $n_k'=10$ | $n_i'=3$ $n_k'=20$ |
| | | | | | | | | | (Contributions to $\sigma^2(\Delta)$) |
| $I$ | 0.13[b] | $n_i'$ | 0.01 | 0.02 | 0.04 | $n_i'n_j'$ | 0.00 | 0.00 | 0.00 |
| $J$ | 0.02[b] | $n_j'$ | 0.02 | 0.02 | 0.02 | $n_j'$ | 0.02 | 0.02 | 0.02 |
| $K$ | 2.65[b] | $n_k'$ | 0.44 | 0.26 | 0.13 | $n_k'$ | 0.44 | 0.26 | 0.13 |
| $pI$ | 0.41 | $n_i'$ | 0.04 | 0.07 | 0.14 | $n_i'n_k'$ | 0.01 | 0.01 | 0.01 |
| $pJ$ | 0.15 | $n_j'$ | 0.15 | 0.15 | 0.15 | $n_j'$ | 0.15 | 0.15 | 0.15 |
| $pK$ | 1.71 | $n_k'$ | 0.28 | 0.17 | 0.09 | $n_k'$ | 0.28 | 0.17 | 0.09 |
| $IJ$ | 0.00[b] | $n_i'n_j'$ | 0.00 | 0.00 | 0.00 | $n_i'n_j'n_k'$ | 0.00 | 0.00 | 0.00 |
| $IK$ | 0.12[b] | $n_i'n_k'$ | 0.00 | 0.00 | 0.00 | $n_i'n_k'$ | 0.00 | 0.00 | 0.00 |
| $JK$ | 0.01[b] | $n_j'n_k'$ | 0.00 | 0.00 | 0.00 | $n_j'n_k'$ | 0.00 | 0.00 | 0.00 |
| $pIJ$ | 0.15 | $n_i'n_j'$ | 0.02 | 0.03 | 0.05 | $n_i'n_j'n_k'$ | 0.00 | 0.00 | 0.00 |
| $pIK$ | 2.01 | $n_i'n_k'$ | 0.03 | 0.03 | 0.03 | $n_i'n_k'$ | 0.03 | 0.03 | 0.03 |
| $pJK$ | 0.07 | $n_j'n_k'$ | 0.01 | 0.01 | 0.00 | $n_j'n_k'$ | 0.01 | 0.01 | 0.00 |
| $IJK$ | 0.01[b] | $n_i'n_j'n_k'$ | 0.00 | 0.00 | 0.00 | $n_i'n_j'n_k'$ | 0.00 | 0.00 | 0.00 |
| $pIJK, e$ | 0.99 | $n_i'n_j'n_k'$ | 0.02 | 0.02 | 0.02 | $n_i'n_j'n_k'$ | 0.02 | 0.02 | 0.02 |
| $\sigma^2(\Delta)$ | | | 1.02 | 0.78 | 0.67 | | 0.96 | 0.67 | 0.45 |
| $\mathscr{E}\sigma^2(\delta)$ | | | 0.55 | 0.48 | 0.48 | | 0.50 | 0.39 | 0.30 |

[a]Values calculated in Table 6.4.

[b]These components do not contribute to the expected observed variance and $\mathscr{E}\sigma^2(\delta)$, and do not reduce the coefficient of generalizability when $i$, $j$, and $k$ are crossed with persons.

to make. It would be unwise to generalize from the absolute level of scores on one mode of response to the absolute level on other modes. The large patient-subtest interaction confirms that the subtests call upon somewhat different functions, hence differences within the profile are likely to have meaning. The *pik* interaction is made large by momentary fluctuations of responsiveness which are probably greater for aphasics than for normal subjects.

### Reporting the study

Having approached the Porch data dealing with object and scorer variation in various ways, we may reflect on what would be recommended to the developer of such a test who wants to make as straightforward a study as possible and to report it simply. The developer should design a G study to systematically represent as many important facets as he can. He may then consider the effects on generalizability of many possible D-study designs. If he is prepared to recommend one single design to all users of the procedure, he can make a fairly straightforward report. It also makes the report simpler if only one universe of generalization is likely to be of interest to users.

A report along the following lines would appear to be suitable for the study we have been considering:

1. Assume an intention to generalize over objects and scorers, for the score on each subtest separately, the score for each section, and for the overall score. Regard subtests as fixed. Report the experimental design for the G study and, describe in general terms, the scorers used. Assume that the D study has design $i \times j \times p$, and specify $n'_i$ and $n'_j$.

2. Perform a three-way analysis of variance of $X_{pij}$ within each subtest. Report mean squares. Use the random model to estimate components of variance, and $\sigma(\Delta)$, $\mathscr{E}\sigma^2(X)$, and $\mathscr{E}\rho^2$ for each subtest.

3. Perform a three-way analysis of variance of scores $X_{pijK*}$ for each section and report mean squares. Calculate estimates for components of the type $\sigma^2(p \mid K*)$, etc. and report them. For each section score report $\hat{\sigma}(\Delta)$, $\widehat{\mathscr{E}\sigma^2(X)}$, and $\widehat{\mathscr{E}\rho^2}$.

4. Combine within-section data (by methods for composite scores to be presented in Chapter 10) to get estimates of the variance components for the overall score and of its $\sigma(\Delta)$, $\mathscr{E}\sigma^2(X)$, and $\mathscr{E}\rho^2$.

This gives more detail then any one user needs, but it answers a variety of questions and provides information from which the user who has an unusual question can calculate an answer for himself. The report might be extended to give similar estimates for a design with scorers nested within subjects, because some clinical practice and research will mix records obtained by different scorers. Or, if the data support the statement that generalizability with nested scorers is nearly as good as for crossed scorers, the report might say simply that.

Obviously, reports of this sort will require a more sophisticated audience than test manuals now assume. Nevertheless, if generalizability is a complicated matter, research reports must somehow make full findings available and hope that readers can be educated to use them. A report that tells a simple but incorrect story is not to be recommended.

### Analysis 5: Variance associated with occasions

Porch analyzed the data of sample 2 to determine the consistency between tests given on two occasions within a two-week period. Several patients had been tested many times during their treatment; if there are three or more scores for a person, we use only the earliest pair of tests. Since all these patients were tested by Porch, tester and scorer are hidden facets in the design.

There are two ways to regard this study. From one point of view occasions are crossed with patients; from another, occasions are nested. If we consider occasions as differing because they occur on different calendar dates (accompanied by different weather, different degrees of attentiveness of the tester, etc.), occasions are nested; only sporadically are two subjects tested on the same date. If, on the other hand, we entertain the possibility of an order effect, because of the fact that the patient makes some recovery during the two-week interval, or profits from practice on the test, there is good reason to distinguish between the first and second test and to treat occasions as crossed. Every subject does have a score that identifiably belongs in the first or second column. The random-sampling model of generalizability theory applies satisfactorily only if we assume the subjects to be in a steady state or at worst assume that any practice effect is a constant that can be corrected for. The random model cannot give serious attention to an order effect; it might be appropriate to regard "first" and "second" occasion as fixed. We shall analyze the Porch retest data for Subtest III within the random model, with two interpretations of the design.

Organizing the data according to an $i \times o \times p$ design permits a three-way analysis of variance, with results given in Table 6.7. The large effects are those attributable to patients, objects, patient-object interaction, and residual. The sample includes a wide variety of clinical types, and consequently the variance component for patients is large relative to the score range of 16 points. The substantial $pi$ effect implies that a person who scores higher on some objects and lower on others tends to show the same pattern the next week, a fact that may have diagnostic significance. Moreover, this implies that a person's standing on one object, even if determined by repeated measurement, has limited value as an indication of his standing, averaged over occasions, on the universe of objects. The substantial object $i$ component reflects the difficulty of handling some objects correctly in this subtest.

**TABLE 6.7.**   *Estimates of Variance Components for Subtest III Over Occasions*

| Source of variance | Sum of squares | Degrees of freedom | Mean square | Estimate of variance component |
|---|---|---|---|---|
| Patients $p$ | 3890.3744 | 39 | 99.75319 | 4.65 |
| Objects $i$ | 733.8747 | 9 | 81.54163 | 0.95 |
| Occasion $o$ | 14.5800 | 1 | 14.57997 | 0.03 |
| $pi$ | 1891.6162 | 351 | 5.38922 | 1.17 |
| $po$ | 170.9198 | 39 | 4.38256 | 0.13 |
| $io$ | 16.3702 | 9 | 1.81891 | —[a] |
| Residual | 1073.1228 | 351 | 3.05733 | 3.06 |

[a] Negative value, treated as zero.

The *pio,e* component of variance is quite large. This reflects a trial-to-trial inconsistency of performance (which contributed to the sizeable *pi* component of our earlier analysis—see p. 167—within subtests on a single occasion). The small *io* mean square generates a negative estimate of the *io* component of variance. To treat this component as zero seems entirely reasonable, because there is no reason to think that certain objects will be systematically easier to respond to on the first occasion than on the second, and others harder. The occasion effect is quite small, implying that these patients did not make appreciable progress during the period between tests and that there was little practice effect. The *po* component is also rather small, implying that day-to-day fluctuations in responsiveness are not a particularly significant source of error of measurement.

Patients are not tested on the same two occasions, therefore the G study may be considered to have the design $i \times (o:p)$. Then, as can be inferred from information on Design V-B in Figure 2.4 and Table 2.1, the *io* component is confounded with *pio,e* and the *o* component is confounded with *po*. Combining sums of squares and degrees of freedom as suggested in Chapter 2 gives these values for the confounded components:

| | Sum of squares | Degrees of freedom | Mean square |
|---|---|---|---|
| *o,po* | 185.4998 | 40 | 4.637 |
| *io,pio,e* | 1089.4930 | 360 | 3.026 |

The variance components are estimated to be as follows:

| | |
|---|---|
| $p$ | 4.64 |
| $i$ | 0.95 |
| $pi$ | 1.18 |
| *o,po* | 0.16 |
| *io,pio,e* | 3.03 |

**TABLE 6.8.** *Estimation of $\sigma^2(\Delta)$ and $\mathscr{E}\sigma^2(X)$ for Subtest III for a D Study with the Design $i \times (o:p)$ $(n_i' = 10, n_o' = 1;$ Generalization to $\mu_p)$*

| $X_{pIo}$ | $\Delta_{pIo}$ | Estimate of variance component | Frequency within $p$ | $\sigma^2(\Delta)$ | $\mathscr{E}\sigma^2(X)$ |
|---|---|---|---|---|---|
| $p$ | | 4.64 | | | 4.64 |
| $I$ | $\mu_I \sim$ | 0.95 | $n_i' = 10$ | 0.10 | |
| $pI$ | $\mu_{pI} \sim$ | 1.18 | $n_i' = 10$ | 0.12 | 0.12 |
| $o, po$ | $\mu_{o,po} \sim$ | 0.16 | $n_o' = 1$ | 0.16 | 0.16 |
| $Io, pIo, e$ | $\mu_{Io,pIo,e} \sim$ | 3.03 | $n_o' n_i' = 10$ | 0.30 | 0.30 |
| | | | | 0.68 | 5.22 |

While the components for $p$ and $pi$ (and in the third decimal place $i$) have been altered, with these data the change in analysis makes no difference in the interpretation.

We shall consider a nested D study, since it conforms more satisfactorily to our model, and there is little evidence of systematic $o$ effects over a limited time span. The resulting estimates are listed in Table 6.8. Using these values we obtain $\mathscr{E}\hat{\rho}^2 = 4.64/5.22 = 0.89$.

We now have the following distinct kinds of coefficient (among others) for Subtest III: over objects (Table 6.2), 0.95; over objects and scorers (Table 6.3), 0.90; over objects and occasions (Table 6.8), 0.89. Insofar as we can judge from these data, which arise from two distinct samples, scorer disagreement is as much a source of inexactness as is variability of response and scoring from one occasion to another. One might well ask for a further study that allows for a combined estimate for all three sorts of variability, and, indeed, that goes on to investigate effects associated with the person who administers the test.

### B. A Two-Facet Anxiety Inventory

*Description of instrument and basic data*

The *S–R Inventory of Anxiousness* (Endler, *et al.*, 1962) is similar to PICA in that items are defined according to a two-facet design. Every item asks the subject to indicate on a 1–5 scale how strongly he experiences a certain response (e.g., "gets an uneasy feeling") when confronted with a certain situation (e.g., "starting off on a long automobile trip"). There is, presumably, a universe of situations and a universe of modes of response, and an item can be formed for any situation-response pair. Data are collected by Design

VII, with $n_i = 11$ situations crossed with $n_j = 14$ modes of response; the subject responds to 154 items. [Additional forms of the test have been prepared, and there is an inventory of hostility of the same type. (Endler & Hunt, 1968)].

The 1962 report on the test included an analysis of variance. At that time, mean squares were erroneously used as a basis for inferring the magnitude of the several effects. That is to say, mean squares were interpreted as we have interpreted variance components. In 1966, Endler estimated components of variance from these same data, incidentally describing how components change when one assumes a mixed model rather than the random model. A substantive interpretation of the variance components was presented by Endler and Hunt (1966). These interpretations of course differ from those originally derived from the mean squares, because the mean squares reflect the arbitrarily chosen values of $n_p$, $n_i$, and $n_j$.

Three samples are treated in the 1966 paper; we confine attention to a sample of 169 Pennsylvania State University freshmen. This analysis is reprinted because it is a clear and useful illustration, and because we can add to the interpretation by considering designs other than the one employed in the G study.

### Summary of the Endler–Hunt results

Table 6.9 presents the chief results for students tested at Pennsylvania State University. The analysis assumes an indefinitely large number of admissible modes of response. It is reasonable enough to think of the universe of anxiety-inducing situations as very large, but perhaps the number of kinds of response indicative of anxiety is rather limited. If the universe were restricted to a finite but large number of conditions, however, the estimates would not change greatly.

**TABLE 6.9.** *Estimates of Variance Components for the* S–R Inventory of Anxiousness (*after Endler & Hunt, 1966*)

| Source of variance | Degrees of freedom | Mean square | Estimate of variance component | Percentage |
|---|---|---|---|---|
| Subjects $p$ | 168 | 21.26 | 0.10 | 5.6 |
| Situations $i$ | 10 | 244.37 | 0.09 | 5.0 |
| Modes of response $j$ | 13 | 836.51 | 0.44 | 24.6 |
| $pi$ | 1680 | 3.16 | 0.18 | 10.1 |
| $pj$ | 2184 | 2.86 | 0.20 | 11.2 |
| $ij$ | 130 | 20.62 | 0.12 | 6.7 |
| Residual | 21,840 | 0.66 | 0.66 | 36.8 |
| | | | | 100.0 |

**TABLE 6.10.**   Estimates[a] of Variance Components in Seven Male Samples Taking
S–R Inventory of Anxiousness *Form 0*

| Source of variance | Range | Median[b] |
|---|---|---|
| $p$ | 0.08–0.13 | 0.11 |
| $i$ | 0.06–0.09 | 0.07 |
| $j$ | 0.34–0.44 | 0.41 |
| $pi$ | 0.16–0.21 | 0.19 |
| $pj$ | 0.13–0.23 | 0.20 |
| $ij$ | 0.07–0.14 | 0.10 |
| $pij, e$ | 0.61–0.75 | 0.66 |
| Number of cases | 30–206 | 93 |

[a] From Table C of supplementary materials for Endler and Hunt, 1969.
[b] Corresponding medians for females from the same schools: 0.11, 0.15, 0.48, 0.23, 0.25, 0.12, and 0.68; the median $N$ is 55.

In interpreting the components, it must be borne in mind that what has been analyzed is the variance of item scores (i.e., of single $pij$ combinations). The components are not on the scale of the *test* scores.

Data were collected with the same form on seven samples of males. The information in Table 6.10 gives some indication of the sampling error of components. There are differences from sample to sample, but they are remarkably small, considering that samples of 30, 41, and 53 cases are included. Even for the two largest samples ($n_p = 206$ and 125), there are disagreements. That the two values of $\sigma^2(p)$ should be 0.13 and 0.08 very likely is explained by differences in selection of cases. That $\widehat{\sigma^2}(ij)$ should be 0.12 in one of the large samples and 0.07 in the other is difficult to understand. Nevertheless, the full set of results suggest that estimates of components are reasonably stable, where $n_i$ and $n_j$ are as large as they are in these G studies.

The Endler–Hunt (1966) interpretation of Table 6.9 made the following substantive points:

There is no single major source of behavioral variance, at least so far as the trait of anxiousness is concerned. Human behavior is complex. In order to describe it, one must take into account not only the main sources of variance (subjects, situations, and modes of response) but also the various simple interactions (Subjects with Situations, Subjects with Modes of Response) and, where feasible, the triple interaction (Subjects with Situations with Modes of Response). Behavior is a function of all these factors in combination.

The fact that very substantial portions of the total variance [of single observations $X_{pij}$, over $p$, $i$, and $j$] come from the interactions of subjects with situations, of subjects with modes of response, and of situations with modes of response, and from the triple interaction, has importance for personality description and for the design of inventories to predict either behavior or feelings. First, it implies that accuracy of personality description in general calls for statements about the modes of response that individuals manifest in various kinds of situations as well as statements about their general proneness to make certain responses rather than others, and about their proneness to be responsive rather than unresponsive.

Second, the fact that substantial portions of variance come from the interactions suggests that the validity of predictions of personal behavior should be substantially improved by asking the individuals concerned to report the trait-indicating responses of interest in the specific situations, or at least in the specific kinds of situations, concerned.

Endler and Hunt cite studies indicating that subscores for particular situations have substantial correlations with behavior in those situations. Evidently, group factors or factors specific to the situation ($pi$ components) are worth interpreting even though a general factor ($p$ component) is present.

Endler and Hunt did not use the components as a basis for discussing the generalizability of scores. They had reported conventional one-facet reliability studies in the original monograph, calculating for each $i$ in turn a coefficient $\alpha$ that indicates how well one can generalize from $X_{piJ}$ to $\mu_{pi}$. They also gave a coefficient for each $j$. Let us now see what a multifacet interpretation can offer.

### Interpretation in the light of generalizability theory

*Generalization to $\mu_p$.* Assume for the present an intention to generalize to $\mu_p$, that is, to interpret the overall score as a measure of general anxiousness. Then all components of variance save that for $p$ contribute to $\sigma^2(\Delta)$. How great is the error of measurement? In Table 6.11, look first at the columns to the right, where it is found that for the 11 × 14-item instrument $\hat{\sigma}(\Delta) = 0.26$. This error is to be judged relative to the scale range of 4.0.

The expected value $\widehat{\mathscr{E}\sigma^2(\delta)}$ is estimated by adding just the components [other than $\widehat{\sigma^2(p)}$] that contribute to observed-score variance in the crossed design. We note that $\hat{\sigma}(\delta)$ is 0.17; since $\hat{\sigma}(p) = 0.10^{1/2} = 0.32$, the test does not discriminate well between persons. This is confirmed by the estimate of 0.77 for $\mathscr{E}\rho^2$, a low value for a questionnaire with 154 responses. However, it is to be realized that the scale was constructed by 11 + 14 acts of sampling, not 154.

**TABLE 6.11.**   Estimation of $\sigma(\Delta)$, $\sigma(\delta)$, and $\mathscr{E}\rho^2$ for D Studies with the S–R Inventory of Anxiousness (*Generalization to* $\mu_p$)

| Source of variance | Frequency within $p$ | Estimate of variance component[a] $(n'_i = 1, n'_j = 1)$ | Contribution to $\sigma^2(\Delta)$ | | | | | |
|---|---|---|---|---|---|---|---|---|
| | | | $n'_i =$ 1 $n'_j =$ 154 | 7 22 | 11 14 | 14 11 | 22 7 | 154 1 |
| $i$ | $n'_i$ | 0.09[b] | 0.09 | 0.01 | 0.01 | 0.01 | 0.00 | 0.00 |
| $j$ | $n'_j$ | 0.44[b] | 0.00 | 0.02 | 0.03 | 0.04 | 0.06 | 0.44 |
| $pi$ | $n'_i$ | 0.18 | 0.18 | 0.03 | 0.02 | 0.01 | 0.01 | 0.00 |
| $pj$ | $n'_j$ | 0.20 | 0.00 | 0.01 | 0.01 | 0.02 | 0.03 | 0.20 |
| $ij$ | $n'_i n'_j$ | 0.12[b] | 0.00 | 0.00 | 0.00 | 0.00 | 0.00 | 0.00 |
| $pij, e$ | $n'_i n'_j$ | 0.66 | 0.00 | 0.00 | 0.00 | 0.00 | 0.00 | 0.00 |
| Est $\sigma^2(\Delta)$ | | 1.69 | 0.27 | 0.07 | 0.07 | 0.08 | 0.10 | 0.64 |
| Est $\mathscr{E}\sigma^2(\delta)$ | | 1.04 | 0.18 | 0.04 | 0.03 | 0.03 | 0.04 | 0.20 |
| Est $\sigma(\Delta)$ | | | 0.52 | 0.26 | 0.26 | 0.28 | 0.32 | 0.80 |
| Est $\sigma(\delta)$ | | | 0.42 | 0.20 | 0.17 | 0.17 | 0.20 | 0.45 |
| Est $\mathscr{E}\sigma^2(X)$ | | | 0.28 | 0.14 | 0.13 | 0.13 | 0.14 | 0.30 |
| Est $\mathscr{E}\rho^2$ | | | 0.36 | 0.71 | 0.77 | 0.77 | 0.71 | 0.33 |

[a] Values calculated in Table 6.9.
[b] Components not entering $\mathscr{E}\sigma^2(\delta)$ and $\mathscr{E}\sigma^2(X)$.

The reader will recall that the Spearman–Brown concept of reliability as a simple function of "length of test" is not applicable to a multifacet instrument. While an investigator improves generalizability by increasing $n'_i$ and $n'_j$, there is no simple formula relating $\overset{\frown}{\mathscr{E}\rho^2}$ or $\overset{\frown}{\sigma^2(\Delta)}$ to the total number of observations. Accuracy of generalization is a function of the sample size for each facet.

Consider the efficiency of various designs for an S–R inventory, all eliciting 154 responses per subject, but with $n'_i$ ranging from 1 to 154 and $n'_j$ from 154 to 1. Table 6.11 estimates $\sigma(\Delta)$, $\sigma(\delta)$, and $\mathscr{E}\rho^2$ for several values of $n'_i$ and $n'_j$. Each change in the $n'_i$, $n'_j$ balance alters these indices even though the "length" of the instrument remains constant. For this example, the optimum $\mathscr{E}\rho^2$ and $\sigma^2(\delta)$ are obtained for $n'_i$ and $n'_j$ nearly equal. In another instrument one might have another result; the optimum balance depends on the relative size of the interaction components. The minimum of $\sigma^2(\Delta)$ is achieved with $n'_i$ around 8 and $n'_j$ around 19. There are different "best designs" for different purposes.

*Generalization to* $\mu_{pi}$.   The investigator concerned with situation-specific anxiety will generalize over modes of response, taking $\mu_{pi}$ as his universe score. Then the $pi$ component is added to the $p$ component in the universe-score variance; this variance is estimated to be 0.28. This value refers to the average variance over all situations, and, therefore, is an unbiased estimate of universe-score variance in any one situation. This leads to $\widehat{\mathscr{E}\rho^2}$ of 0.82 for $n_j' = 14$ and $n_i' = 1$, where generalization is over modes of response only.

A more informative analysis would be to treat scores on each situation separately. This, of course, is what Endler *et al.* did in calculating $\alpha$ for each situation. Their $\alpha$'s range from 0.55 to 0.90 (average 0.76). Each of these is a coefficient of generalizability indicating how accurately one can generalize from an observed score with 14 modes of response and a fixed situation, to the universe score for that situation, over all modes of response.

Because $i$ is fixed, the $i$ and $pi$ components no longer enter $\sigma^2(\Delta)$. Using the analysis for all situations together, $\widehat{\sigma^2}(\Delta)$ for a one-item test equals 1.42 ($= 1.69 - 0.09 - 0.18$) and $\hat{\sigma}(\Delta) = 1.19$. $\widehat{\mathscr{E}\sigma^2}(\delta) = 0.86$. All components contributing to $\widehat{\sigma^2}(\Delta)$ and $\widehat{\mathscr{E}\sigma^2}(\delta)$ are proportional to $n_j'$ for the one-situation test. To appraise the effect of any $n_j'$, one merely divides $\widehat{\sigma^2}(\Delta)$ or $\widehat{\mathscr{E}\sigma^2}(\delta)$ by $n_j'$. To estimate $\mathscr{E}\sigma^2(X)$ for any $n_j'$, we add $\widehat{\sigma^2}(\mu_{pi}) = 0.28$ to $\widehat{\mathscr{E}\sigma^2}(\delta)$; then $0.28/\widehat{\mathscr{E}\sigma^2}(X)$ estimates $\mathscr{E}\rho^2$. As there is just one variable facet, the result is consistent with the Spearman–Brown formula. We arrive at these values:

| $n_j'$ | 1 | 5 | 10 | 20 | 30 |
|---|---|---|---|---|---|
| Est $\mathscr{E}\sigma^2(\delta)$ | 0.86 | 0.17 | 0.09 | 0.04 | 0.03 |
| Est $\mathscr{E}\sigma^2(X)$ | 1.14 | 0.45 | 0.37 | 0.32 | 0.31 |
| Est $\mathscr{E}\rho^2$ | 0.25 | 0.62 | 0.76 | 0.87 | 0.90 |

**Design recommendations.** Now the tester can begin to think about the desired design of his instrument. Suppose he intends to generalize to $\mu_p$, will use the crossed Design VII, and has a fixed total number of items ($= n_i' n_j'$) in mind. If he wants to keep $\sigma(\Delta)$ at a minimum, this is achieved by taking a value of $n_i'$ smaller than $n_j'$ (Table 6.11). If he wants to maximize $\mathscr{E}\rho^2$ and to minimize error in the point estimate of $\mu_p$, he will make $n_i'$ nearly equal to $n_j'$. This is not the whole answer, however. Perhaps he should nest $j$ within $i$ so that he would sample the large $j$ component (and $pj$) $n_i' n_j'$ times. If he retained $n_i'$ at 11 and chose 14 different $n_j'$ for each $i$, $\hat{\sigma}(\Delta)$ would drop to 0.18 and $\hat{\sigma}(\delta)$ to 0.14, and $\widehat{\mathscr{E}\rho^2}$ would rise to 0.83.

It becomes obvious that, if estimating $\mu_p$ is our only concern, the best design is III-B: 154 $ij$ pairs in the test with *no i or j* repeated. For that design, $\widehat{\mathscr{E}\rho^2}$ is 0.94! This mathematically ingenious solution, unfortunately, is almost certainly impractical, because it would take considerable psychological ingenuity to prepare a list of 154 relevant modes of response.

What if the score for each specific situation is of interest? This introduces the bandwidth-fidelity dilemma: as more $i$ are covered in an instrument of fixed overall length, the information on each one becomes less dependable (Cronbach & Gleser, 1965, p. 97 ff.). Consequently, there is some limit on the number of facts one should try to collect. If we inquire about 154 situations with one mode of response for each, and generalize to each of the

$\mu_{pi}$, $\hat{\sigma}(\delta) = 0.86^{1/2} = 0.93$; with more responses $\widehat{\mathscr{E}\sigma^2}(\delta)$ is inversely proportional to $n'_j$. It is hard to say just what precision the tester should want in a score, but if he decided that $\sigma(\delta)$ should not exceed 0.19, for example (so that $\mathscr{E}\rho^2$ in generalizing to $\mu_{pi}$ is 0.88), $n'_j$ would have to be 23, and this would set the limit at 7 situations. (The number of modes of response would have to vary from situation to situation, to bring each specific coefficient to 0.88.) It must be remembered that the optimum design for estimating several $\mu_{pi}$ is not optimum if one is interested also in simultaneously estimating various $\mu_{pj}$ for specific modes of response.

The decision about bandwidth can be improved by going on to a factor analysis of some type. The presence of a variance component for $pi$ does not tell us whether that interaction arises from just one factor that divides the items into two clusters, or from 11 situation-specific factors or something between. For the S–R inventory a conventional factor analysis (Endler, *et al.*, 1962) and a three-mode (multifacet) factor analysis (Tucker, 1964) have been reported. Many of the methods developed in Chapters 9 and 10 could profitably be applied to this instrument.

The investigator planning a D study will not find it hard to draw a conclusion about the proper design, but the conclusion will depend on his purpose. Once he conceptualizes:

1. The relative importance of $\mu_p$, $\mu_{pi}$, and $\mu_{pj}$, or $\mu_P$, $\mu_{Pi}$, and $\mu_{Pj}$ as targets of generalization
2. Whether he wants to interpret the observed score, or observed individual differences, or the point estimate of the universe score
3. A preliminary estimate of the total number of observations per person he can afford

the investigator can quickly arrive at an appropriate design.

## EXERCISES

**E.1.**   In Table 6.1 and 6.2, residual components of variance for PICA subtests are estimated. What kinds of variation in performance contribute to the residual component in each analysis?

**E.2.** Porch analyzed $X_{pij}$ at the item level $i$, obtaining the mean squares listed below, among others. $n_p = 30$ and $n_j = 3$. Determine components of variance as in Table 6.11. Interpret the findings.

| Subtest III | | $p$ | $j$ | Residual |
|---|---|---|---|---|
| **Item** | | | | |
| 1 | toothbrush | 23.539 | 17.544 | 1.740 |
| 2 | cigarette | 16.757 | 14.011 | 2.563 |
| 3 | pen | 25.278 | 6.633 | 2.622 |

**E.3.** Calculate $\sigma^2(\Delta)$ for generalization over objects, scorers, and subtests, using the components estimated for PICA in Table 6.4. Assume $n_i' = 10$, $n_j' = 1$, $n_k' = 6$.

**E.4.** Calculate $\sigma^2(p \mid K^*)$ and the coefficient of generalizability for 6 fixed subtests, 10 random objects, and 1 random scorer. Use the data from Tables 6.4 and 6.5.

**E.5.** Silverstein and Fisher (1968) administered the *S–R Inventory of Anxiousness* twice to the same subjects with a one-month interval between tests. The design was treated as having occasions nested within *pij*, though it could have been treated as $o \times i \times j \times p$ or as $i \times j \times (o:p)$. Discuss the pros and cons of this choice, knowing that the subjects were prisoners assigned to a "guidance center," and that the first test was part of a general battery of measures given to the prisoner individually about a month after arrival.

**E.6.** The analysis of the Silverstein–Fisher data gave these mean squares:

| $p$ | $i$ | $j$ | $pi$ | $pj$ | $ij$ | $pij$ | within $pij$ |
|---|---|---|---|---|---|---|---|
| 39.92 | 144.24 | 985.31 | 3.84 | 5.62 | 10.21 | 0.73 | 0.56 |

$n_p = 100$, $n_i = 11$, $n_j = 14$. Estimate the components and indicate what is learned from this study that adds to, or contradicts, the Endler–Hunt results.

**E.7.** Leler (see Chapter 3, Exercise 4) collected data on preschool children interacting with their mothers as they performed various tasks. Assuming generalization over observers and tasks, $\hat{\sigma}(\delta)$ and $\hat{\mathscr{E}}\rho^2$ were calculated for the design with tasks ($n_i' = 6$) and raters ($n_j' = 2$) crossed with mother–child pairs. In the light of Table 6.E.1, what recommendations seem sensible for improving subsequent data collection? Consider the costliness of increasing the demands upon subject time that would be involved if more tasks were added. Because all scales are rated from the same tape, there is little saving in cost by allowing smaller $n_j'$ for some scales. The alternatives open include dropping a scale, increasing the number of conditions of $i$ or $j$, shifting to some kind of partially nested design, and perhaps others. You may be able to suggest specific information an investigator could seek in the data that would help in deciding on the revisions.

**TABLE 6.E.1.**   Information from G Study of Mother–Child Interactions (after Leler, 1970)

| Scale | Mean | $\hat{\sigma}(\delta)$ | $\mathscr{E}\hat{\rho}^2$ | Component of variance | | | | | | |
| | | | | $p$ | $i$ | $j$ | $pi$ | $pj$ | $ij$ | $pij,e$ |
|---|---|---|---|---|---|---|---|---|---|---|
| D. Mother's use of criticism | 2.9 | 0.56 | 0.00 | 0.0 | 0.1 | 1.3 | 0.5 | 0.3 | 0.1 | 1.1 |
| I. Mother's use of reasoning with child | 3.0 | 0.71 | 0.45 | 0.4 | 0.7 | 1.2 | 0.6 | 0.6 | 0.1 | 1.1 |
| K. Mother's discouragement of child's verbalizations | 2.4 | 0.68 | 0.11 | 0.1 | 1.1 | 0.2 | 0.3 | 1.0 | 0.4 | 1.6 |
| L. Child's independence | 4.7 | 0.63 | 0.57 | 0.5 | 0.1 | 0.0 | 0.7 | 0.3 | 0.1 | 1.3 |
| Approximate median for 14 scales | | 0.55 | 0.50 | 0.4 | 0.6 | 0.1 | 0.5 | 0.2 | 0.1 | 1.1 |

## Answers

**A.1.**   The residual in Table 6.1 includes any scorer–patient interaction, and unsystematic variation in the scoring operation that might arise from the scorer's inattention or shifting standards.

In Table 6.2, patient–item variation enters the residual. This includes any effect of momentary inattention or blocking by the patient, as well as any systematic difficulty the patient has with a certain word.

**A.2.**   Universe scores appear to vary over the entire 16-point scale. Scorer standards differ very little. The residual component shows a moderate amount of scorer inconsistency that is probably irreducible since the scorer has to judge a single brief response.

| | $\widehat{\sigma^2}(p \mid i)$ | $\widehat{\sigma^2}(j \mid i)$ | $\widehat{\sigma^2}(pj,\bar{e} \mid i)$ |
|---|---|---|---|
| Toothbrush | 7.3 | 0.5 | 1.7 |
| Cigarette | 4.7 | 0.4 | 2.6 |
| Pen | 7.6 | 0.1 | 2.6 |

**A.3**   $\frac{1}{10}(0.13) + (0.02) + \frac{1}{6}(2.65) + \frac{1}{10}(0.41) + (0.15) + \frac{1}{6}(1.71)$
$+ \frac{1}{10}(0.00) + \frac{1}{60}(0.12) + \frac{1}{6}(0.01) + \frac{1}{10}(0.15) + \frac{1}{60}(2.01)$
$+ \frac{1}{6}(0.07) + \frac{1}{60}(0.01) + \frac{1}{60}(0.99) = 1.031$

**A.4.**   $\widehat{\sigma^2}(p \mid K^*) = 5.59$

$\widehat{\mathscr{E}\sigma^2}(\delta) = 0.07 + 0.16 + 0.03 = 0.26$

$\widehat{\mathscr{E}\rho^2} = \frac{5.59}{5.85} = 0.96$

**A.5.** Treating occasion as nested within *pij* implies that each "question" (stimulus paired with mode of response) is answered by the person on a different occasion. Operationally, the questions are reached at different instants in time, so that if occasion variation is interpreted as moment-to-moment variation, the analysis is pertinent. But this analysis lumps into one residual term the occasion, occasion × stimulus, occasion × stimulus × mode of response, and other components that might be separated in a different analysis.

A crossed analysis emphasizes the two distinct occasions a month apart (i.e., occasion is interpreted as primarily reflecting effects associated with first-test-vs-second-test differences). Momentary variation will remain in the residual term, but there will be separate estimates of components such as *pio* (the tendency of the person to respond differently to a stimulus on the first and second occasions). One can imagine that faking on certain questions might be different if the person believes at the time of the first testing that his responses will affect his treatment in prison.

Treating occasions as nested within persons implies that the occasion of the first test is different for each person. That is, occasion is identified with the calendar date on which the person is tested, rather than with "first" or "second" test. This analysis will leave confounded certain components that the crossed analysis separated (e.g., *o* and *po*).

This decision about analysis cannot be made on purely statistical grounds; it is essentially a matter of selecting the definition of "occasion" that the investigator considers most meaningful. If one needed unambiguous information about effects related to time, a more elaborate experiment would be designed involving some same-day retests and some crossing of days with persons, to evaluate the strength of effects associated with calendar date, order of testing, and momentary variation.

**A.6.** There is close agreement between the results tabulated below and those of Endler and Hunt (p. 179) for the *p, j,* and *pj* components. The other four components estimated by Endler and Hunt are larger than their counterparts in this analysis; variance has been reassigned to the within-*pij* component. The variance shifted from the Endler–Hunt *i* component to "within *pij*" must be *io* variance. Likewise, there is evidence that the *pio* and *ijo* components are not negligible. (In the Endler–Hunt analysis the hidden facet of occasion left these confounded with *pi* and *ij* respectively.) There is evidently some shift from occasion to occasion in the stimuli the person identifies as arousing anxiety, but little shift in his report on modes of response.

|  |  |  |  |  |  |  | within |
| --- | --- | --- | --- | --- | --- | --- | --- |
| *p* | *i* | *j* | *pi* | *pj* | *ij* | *pij* | *pij* |
| 0.10 | 0.05 | 0.44 | 0.12 | 0.22 | 0.05 | 0.08 | 0.56 |

**A.7.** There is no certain answer to this question. The decisions actually made by Leler were reasonable; perhaps other suggestions could profitably be made.

To begin, Leler discarded none of these scales. While many coefficients were low, none was too low to permit her to investigate whether the variables correlated significantly with language, in a D study using about 60 cases,

It was not feasible to add tasks or raters under the cost limitations of her study. An extra rater would have reduced the generally large *pj* and *pij* components of variance.

Because the *i* and *j* components are occasionally large, a nested design is not recommended.

With regard to Scale D, the key problem is absence of person variance rather than extremely large error. The exceptional component for raters *j* led to further inquiry, which disclosed that one rater was taking only the mother's words into account while the other was also considering tone of voice. Directions for raters were modified to increase emphasis on tone of voice. If the trait is better defined, generalizable data may be expected.

With regard to Scale I, the *j* and *pj* components are larger than for other scales. Conferences with raters showed a difference in their definitions for "reasoning"; the directions were revised. A task-by-task study showed deviant scores on the sixth task, where the mother was required to fill out a form while the child had opportunities to distract her. The mother had little occasion to reason with the child while preoccupied; this task was dropped from the scoring of Scale I (and also from Scale D).

On Scale K, the mean for the sixth task was high. In retrospect, it was realized that discouraging the child's verbal interruptions was appropriate here, and therefore had a different psychological significance than elsewhere. The sixth task was ignored in subsequent use of this scale. This constitutes redefinition of the universe. No way of reducing *pj* effects was discovered.

On Scale L, the only improvement was to revise wording of some scale items. After these revisions, on a larger sample of subjects, the coefficients became: for D, 0.61 (vs 0.00 originally!); for I, 0.66 (vs 0.45); for K, 0.64 (vs 0.11); for L, 0.68 (vs 0.57).

# CHAPTER 7
# Illustrative Analyses of Partially Nested Designs

## A. Observations of Classrooms

We have referred earlier to the pioneering expositions of the multifacet approach by Medley and Mitzel (1963; with Doi, 1956). Their theoretical statement gives an excellent introduction to the basic crossed design. We propose to work with the numerical example of their 1963 paper.

### Design and basic data

Medley and Mitzel were concerned with characteristics of classrooms. The "class" is the subject under study: data are collected to learn about the tone of the classroom rather than the traits of particular pupils. To evaluate the technique, 2 recorders simultaneously observed each of 24 teachers, each on 5 occasions. This is Design V-B $[r \times (o:t)]$, occasions nested within teachers and crossed with recorders.

A statement in which the authors emphasize the significance of a multifacet analysis is worth quoting (notation has been adapted to conform to that in this work), to reinforce points made in our own theoretical chapters (Medley & Mitzel, 1963, p. 310):

> Most observational studies in the past have studied reliability either in terms of per cent of observer agreement or in terms of an interclass correlation (usually the product-moment, but occasionally the rank-order, coefficient) between two sets of observations.
>
> A per cent of observer agreement tells almost nothing about the accuracy of the scores to be used, mainly because the per cent of agreement

**189**

between observers is relevant to only a part—and, the evidence indicates, a small part at that—of the reliability problem. The experience with observational studies summarized in this chapter clearly bears out a fact pointed out by Barr in 1929: that errors arising from variation in behavior from one situation or occasion to another far outweigh errors arising from failure of two observers to agree exactly in their records of the same behavior. It is not impossible to find observers agreeing 99 per cent in recording behaviors on a scale whose reliability . . . over occasions does not differ significantly from zero.

Reliability can be low even though observer agreement is high for a number of reasons. For example, observers might be able to agree perfectly on the number of seats in a room, yet if the number of seats in all rooms is equal, or nearly so, the reliability of seat counts as a measure of differences between classes will be zero. Near-perfect agreement could also be reached about the number of boys in a room wearing red neckties; but if every boy changed the color of his tie every day, the reliability of these counts would be zero. So long as an interclass (product-moment) correlation is based on scores obtained on two different occasions by two different observers, it does estimate $\rho_{XX'}$. But it is not likely to be a very accurate estimate because the number of classrooms $n_t$ is usually small in observational studies, and the size of $n_t$ determines the precision of a product–moment correlation coefficient (its standard error varies inversely as the square root of $n_t$). In even a rather ambitious study, using 100 classrooms, the 90 per cent confidence interval of $\rho_{XX'}$ estimated in this way would be about 0.33 points wide! If the number of situations or occasions per teacher $(n_o)$ is increased to more than two, several correlations can be calculated, one between each pair of situations; but since they are not independent, it is difficult to combine all of the correlations into a single best estimate.

A single intraclass correlation can be calculated from an analysis of variance of a set of data collected according to the plan suggested above. Such a single coefficient combines all of the information in the $n_o n_r$ independent measurements of each of the $n_t$ classes. The estimate of $\rho_{XX'}$ so obtained is unbiased and also more precise than any combination of interclass correlations . . . . Moreover, the different reliability coefficients appropriate to the various uses to which the scores might be put can all be estimated from the one analysis of variance.

*Review of original analysis*

The design of greatest all-around importance in generalizability studies is the completely crossed Design VII, and for their pedagogical aims Medley and Mitzel develop the example as if such a design had been used, teachers $t$ being crossed with observers $r$ and occasions $o$. They speak of "situations"

**TABLE 7.1.** *Estimates of Variance Components from a G Study with the Design* $r \times o \times t$ *(after Medley & Mitzel)*

| Source of variance | Sum of squares | Degrees of freedom | Mean square | Estimate of variance component when $N_o$ is considered to be | |
|---|---|---|---|---|---|
| | | | | Very large | 5 |
| Teacher $t$ | 203 | 23 | 8.83 | 0.53 | 0.81 |
| Recorder $r$ | 13 | 1 | 13.00 | 0.10 | 0.10 |
| Occasions $o$ | 19 | 4 | 4.75 | 0.03 | 0.03 |
| $tr$ | 17 | 23 | 0.74 | 0.01 | 0.15[b] |
| $to$ | 321 | 92 | 3.49 | 1.41 | 1.41 |
| $ro$ | 1 | 4 | 0.25 | —[a] | —[a] |
| Residual | 62 | 92 | 0.67 | 0.67 | 0.67 |

[a] Negative value, treated as zero. Because of this procedural decision, the numerical values for $r$ and $o$ components disagree with the values given by Medley and Mitzel.
[b] The formula used by Medley and Mitzel to estimate the $tr$ component obtains a different result because they, in effect, assume the $tro$ interaction to be negligible.

where we speak of "occasions," and analyze the data as if each teacher had been observed "in the same situations." This might be a reasonable interpretation, if each teacher taught certain standard lessons and the recorders observed each teacher with each lesson. Although this is not the study actually made, we review the crossed analysis both as an additional demonstration of the method and to show what difference a change to the strictly correct analysis makes.

Table 7.1 (see Medley & Mitzel, 1963, pp. 314–316) tabulates (in our notation) the main results from the crossed three-way analysis of variance components. Medley and Mitzel made two analyses for illustrative purposes; one that assumes an indefinitely large number of occasions, and one that regards the universe as limited to five fixed occasions (situations). The latter model would be especially reasonable if there were one observation on each of the five school subjects in the school program, though one would then like to have two observations per school subject (crossed with teachers and recorders) in order to disentangle occasion effects.

The notable difference in the teacher component between the analyses with $N_o$ large and $N_o$ small is readily understood. If situations are fixed, generalization is over recorders and the universe score is $\mu_{tO*}$ ($O*$ referring to the set of situations). Therefore, the $to$ component contributes to $\sigma^2(t \mid O*)$. The $to$ component is 1.41 when $N_o$ is assumed large. For five situations from this large universe, the $tO$ component is one-fifth of 1.41, or 0.28; $0.81 = 0.53 + 0.28$. Similarly, $\widehat{\sigma^2}(tr \mid O*)$ is increased by one-fifth of the $tro,e$ component.

For the D study, Medley and Mitzel note, the universe-score variance for generalization to $\mu_p$ will be $0.53(n'_r n'_o)^2$. (They treat the observed score as a sum rather than as an average as we do.) Assuming a crossed design for the D study, the corresponding expected observed-score variance will be $n'_r n'_o (0.53 n'_r n'_o + 0.01 n'_o + 1.41 n'_r + 0.67)$. "Inspection of this formula shows clearly that increasing the number of visits $n'_o$ will [decrease observed variance relative to universe-score variance and, hence, will] increase reliability much more rapidly than increasing the number of recorders $n'_r$. . . .

When $n'_r = 1$ and $n'_o = 5$, for example, $\widehat{\mathscr{E}\rho^2} = 0.55$; when $n'_r = 5$ and $n'_o = 1$, $\widehat{\mathscr{E}\rho^2} = 0.26$ [notation altered]." Clearly, the first requirement for improving generalizability is to sample the large *to* component repeatedly.

Medley and Mitzel (1963, pp. 315–316; notation altered) expand:

The largest components of variation in these observational records are three: $\sigma^2(t)$, $\sigma^2(to)$, $\sigma^2(tro,e)$. Variation from situation to situation within the same class $\sigma^2(to)$ appears greater than variation in average behavior from one class to another $\sigma^2(t)$. In order to measure differences between classes reliably, therefore, it is necessary to observe each class in a number of situations, so that the fluctuations measured by $\sigma^2(to)$ can cancel one another out.

The large contribution of unexplained sources of variance indicated by the magnitude of $\sigma^2(tro,e)$ shows that there are sizable influences affecting behavior records that were not isolated in this experiment.

There is no variation at all which can be attributed to interaction between recorder and situation $\sigma^2(ro)$, indicating that the observers are not biased in favor of any one situation over any other. The fact that $\sigma^2(tr)$ is estimated to be only .01 reflects the fact that "observer errors" are very slight.

The fact that the estimates of $\sigma^2(tr)$, $\sigma^2(o)$, and $\sigma^2(r)$ are all relatively small, but not zero, makes one wonder whether or not they could be neglected, i.e., whether true values could be assumed to be zero . . . .

It would be quite satisfactory in using this scale to employ only one recorder per visit; if more than one competent observer were available, it would be advisable to send them to visit the classes one at a time, so that the number of different situations recorded would be as large as possible.

Prior to the last paragraph quoted, the authors recommend use of the $F$ test to decide the significance of each component, with the idea of replacing any nonsignificant component with zero. This we do not recommend. A facet is represented in the G-study design only if it is thought to be a source of variation, whether large or small. If the model is soundly chosen, the analysis gives an unbiased estimate of the components involving that facet. When the occasion component of 0.05 is "nonsignificant," 0.05 is nonetheless likely to be a better estimate than 0.00. Indeed, because the "significance" of a component is a function of the number of times that component was sampled in the design, as well as its magnitude, there is a risk of discarding a meaningfully large component when only a few degrees of freedom went into its estimate.

### Analysis as a partially nested design

To arrive at the proper mean squares, it is recognized that the Medley–Mitzel design is of type V-B, $r \times (o:t)$. Figure 2.5 illustrates the design (though it uses the symbols $p$, $i$, and $j$ in place of the present $t$, $r$, and $o$). Within $t$, the components for $o$ and $to$ are confounded. Consequently, one pools the component sums of squares and degrees of freedom, arriving at these within-$t$ results: sum of squares = 340, degrees of freedom = 96, mean square = 3.54. For $ro$, $tro$, $e$ a similar procedure yields sum of squares = 63, degrees of freedom = 96, mean square = 0.66.

The mean squares from Table 7.1, along with those calculated above for within $t$ and within $tr$, are taken as estimates of the expected mean square. The five components estimated from (2.3) are as follows:

$$\widehat{\sigma^2}(\text{within } tr) = \widehat{\sigma^2}(ro,tro,e) = 0.66$$

$$\widehat{\sigma^2}(tr) \qquad\qquad\qquad = 0.02$$

$$\widehat{\sigma^2}(r) \qquad\qquad\qquad\quad = 0.10$$

$$\widehat{\sigma^2}(\text{within } t) = \widehat{\sigma^2}(o,to) \qquad = 1.44$$

$$\widehat{\sigma^2}(t) \qquad\qquad\qquad\quad = 0.52$$

The estimate for the teacher component is almost the same as the estimate of Table 7.1 with $N_o$ large. The recorder component is not changed. The new within-teacher component replaces the $to$ component. This estimate, like the original one, implies the necessity of observing the teacher on many occasions, but now a systematic teacher-occasion interaction is not implied. The design used for the G study could give no information on the comparative size of $o$ and $to$ effects.

For $n'_r = 1$, $n'_o = 5$, $\mathscr{E}\rho^2$ is estimated at 0.56. While shifting to the correct scheme for analysis of *these* data makes no great difference in conclusions, the reader should not conclude that that will usually be the case. The estimation formulas for a design not actually used can give misleading estimates.

### B. Classroom Observations as Dependent Variable in an Experiment

*Description of study*

Dwight Goodwin (1966) carried out a study (also distributed as a contract report under the authorship of Krumboltz & Goodwin, 1966) of a special teacher-training technique that was intended to promote, as an end result, better conduct and attentiveness among the pupils directed by those teachers. Unlike the study of Medley and Mitzel, which was concerned with scores for single classrooms, this is an experiment where conclusions rest on comparisons of sets of pupils whose teachers have been treated differently. The effectiveness of the training procedure had to be judged at two levels: by determining whether the teachers adopted the prescribed practices, and by determining whether the pupils acted in the desired way. Both questions were investigated by sending trained observers into the classroom according to a time-sampling schedule.

Generalizability studies were made during Goodwin's preliminary testing of the procedure, and again during "baserate" studies that investigated the frequency of significant kinds of behavior prior to training. Similar analyses might also have been made of post-training data, but the G studies were carried out primarily to aid in the planning of final data collection.

Goodwin's observer watched a particular pupil during a 5-second period and recorded a numeral to represent teacher and pupil behavior. There were five spaces within which the numeral could be written, representing five explicitly defined "degrees of task orientation of the pupil." The numeral (1–9) described the extent to which the teacher encouraged what the child did; for example, 1 indicated individual reward for the child being observed, and 8 indicated punishment or scolding of a group of which the child is part. The data in effect constitute a pair of scores, one for the pupil's action and one for the teacher's response.

The 5-second intervals of observation were made consecutively, during a 1-minute period. During the next minute, the observer made 12 observations of a second child. He continued to alternate between the two children for 10 minutes, after which he rested. During a single class hour there would be three 10-minute intervals of active observation, 5 of the 10 minutes being devoted to each child. All data were collected in second- or third-grade classrooms. We shall discuss two analyses of partially nested designs employing Goodwin's data.

*Baserate study of pupil scores: intervals within days within pupils*

In the baserate study, with one observer per pupil, the primary question of generalizability was the adequacy of the time sampling. The pilot study (discussion following) had by this time given satisfactory evidence of observer accuracy, and only one observer was used to collect baserate data on any child. However, not all children were observed by the same persons, therefore any variance arising from observer differences is confounded with pupil components. Components of variance associated with the observer are thought to be negligible, however.

The baserate study provided both G and D data. The design of the study is intervals $i$ within days $d$ within pupils $p$ (i.e., Design III-A, $i:d:p$). Attention in this analysis is entirely on the pupil's score for each 5-second observation. A group of 60 scores (1,2, . . . , 5) were recorded for a child during a 10-minute interval. (In this analysis the companion score for the teacher is ignored.) There were three intervals of observation on one day and three intervals on another ($n_p = 28, n_d = 2, n_i = 3$). Days were treated as randomly and independently sampled within pupils, even though several pupils were observed on the same day.

A score was available for each 5-second interval, but there seemed to be no major purpose in the present study in determining components for units of behavior smaller than the 10-minute interval. The 60 observations for any 10-minute period were therefore entered in the computer simply as replicates. Analysis within the 10-minute period might be useful if one were considering the possibility of taking fewer samples of behavior within the 10-minute period.

As usual in a nested design, the estimates for seven sums of squares and mean squares are obtained and grouped to arrive at the correct mean squares (Table 7.2). The additional step of dividing mean squares by 60 is required to correct for the fact that 60 observations enter the analysis for each interval.

The component of variance for pupils is quite small because the sample was selected to represent inattentive pupils, and therefore has a restricted range on task orientation. In a correlational study, pupil variance is usually "wanted" variance, but not in the pretest for an experiment. The small pupil component here is encouraging, as variation among pupils contributes to the standard error of the means on which the conclusions of the experiment ultimately rest.

The component for intervals is considerably larger than that for days. This implies that variation in attention is associated with shifts in the classroom activity or fatigue more than with the pupil's mood on a particular day or the general level of excitement in the class on a particular day.

The Venn diagrams shown in Figure 7.1 may be used for practice in deriving the estimation equations; results can be checked by entering the mean squares from Table 7.2 in the equations and comparing the estimates of variance components with those of the table.

These baserate data were used to draw conclusions about the initial

**TABLE 7.2.** *Estimates of Variance Components from a G Study with the Design* $i:d:p$

| | Analysis of variance as if crossed | | | Analysis as nested | | | |
|---|---|---|---|---|---|---|---|
| Source of variance | Sum of squares | Degrees of freedom | Sum of squares | Degrees of freedom | Mean square | Mean square rescaled[a] | Estimate of variance component |
| $p$ | 1149 | 27 | 1149 | 27 | 42.55 | 0.71 | 0.04 |
| within $p$ | | | | | | | |
| $d$ | 71 | 1 ⎤ | 762 | 28 | 27.20 | 0.45 | 0.08 |
| $pd$ | 691 | 27 ⎦ | | | | | |
| within $pd$ | | | | | | | |
| $i$ | 48 | 2 ⎤ | 1400 | 112 | 12.50 | 0.21 | 0.21 |
| $di$ | 1.69 | 2 | | | | | |
| $pi$ | 824 | 54 | | | | | |
| $pdi, e$ | 526 | 54 ⎦ | | | | | |

[a] Dividing by 60 reduces data to the basic 1–5 scale.

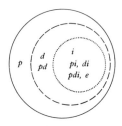

(a)  Components in the design

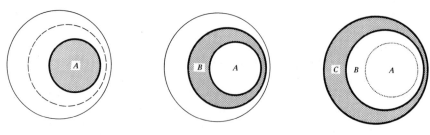

(b) the $i$ circle          (c)  The $d$ circle          (d)  The $p$ circle

**FIGURE 7.1.**  *Schematic Analysis of the $i:d:p$ Design.*

similarity of experimental and control children and as a base for estimating changes during treatment. The error in the individual score is the discrepancy $\Delta_{pDI}$ between observed score (two days combined, six intervals combined) and universe score. Therefore $\widehat{\sigma^2}(\Delta) = \widehat{\sigma^2}(d,pd)/2 + \widehat{\sigma^2}(\text{within } pd)/6 = 0.075$ and $\hat{\sigma}(\Delta) = 0.28$. Judged against the possible range of 4.00, this was considered to indicate adequate precision. The technique is inadequate for determining individual differences within the inattentive group, but that is irrelevant to the study.

For the main study, interest centered on a comparison of the means for groups of children. The means of greatest importance to Goodwin were to be calculated for groups of seven pupils (there being just seven trained and seven control teachers, each with one "target" pupil). Sampling errors arising from intervals and days are independent for the seven children. The problem of generalizability for Goodwin was to compare $X_{PDI}$, the average for the group of seven, with $\mu$, the mean for the population and universe.

As conclusions were not to be drawn about individual pupils, Goodwin's tentative plan for the posttest study called for $n'_p = 7$, $n'_a = 2$, and $n'_i = 3$ in an $i:d:p$ design. Then, for the discrepancy of the observed mean from $\mu$,[1]

$$\widehat{\sigma^2}(\Delta) = \tfrac{1}{7}\widehat{\sigma^2}(p) + \tfrac{1}{14}\widehat{\sigma^2}(d,pd) + \tfrac{1}{42}\widehat{\sigma^2}(\text{within } pd)$$

$$= 0.006 + 0.006 + 0.005 = 0.017$$

The standard error of the mean is therefore 0.13, and that for a difference between groups is $2^{1/2}(0.13)$ or 0.18. While this does not seem to be large, relative to the range, Goodwin decided to obtain somewhat greater precision by observing on a third day. This reduced $\widehat{\sigma^2}(\Delta)$ to 0.013 and $\hat{\sigma}(\Delta)$ to 0.11. Adding more pupils, which would have increased the teacher's effort, was judged to be impractical. In the main comparisons on the posttest data, the variation within groups from the pupil component was reduced by means of analysis of covariance, using the baserate observation as covariate.

Another feature of Goodwin's investigation called for comparing matched pairs of pupils, one pair in each experimental class. Within a pair, one child had been randomly selected as a "target" child, and the teacher had been told that the child would be observed in order to judge the teacher's success. On the posttest the second child was observed without the teacher's awareness, to guard against the possibility that the teacher was using the techniques only on the target child, to impress the observer. The observer was instructed to observe the target child and the non-target child during alternate minutes of an

---

[1] There is no reason to think that $\sigma^2(p)$ in the experimental group at the posttest will equal the value found in the pretest. But the pretest value must be used at the time of planning, unless the design can be changed after some posttest data are collected.

interval. This eliminates, from the observed difference between these children, components associated with $d$ and $i$ (but not the $pd$ and $pi$ interactions), and makes the comparison more precise. We have no measure of the $pd$ and $pi$ components of variance by themselves, because no generalizability analysis was made from posttest data.

(The reader may be tempted to make inferences from the original sums of squares that were compounded to form the within-$p$ and within-$pd$ estimates. He might, for example, conclude that SS $i$ of 48 with 2 degrees of freedom suggests a large $i$ component. No such inference is warranted. The sum of squares for $i$ is simply a calculation based on averages, each a combination of many different $i$ arbitrarily assembled in the same column of a matrix. A similar remark is to be made about any other entry in the "analysis as if crossed.")

An analysis was made for pretest teacher scores. The estimated variance components for $tp$ and days within $tp$ were less than 0.01, indicating marked similarity among teachers prior to the treatment. The within-$tpd$ component was estimated at 0.13, leading to the conclusion that the technique was sensitive enough to detect effects of the training.

This study is not really adequate, however, because it deals only with the numerals on the 9-point scale. The psychological significance of that score depends on what the pupil is doing. That is, a response of 1 (rewards) is desirable teacher behavior when the pupil is attentive but not when the pupil is creating a disturbance. In the main study, teacher and pupil data were considered in combination and rescored—in effect on a 3-point scale, by counting certain score-pairs as desirable, certain others as undesirable, and assigning zeros to the remainder. For this ad hoc method of recombining data, an equally ad hoc method of appraising generalizability could be developed.

### Pilot study of pupil scores: intervals within days, days within pupils, these crossed with observers

Prior to the baserate study, Goodwin made various pilot studies. One of these illuminated the plan for later observing. The design was $(i:d:p) \times r$. For each pupil, two days were sampled, not the same for every pupil. For each pupil-day combination, two intervals (observation periods) were scheduled. Within each interval, the observer recorded 48 ratings (not 60 as in the study treated above). There were two observers who worked simultaneously; observers, then, are crossed with $p$, $d$, and $i$ (Figure 7.2).

A group of 12 children were studied from 6 classrooms; 6 had been nominated as habitually inattentive and 6 as habitually attentive. Neither of these breakdowns enters the pilot-study analysis.

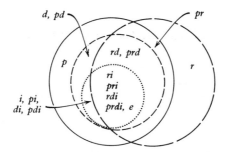

*FIGURE 7.2. Schematic Representation of the (i:d:p) × r Design.*

A 4-way analysis generates 15 sums of squares that recombine as shown in Table 7.3. For two effects, the estimates are negative. The apparently small recorder main effect confirms that the raters were well instructed in their task. This is further supported by the very small components for *pr*, within *pr*, and within *prd*.

It is more surprising to find a small estimate for the within-*p* component, which implies that pupil behavior does not vary much from day to day. This is in contrast with the large within-*pd* component, implying large variation between periods observed on the same day. According to these data, it is inappropriate to say of a pupil, "He's having an off-day today." Attentiveness shifts a good deal from period to period within the day.

It may be worthwhile to examine the estimation equations for this complex design. Figure 7.3 makes the equations easier to grasp intuitively. As usual, the finest division of scores in the design gives a residual mean square that estimates the residual variance component.

(7.1)
$$\text{EMS } i \text{ within } prd = \sigma^2(i \text{ within } prd) = A$$
$$\widehat{\sigma^2}(\text{within } prd) = 0.008$$

This is a bundle of effects—*ri*, *pri*, *rdi*, *prdi*, *e*—representing rater disagreement in viewing the same series of incidents within an interval.

Diagram (b) of Figure 7.3 shows the *d,r* intersection, representing EMS *d* within *pr*, as containing the kernel identified as *A* in diagram (a), plus the shaded area that represents the component $\sigma^2(d$ within $pr)$. Adding the needed multiplier, the diagram suggests the equation:

(7.2)
$$\text{EMS } d \text{ within } pr = A + B$$
$$= A + n_i\sigma^2(rd,prd)$$
$$\widehat{\sigma^2}(rd,prd) = 0.001$$

**TABLE 7.3.** *Estimates of Variance Components from a G Study with the Design* $(i:d:p) \times r$

| Source of variance | Analysis of variance as if crossed | | Analysis as nested | | | | Estimate of variance component |
|---|---|---|---|---|---|---|---|
| | Sum of squares | Degrees of freedom | Sum of squares | Degrees of freedom | Mean square | Mean square rescaled[a] | |
| $p$ | 337.92 | 11 | 337.92 | 11 | 30.72 | 0.640 (A + B + C + D + E + F) | 0.001 |
| within $p$ | | | | | | | |
| $d$ | 0.92 | 1 ⎫ | 258.54 | 12 | 21.54 | 0.449 (A + B + D + E) | 0 |
| $pd$ | 257.62 | 11 ⎭ | | | | | |
| within $pd$ | | | | | | | |
| $i$ | 1.29 | 1 ⎫ | 720.44 | 24 | 30.02 | 0.625 (A + D) | 0.309 |
| $pi$ | 478.00 | 11 ⎪ | | | | | |
| $di$ | 1.03 | 1 ⎬ | | | | | |
| $pdi$ | 240.12 | 11 ⎭ | | | | | |
| $r$ | 0.11 | 1 | 0.11 | 1 | 0.11 | 0.003 | 0 |
| $pr$ | 7.26 | 11 | 7.26 | 11 | 0.66 | 0.014 (A + B + C) | 0.001 |
| within $pr$ | | | | | | | |
| $rd$ | 0.21 | 1 ⎫ | 6.12 | 12 | 0.51 | 0.011 (A + B) | 0.001 |
| $prd$ | 5.91 | 11 ⎭ | | | | | |
| within $prd$ | | | | | | | |
| $ri$ | 0.33 | 1 ⎫ | 9.61 | 24 | 0.40 | 0.008 (A) | 0.008 |
| $pri$ | 4.20 | 11 ⎪ | | | | | |
| $rdi$ | 0.33 | 1 ⎬ | | | | | |
| $prdi, e$ | 4.75 | 11 ⎭ | | | | | |

[a] Dividing by 48 reduces data to the basic 1–5 scale. Capital letters refer to Figure 7.3.

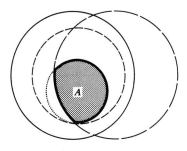

(a) The *p,j* intersection

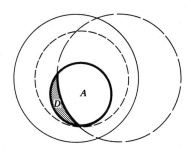

(b) The *d,r* intersection

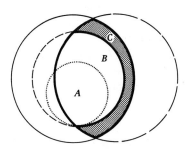

(c) The *p,r* intersection

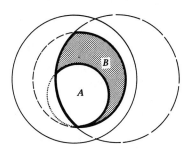

(d) The *i* circle

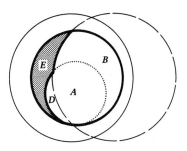

(e) The *d* circle

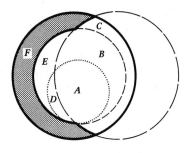

(f) the *p* circle

FIGURE 7.3.   Schematic Analysis of the (i:d:p) × r Design.

201

There is little or no tendency for raters to vary their standards from day to day. A similar argument, applied to diagram (c), gives:

(7.3) $$\text{EMS } pr = A + B + n_d n_i \sigma^2(pr)$$

$$\widehat{\sigma^2}(pr) = 0.001$$

The negligible pupil–observer interaction is all to the good; it denies that raters show favoritism.

The *i*-within-*pd* mean square includes all the information within the *i* circle, as shown in diagram (d):

(7.4) $$\text{EMS } i \text{ within } pd = A + n_r \sigma^2(i,pi,di,pdi)$$

$$\widehat{\sigma^2}(i,pi,di,pdi) = 0.309$$

This is the large component reflecting hour-to-hour pupil variability. There is a temptation to look back at the analysis as if crossed and to emphasize the magnitude of the *pi* mean square. But the design thoroughly confounds *i* and *pi*. If the hour-to-hour shift in behavior were entirely due to an interval main effect (i.e., to a tendency of the whole group to be disorderly during some periods) one would observe this in large *pi* and *pdi* mean squares, as long as a different pupil is seen in each period. The low *i* mean square is a misleading value; it is a composite based on many intervals whose main effects tend to cancel out.

The equation corresponding to diagram (e) is:

(7.5) $$\text{EMS } d \text{ within } p = A + B + D + n_r n_i \sigma^2(d,pd)$$

$A + B = 0.011$ and $D$ was just estimated as 0.617. This would imply that $n_r n_i \sigma^2(d,pd)$ equals $0.449 - 0.628$. Because this is negative, we estimate $\sigma^2(d,pd)$ to be zero. For the *p* component (diagram f):

(7.6) $$\text{EMS } p = A + B + C + D + E + n_r n_d n_i \sigma^2(p)$$
$$= \text{EMS within } p + n_d n_i \sigma^2(pr) + n_r n_d n_i \sigma^2(p)$$

The calculation must use a value of zero for $E$. Then $\widehat{\sigma^2}(p) = 0.001$. [The computation of the $r$ component is left as an exercise (Exercise 3, this chapter).]

The implications for design of Goodwin's D study lay in the large component for intervals within *pd* and the small components for *r* and its interactions. The decision was made to raise the number of scores within the interval to 60 and the number of intervals within the day to 3. Also, it was decided that one observer for any session was sufficient, and that no pains need be taken to cross observers with children or teachers. One might have decided to observe at six intervals on the same day rather than at three on each of two days, in view of the small component for days within *pr*. However, this offered no practical advantage. This set of recommendations is an especially clear example of the usefulness of a multifacet G study. The baserate study demonstrated that they lead to satisfactory data.

## C. *Pupil Achievement as a Measure of Teacher Effectiveness*

Belgard, Rosenshine, and Gage (1972) collected data on the effectiveness of teachers in presenting informational lessons by the following set of operations:

1. A textual lesson on Yugoslavia, and 10 pertinent test items, were prepared.
2. The teacher was given the text and five of the test items to define what he was to teach.
3. He presented the lesson to his own regular class as a lecture (under certain constraints).
4. The 10-item test was given to the pupils. The average score earned by his pupils was taken as the teacher's unadjusted score.

This procedure was repeated for a lesson on Thailand. For purposes of this discussion we shall ignore the difference between the five exposed items and the five secure items. We shall also ignore an adjustment of scores made to allow for variation in pupil ability.

### *Design and basic data*

The G study collected data on 43 classes, each of which received both lessons. The teachers $t$ are the objects of investigation; the facets are pupils $p$, lessons $j$, and items $i$. The design is lessons crossed with teachers and pupils, pupils nested within teachers, items nested within lessons and crossed with teachers and pupils $[(i:j) \times (p:t)]$.

Here we encounter a facet nested *in the universe*. The structure of the universe is $(i:j) \times t \times p$. While in principle, any teacher might be assigned any pupil and any lesson, the items accompany one and only one lesson. The model underlying the observed score takes the following form:

$$(7.7) \quad X_{tpji} = \mu + (\mu_t - \mu) + (\mu_p - \mu) + (\mu_j - \mu)$$
$$+ (\mu_{tp} - \mu_t - \mu_p + \mu) + (\mu_{tj} - \mu_t - \mu_j + \mu)$$
$$+ (\mu_{pj} - \mu_p - \mu_j + \mu) + (\mu_{ji} - \mu_j) + (\mu_{tji} - \mu_{tj}$$
$$- \mu_{ji} + \mu_j)$$
$$+ (\mu_{pji} - \mu_{pj} - \mu_{ji} + \mu_j) + (\mu_{tpj} - \mu_{tp} - \mu_{tj} - \mu_{pj}$$
$$+ \mu_t + \mu_p + \mu_j - \mu)$$
$$+ (\mu_{tpji} - \mu_{tji} - \mu_{pji} - \mu_{tpj} + \mu_{tj} + \mu_{pj} + \mu_{ji} - \mu_j)$$
$$+ e_{tpji}$$

The usual $\mu_i$, $\mu_{ti}$, $\mu_{pi}$, and $\mu_{tpi}$ are undefined. Because one can average over $i$ only within lessons $j$, any $i$ effect is part of such a score component as $ji$,

**TABLE 7.4.** *Summary of G Studies Pertinent to Teacher's Score on a One-Lesson D Study with About 19 Pupils*

| Analysis | Components allocated to universe score | Sources allocated to the error $\delta$ | Estimate[a] of variance for Universe score | Estimate[a] of variance for Error | Coefficient of generalizability |
|---|---|---|---|---|---|
| One-facet analyses | | | | | |
| 1. Corrected correlation of half-test scores for each lesson | $t; P_t, tP_t; tj^*, P_{tj}^*, tP_{tj}^*$ | Interactions of $t$ and $p$ with items within the lesson | Y 0.579<br>T 0.495 | 0.236<br>0.156 | 0.71<br>0.76 |
| 2. Correlation between lessons | $t; P_t, tP_t$ | Interactions of $t$ and $p$ with lessons and items | Y 0.514<br>T 0.410 | 0.301<br>0.241 | 0.63 |
| 3. One-way anova for each lesson, pupil scores nested within the teacher | $t; tj^*, tj^*I_j^*$ | Pupils; pupil-teacher interaction within the lesson | Y 0.625<br>T 0.503 | 0.199<br>0.174 | 0.76<br>0.74 |
| Three-facet analysis | | | | | |
| 4. Anova over teachers, pupils, lessons, and items | $t$ | Pupils; all interactions except that for items with lessons | 0.380 | 0.395 | 0.49 |

[a] Where two entries appear, one applies to the lesson on Yugoslavia and one to the lesson on Thailand. All variances are expressed in terms of the score scale of the test (0–10 scale). Scores were added over items, not averaged; but scores were averaged over pupils.

*tji*, etc. (For convenience we shall refer to the item-within-lesson effect as a *ji* interaction. The other "interactions" also are unconventional in form.)

### Summary of results

The original investigators analyzed the data by several conventional techniques before multifacet analysis was considered. The analyses slice through the data in many ways; there are confusing shifts in the definition of the universe score and in the composition of the "error" variance that are unavoidable in one-facet studies. We therefore start with a summary table. For purposes of this table all results are stated in terms of a score based on just 1 lesson, where the 10 items are *summed* to provide a pupil score and the teacher's score is an average over pupils. The number of pupils varies from 10 to 31, with a mean of 21. Throughout these studies, interest centers on differences among teachers. Because all teachers teach the same lessons, components *j* and *ji* can be ignored. In effect, we discuss the teacher's deviation from the mean of the group of teachers who taught the same lesson and gave the same test.

In the first and second one-facet procedures, two scores were obtained for each teacher, and two observed-score variances and the inter-correlation were calculated. The estimate of the universe-score variance is the product of the observed-score variance and the correlation; the remainder of the observed-score variance constitutes the "error."

Lesson *j* is a hidden facet in analysis 1, treated as fixed in the universe. In analysis 2, lessons vary; pupils are in effect fixed, though the fixed set of pupils $P_t$ varies with the teacher. Items are fixed within lessons but there is nevertheless generalization over items, as the number of items in the universe of generalization (all lessons) is large. In analysis 3, pupils are treated as the source of error; lesson and items within the lesson are fixed. The one-facet procedures generate separate estimates of variances for each lesson. The multifacet study reports one estimate, applying to the expected variance of teacher's mean scores (any lesson, any set of pupils).

The most obvious fact in Table 7.4 is the variation in rationale and results from one procedure to the next. One-facet analyses produce coefficients ranging from 0.63 to 0.76. The multifacet study, which attends to all types of error simultaneously, reports a coefficient of only 0.49 (cf. p. 181). The universe-score variance changes similarly. When the universe score is defined as the teacher score averaged over all lessons and all pupils, its variance is estimated as 0.38. In analyses 1–3, where averaging is over only pupils or lessons or item-sets, the estimate is larger. Each such average leaves a facet fixed, and interaction components and the main effect for the fixed facet are then counted in the universe score. If the investigator wants to generalize over lessons *and* items within lessons *and* pupils, so that only the component for $t$ enters the universe score, the only usable one-facet correlational procedure is to correlate a score on one lesson for one group of pupils with the score on another lesson in a second group of pupils. This confounds lessons and pupils as sources of variance, but both appear within the error and reduce the correlation.

### Analysis 1: Split-half study of generalizability over items

Analysis 1 is a conventional split-half study. This type of analysis investigates generalizability over items, assuming pupils and the lesson fixed.

For each five-item half-test, a class mean was calculated for each teacher. The split-half correlation, after correction by the Spearman–Brown formula, was 0.71 for the lesson on Yugoslavia and 0.76 for Thailand. Such a coefficient indicates the extent to which administering different item sets alters the comparative standings of the teachers. Any moment-to-moment inconsistency of performance (e.g., pupil's careless reading of any item) also reduces the coefficient. Stable pupil effects are confounded with teacher effects, and contribute to the estimated universe-score variance.

At the risk of notational indigestion, we shall try to identify components precisely by letting $P_t$ represent the particular set of pupils associated with teacher $t$ (i.e., the pupil sample for the teacher), and letting $I_j$ represent the item-set employed for lesson $j$. The universe score implied by this split-half analysis is specific to the teacher, the set of pupils for this teacher, and the given lesson; it is an average over the universe of items for that lesson. It could be denoted $\mu_{tP_{t}j*}$. There is no asterisk on $P$ because in this analysis a different set of pupils defines the universe of generalization for each $t$. The very fact that our symbol must be so complicated highlights the inadequacy of the classical theory, which looks on the corrected split-half correlation as the squared correlation of observed score with "the" true score of the teacher, which is denoted by some simple expression such as $\mu_t$.

It is instructive to ask what the split-half study indicates about variance components. In the basic model for decomposing $X_{tpji}$ (7.10) there are three main effects, four first-order interactions, three second-order interactions, and one final $tpji,e$ component. The correlation is determined from deviation scores. The $j$ and $ji$ components drop out of consideration, because they appear in the mean $X_{TPjI}$ and therefore do not enter the deviation score. The "error" variance determined by the split-half analysis embraces all the remaining components that involve $i$. The analysis allocates the components as follows:

|  | To universe score | To error |
|---|---|---|
| Main effects | $t, P_t$ | |
| First-order interactions | $tP_t, tj*, P_tj*$ | |
| Second-order interactions | $tP_tj*$ | $tj*I_{j*}, P_tj*I_{j*}$ |
| Residual | | $tP_tj*I_{j*}, e$ |

The universe-score variance (six components combined) is estimated by multiplying the observed-score variance (0.815 for Yugoslavia) by the coefficient (0.71). Therefore, the estimated universe-score variance is 0.579. The remainder is the "error" variance (0.236). An alternative estimate is made from the Thailand data (see Table 7.4).

### Analysis 2: Correlation of one lesson with another

Analysis 2 obtained an interclass correlation of 0.63 between the Thailand and Yugoslavia scores. This analysis compares two scores for a teacher-class combination; it gives information on how closely the score of the class on one lesson is likely to agree with the universe score that could be determined from very many lessons. In generalizing over lessons, one inevitably generalizes over items also. The use of deviation scores in computing the correlation eliminates the $j$ and $ji$ components from consideration.

The universe score implied by this analysis is $\mu_{tP_t}$, which embraces components for $t$, $P_t$, and $tP_t$. The universe-score variance is smaller than that from the split-half study, as it embraces fewer components. For Yugoslavia, the estimated universe-score variance in the split-half study is 0.579 and in this analysis is 0.514. The difference of 0.065 is a rough indication of the combined magnitude of the three $j*$ interactions that contributed to the universe-score variance in analysis 1.

Analysis 2 provides only one coefficient (see Table 7.4), not a separate coefficient for each lesson. The coefficient is lower than in analysis 1 because generalization is to a broader (and more significant) universe. The split-half study, treating the lesson as fixed, did not examine how well the observed score represents the teacher's general power to teach all lessons of this type. But analysis 2, like analysis 1, ignores pupil characteristics as a source of error in evaluating the teacher.

### Analysis 3: One-way anova over pupils

Analysis 3 produced an intraclass correlation for teacher scores with pupils as the variable facet. The coefficient is like the one Horst (1949) recommended for "reliability over raters," except that here the correlation was computed from unbiased variance estimates. Since the analysis was made on each lesson separately, the universe score has the form $\mu_{t_j*I*}$.

A one-way analysis of variance (pupils within teachers, Design II) was carried out for the lesson. This is not entirely a routine matter, because the number of pupils per class varies. Two alternatives are available for coping with this (apart from the device of discarding cases to make the number of pupils the same for every teacher). The procedure suggested by Horst is to estimate the within-teacher variance for each teacher separately, and average over teachers. The alternative is to estimate the within-teacher variance from the mean square within teachers (i.e., data for all teachers pooled).

The former procedure weights teachers or classes equally, the latter weights individual pupils equally. If it may be assumed that variance within the teacher is uniform in the population, or that these variances—while not uniform—are uncorrelated with class size, then the calculation from the pooled mean square is to be preferred. (However, the two results are unlikely to differ appreciably.)

The one-facet analysis of variance gives a mean square for teachers, but because class sizes vary, a special treatment is needed to estimate the universe-score variance. Let us define symbols as follows:

$n_t$    number of teachers in study

$n_p \mid t$    number of pupils in class of teacher $t$

$n$    number of pupils in study $\left( = \sum_t n_p \mid t \right)$

$\sigma^2(a)$    sum of within-teacher components of variance, weighted as in G study

$\sigma^2(b)$    universe-score variance

(7.8)
$$k_0 = \frac{n^2 - \sum_t (n_p^2 \mid t)}{n(n_t - 1)} .$$

The expected mean square takes the form (Graybill, 1961, p. 353):

(7.9)    $\text{EMS } p = \sigma^2(a) + k_0\sigma^2(b)$

The results are given in Table 7.4 and need not be repeated here.

The within-teacher mean square is an unbiased estimate of the variance arising from differences among pupils and from inconsistency in the pupil's performance (e.g., inattention). The estimate of this variance embraces components $p$, $pj^*$, $pj^*I_{j*}^*$, $tp$, $tpj^*$, and $tpj^*I_{j*}^*$, $e$.

The fact that the estimate of this error variance is considerably lower than the error variance in analysis 2 is suggestive. Because shifting the components involving $tj^*$ from error to universe score has reduced the error variance by about one-third, teacher-lesson or teacher-lesson-item interaction must be substantial. Comparisons from one single-facet study to another can shed light on the magnitude of variance components. It is difficult to disentangle the effects, however, and a multifacet study is more directly informative.

The intraclass correlation estimated from the variances counts the pupil main effect as error. In analyses 1 and 2, where generalization was over items and lessons, respectively, the main effects for items and lessons were disregarded. In analysis 3, where generalization is over pupils, the pupil main effect contributes to the observed-score variance and the error variance. The intraclass correlation recognizes that each teacher is measured by different pupils; pupil main effects add to observed differences among teachers.

## Analysis 4: The multifacet study

A G study along the lines suggested in this monograph proceeded from the scores of individual pupils on individual items. To avoid complications in data processing, the scores were reduced to a "box" design by discarding all classes with fewer than 19 pupils, and discarding at random within the remaining classes to reduce the number to 19. This means that the data are not identical to those used in the one-facet studies (analyses 1–3).

In Chapter 2, a two-facet analysis from this study was displayed in Table 2.4, which was not interpreted. To avoid substantive considerations irrelevant at that point, the design was described as $i \times (j{:}p)$. In our present notation, the design for that three-way analysis is $j \times (p{:}t)$. Items were ignored, and analysis proceeded from the test score for each lesson. Moreover, the test scores used had been standardized within lessons. We shall not discuss the three-way analysis; the results are not inconsistent with the four-way analysis.

*Estimation and interpretation of components.* The four-way design is diagrammed in Figure 7.4. The analysis of variance was performed as if the data had been collected under a crossed design, and sums of squares and degrees of freedom were then combined to reflect the nesting of pupils within teachers and of items within lessons (Table 7.5). We have given the equations for expected mean squares to enable the reader to trace the complex calculations, and, in Figure 7.5, the diagrams corresponding to the equations.

Table 7.5 separates many of the components that previously were confounded. The components are stated on the scale of the scores that entered the analysis (i.e., for a single pupil on a single item). Elsewhere, results have been reported for a sum of 10 items; to shift to that scale, components of variance will be multiplied by 100 in Table 7.6.

The size of the components indicates something about the structure of behavior and about the sources of error to be brought under control. As

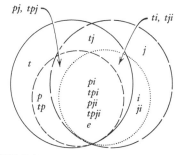

**FIGURE 7.4.** *Schematic Representation of the* $(i{:}j) \times (p{:}t)$ *Design.*

TABLE 7.5. Estimates of Variance Components from a G Study with the Design $(i{:}j) \times (p{:}t)$

| Source of variance | Analysis of variance as if crossed | | Mean square from analysis as nested | Expected mean square | Estimate of variance component |
|---|---|---|---|---|---|
| | Sum of squares | Degrees of freedom | | | |
| $t$ | 65.46 | 28 | 2.338 | $\sigma^2 + n_p\sigma^2(ti,tji) + n_i\sigma^2(pj,tpj) + n_jn_i\sigma^2(p,tp) + n_pn_i\sigma^2(tj)$ $+ n_pn_jn_i\sigma^2(t)$ | 0.0038 |
| within $t$ | | | | | |
| $p$ | 11.32 | 18⎫ | 0.522 | $\sigma^2 + n_i\sigma^2(pj,tpj) + n_jn_i\sigma^2(p,tp)$ | 0.0148 |
| $tp$ | 261.19 | 504⎭ | | | |
| $j$ | 10.74 | 1 | 10.740 | $\sigma^2 + n_p\sigma^2(ti,tji) + n_i\sigma^2(pj,tpj) + n_tn_p\sigma^2(i,ji)$ $+ n_pn_i\sigma^2(tj) + n_tn_pn_i\sigma^2(j)$ | 0.0002 |
| within $j$ | | | | | |
| $i$ | 93.94 | 9⎫ | 9.365 | $\sigma^2 + n_p\sigma^2(ti,tji) + n_tn_p\sigma^2(i,ji)$ | 0.0163 |
| $ji$ | 74.63 | 9⎭ | | | |
| $tj$ | 17.02 | 28 | 0.608 | $\sigma^2 + n_p\sigma^2(ti,tji) + n_i\sigma^2(pj,tpj) + n_pn_i\sigma^2(tj)$ | 0.0009 |
| within $tj$ | | | | | |
| $pj$ | 6.13 | 18⎫ | 0.224 | $\sigma^2 + n_i\sigma^2(pj,tpj)$ | 0.0047 |
| $tpj$ | 111.01 | 504⎭ | | | |
| $ti$ | 108.01 | 252⎫ | 0.383 | $\sigma^2 + n_p\sigma^2(ti,tji)$ | 0.0108 |
| $tji$ | 85.14 | 252⎭ | | | |
| $pi$ | 29.31 | 162⎫ | | | |
| $pji$ | 32.85 | 162⎭ | 0.1778 | $\sigma^2$ | 0.1778 |
| $tpi$ | 807.42 | 4536⎫ | | | |
| $tpji, e$ | 801.43 | 4536⎭ | | | |

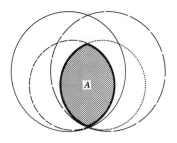

(a)  The *p,i* intersection

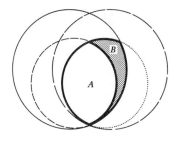

(b)  The *t,i* intersection

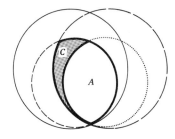

(c)  The *p,j* intersection

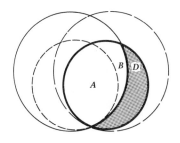

(d)  The *i* circle

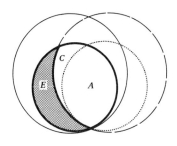

(e)  The *p* circle

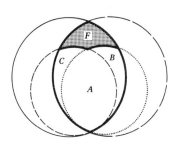

(f)  The *t,j* intersection

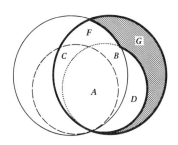

(g)  The *j* circle

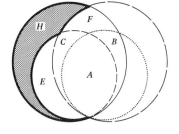

(h)  The *t* circle

FIGURE 7.5.  *Schematic Analysis of the (i:j) × (p:t) Design.*

usual,[2] the largest component is the residual. This is not surprising; whether a pupil will get a particular item correct is difficult to predict from his ability, the item difficulty, and other basic effects. Vagaries of attention, lapses of memory, and chance in guessing all contribute to this residual. As such a component is sure to be sampled many times in any design, its magnitude is not of great concern.

Next largest is the within-lesson component. This component has to do with item difficulty, and can be reduced by using a large number of items. Since the component does not affect differences among teachers (where items are crossed with teachers), it is of no concern in the Belgard study. The same may be said about the component for lessons; as long as lessons are crossed with teachers, they do not affect teacher differences. The tiny size of this component implies that the investigators developed lessons and tests of similar difficulty. This tends to warrant using one lesson as pretest and one as posttest in some experiment. A counterbalance of order would not be critical. The $tj$ component, also small, indicates the extent to which teacher scores depend on the lesson they are teaching. Since the $tj$ variance is small relative to the $t$ component, ability to present lessons of this type appears to be general over topics. This finding is welcome, because a large component would force the investigator to explain how "ability to teach about Yugoslavia" differs from "ability to teach about Thailand," etc.

The third largest component is that within $t$, which arises from pupil variability. Since in operational use of the procedure, different teachers will teach different pupils, this source of error is one to be held under control. This can be accomplished by random assignment, by using many pupils per teacher, and/or by partialling out pupil differences with some covariate such as an ability test.

The value of 0.0108 for $i$ within $tj$ suggests the presence of a teacher–item interaction. This need not cause concern; it is obvious that teachers will emphasize different subtopics, and each will present some more clearly than others. Employing many items will reduce the effect of this unwanted component in the teacher's score.

Finally, the $p$-within-$tj$ component suggests a pupil-lesson interaction of modest size. Most likely, this represents variation in attentiveness from day to day. Its effect on the teacher's score will be reduced by using more pupils or more lessons.

***Generalizability in the D study.*** To appraise any one design, it is necessary to divide the components by the number of observations made on each.

---

[2] Usual, that is, except where one facet is observers. In such studies, momentary variations in subject behavior appear in some component higher in the table, because the several observers concur in reporting that fluctuation.

**TABLE 7.6.**   *Composition of Variance in a D Study with Design* $(i{:}j) \times (p{:}t)$

| Source of variance | Estimate of variance component[a] | Frequency | One-lesson study[b] Assumed frequency | Contribution to $\mathscr{E}\sigma^2(\delta)$[c] | Two-lesson study Assumed frequency | Contribution to $\mathscr{E}\sigma^2(\delta)$[c] |
|---|---|---|---|---|---|---|
| $t$ | 0.38 | 1 | | | | |
| $tj$ | 0.09 | $n'_j$ | 1 | 0.090 | 2 | 0.045 |
| $p$ within $t$ | 1.48 | $n'_p$ | 19 | 0.078 | 19 | 0.078 |
| $p$ within $tj$ | 0.47 | $n'_p n'_j$ | 19 | 0.025 | 38 | 0.012 |
| $i$ within $tj$ | 1.08 | $n'_i n'_j$ | 10 | 0.108 | 20 | 0.054 |
| $i$ within $ptj$ | 17.78 | $n'_p n'_i n'_j$ | 190 | 0.094 | 380 | 0.047 |
| $\widehat{\mathscr{E}\sigma^2(\delta)}$ | | | | 0.395 | | 0.237 |
| $\widehat{\mathscr{E}\sigma^2(X)}$ | | | | 0.775 | | 0.617 |

[a] Estimated from Table 7.5, rescaled to test-score scale.
[b] Result used in Table 7.4.
[c] Expected observed-score variance equals $\mathscr{E}\sigma^2(\delta)$ plus universe-score variance given by $t$ component.

This is done in Table 7.6. We have already mentioned the rescaling employed to make these numbers comparable to those from the one-facet analyses. Because only comparisons between teachers are intended, the components for $j$ and $i$ within $j$ are not carried forward. In this crossed design, components lying outside the $t$ circle of Figure 7.4 are the ones missing from the observed-score variance.

Table 7.6 assumes that the D study will employ a design just like that of the G study, except for possible variation in the number of pupils, items, or lessons. The assumption is that the number of pupils will be the same for all teachers. If this assumption is grossly violated in the D study, one can compute the error variance for each possible $n'_p \mid t$ and use the distribution of $n'_p$ to compute a weighted average error variance.

For comparability to the one-facet studies, Table 7.6 gives results for a one-lesson D study. We observed that in analysis 1, the error variance is the sum of $t$ and $p$ interactions with items (within a lesson). Table 7.6 shows this sum to be $0.108 + 0.094 = 0.202$; this compares well with the values of 0.236 and 0.163 calculated for the lessons separately in Table 7.4.

The two-lesson error variance of 0.237 (Table 7.6) implies a standard error $\hat{\sigma}(\delta)$ of 0.49 on the 10-point scale. From this, the investigator decides whether the two-lesson design is adequate to measure differences in teacher effectiveness. Since the standard error of a difference between teachers will be $2^{1/2}(0.49)$, the measurement is moderately satisfactory. An observed difference between two teachers of one point on the ten-point scale will be inconsistent with the direction of the universe-score difference in about one case out of six. The universe-score variance implies a standard deviation for universe scores of 0.62; evidently, teacher scores are confined to a range of about three scale points. For the two-lesson study, the ratio of universe-score variance to observed-score variance is $0.38/(0.38 + 0.24) = 0.62$. This coefficient of generalizability is large enough to justify using the procedure in studies that correlate teacher differences with other variables. However, the procedure is probably not adequate where a measure of the comparative standing of the individual teacher is desired.

*Alternative D-study designs.* A multifacet study does not end with the report of a coefficient. An important use of the results is to evaluate possible designs for a D study. One may consider not only changes in $n'_p$, $n'_j$, etc., but also a relaxation or tightening of crossing and nesting. Some of this can be done by inspection. For instance, it would be inexpensive to increase the number of test items. Would this have a worthwhile effect? In Table 7.6, $n'_i$ affects two fairly large components. It looks as if doubling $n'_i$ would reduce the error variance for the two-lesson study by about 0.05—a worthwhile gain. (But not as large as routine application of the Spearman–Brown formula would have suggested. That formula assumes that doubling test length cuts the entire error variance in half.) Extending test length would have diminishing returns; even an indefinitely long test would not reduce the error variance below 0.135, at which point the coefficient of generalizability would be 0.38/0.515, or 0.74.

Another alternative would be to have each teacher teach a lesson to two classes. This would not require extra preparation, and might be more practicable than increasing $n'_j$, the number of lessons. We can only assume that teaching a second class is the same as teaching a class twice as large, as we have no data to assess the teacher–presentation interaction (with lesson fixed). We do know that the teacher–lesson effect in Table 7.6 is small; because the lessons were given on different occasions, this encourages us to think that occasion-to-occasion variation in a teacher's effectiveness is small. Consider the effect of doubling $n'_p$; this chops three components in half, reducing the error variance from 0.24 to about 0.17 and raising the coefficient to 0.69. The hypothetical limit, from an infinite number of presentations, is 0.10 for the error variance and 0.79 for the variance ratio. Generalizability improves indefinitely as one increases both the number of lessons and the number of pupils, but this enters the realm of fantasy.

More interesting is the question of the "exchange rate" between lessons and pupils. One may specify a certain number of lessons, three for instance, and ask what total number of pupils per teacher-lesson, if any, will bring the coefficient to, arbitrarily, 0.75. While this problem takes a different form in every study, and therefore no algorithm can be offered, a detailed example may assist the reader to formulate such problems.

Let $r$ equal the desired coefficient, and write $y$ for the corresponding error variance, equal to $[(1 - r)/r]\sigma^2(t)$. Label the five components of the error $a$, $b$, $c$, $d$, and $e$, in order.

$$(7.10) \qquad y = \frac{a}{n'_j} + \frac{b}{n'_p} + \frac{c}{n'_p n'_j} + \frac{d}{n'_i n'_j} + \frac{e}{n'_p n'_i n'_j}$$

Assume for the sake of argument that $n'_i$ is fixed and $n'_j$ is temporarily fixed. Solving for $n'_p$,

$$(7.11) \qquad n'_p = \frac{n'_i n'_j b + n'_i c + e}{n'_i n'_j y - n'_i a - d}$$

For our example, $n'_i = 10$. The desired $r$ is 0.75. Suppose a trial $n'_j$ is set at 3. Since $\widehat{\sigma^2}(t)$ is 0.38, $y = 0.127$.

$$n'_p = \frac{30(1.48) + 10(0.47) + 17.80}{30(0.127) - 10(0.09) - 1.08} = 37$$

If $n'_j$ were 10, to construct a dramatic contrast, the required $n'_p$ is 16. The investigator could get roughly the same precision by using 3 lessons with classes of size 37 or 10 lessons with classes of size 16.

By simply setting the denominator of (7.11) equal to zero and entering $n'_i$, one can determine the lowest value of $n'_j$ that allows the coefficient to reach the desired $r$. Within the given $y$, the minimum $n'_j$ equals 1.6, which means that there is no increase in the number of pupils alone capable of raising the coefficient to 0.75 when $n'_j = 1$ and $n'_i = 10$.

To illustrate one further improvement in design, consider joint sampling of pupils and lessons. Let the teacher teach two classes, one lesson to each class. This, without increasing the total number of manhours of work, doubles the number of observations on $p$ within $t$ (because in *this* design there are $n'_p n'_j$ such observations). The error is reduced by 0.04.

### D. *Item-Sampling Studies of Test and Item Means*

A striking innovation in recent educational testing has been the "item sampling" design (Lord & Novick, 1968, Chapter 11; Sirotnik, 1970). Different test items, selected at random, are given to different pupils or groups of pupils. The design was originally suggested by Turnbull and by Ebel to meet a practical problem in the collection of test norms. A testing program often requires new forms of its test each year, and these forms have to be more or less parallel. In testing for college admission, for example, the new test must be kept secure until the date of use, yet statistical information on items is needed to assemble the new form. Suppose that the pool of tryout items is divided into small subsets, and that these subsets are introduced into the regular forms of the test of the current year. Each pupil would receive only one of the subsets in his test booklet. By this device every item is applied to a representative fraction of the national sample. The resulting data about the experimental items permit selection of items that will give next year's test the desired statistical properties. Since no one has seen much of the new test, it is adequately secure.

Item sampling is sensible in evaluation studies, experiments, and surveys where interest attaches to performance on a domain rather than on a fixed set of items. Because many more items can be administered than in the items × pupils design, the behavioral domain of interest is more adequately sampled and the universe-and-population mean and variance are better estimated. Another advantage appears in questionnaire surveys conducted by mail, where a short list of questions will bring more returns than a long one.

Generalizability theory provides a useful approach to the evaluation of item-sampling designs, because the components of variance indicate just how much is to be gained or lost by changes in the proportion of pupils who receive each item and by other variations in design. To illustrate our analytic methods, we employ data collected in the *National Longitudinal Study of Mathematical Abilities* (NLSMA), made available through the courtesy of E. G. Begle, project director, and Leonard Cahen and Walter Zwirner.[3] The general concern of NLSMA has been to compare groups who studied certain texts with respect to various mathematical accomplishments. The inquiry estimates the mean score of a population of pupils on a universe of items. The comparative study calls for an estimate for each textbook group as a separate population. We shall discuss only a single such group.

Bock and Wiley (1967) investigate how many schools and how many pupils within a school should be used to estimate a mean efficiently; but they use a fixed test form and so deal with just a part of our problem. Lord and Novick restrict themselves to one-facet G studies and use specialized mathematics applicable only to items with 1–0 scoring. For such items, they offer procedures for estimating moments of the universe-score distribution in the population. They assume random distribution of items over subjects, and they introduce corrections for sampling from a finite universe. We ignore these corrections but we treat pupils as sampled within schools and so take school differences into account.

---

[3] For another report on NLSMA item sampling that investigates other aspects of the procedure, see Cahen, Romberg, and Zwirner (1970).

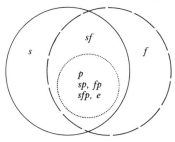

**FIGURE 7.6.** *Schematic Representation of the p:(s × f) Design.*

## A study with Design IV-A

Data are available for ninth graders who took a certain test in the item-sampling manner. The test had been divided into ten 5-item "forms" by random allocation of items. Within a class, the pupils to receive each form were determined at random. Data were available from many classes, large and small. Because class size varied, sometimes as few as four pupils in a school received a certain group of items. Wherever more than four pupils in a school had taken any form, we reduced the number to four by random selection.

The design of the study is type IV-A, $p:(s \times f)$; every form is given in every school, and pupils are nested within $sf$ cells ($n_s = 29, n_f = 10, n_p = 4$; Figure 7.6). It is assumed that schools are a random sample from those using the text, that the forms are random samples from a pool of admissible items, and that pupils are random within schools. Table 7.7 gives the analysis of variance components. The score on a five-item set is totalled, allowing a 0–5 range of scores; components are expressed on that scale.

Since items are randomly assigned to forms, it would be possible to analyze scores at the item level. For any one form, the design is $i \times (p:s)$ so that components of variance for $pi$, $spi$ and $p$, $ps$ can be isolated. These components, which are now confounded in $p:sf$, would be important in

**TABLE 7.7.** *Estimates of Variance Components from a G Study with the Design $p:(s \times f)$.*

| Source of variance | Sum of squares | Degrees of freedom | Mean square | Estimate of variance component |
|---|---|---|---|---|
| Schools $s$ | 225.61721 | 28 | 8.05776 | 0.165 |
| Forms $f$ | 264.21030 | 9 | 29.35670 | 0.241 |
| $sf$ | 363.58319 | 252 | 1.44279 | 0.018 |
| $p:sf$ | 1194.00000 | 870 | 1.37241 | 1.372 |

**TABLE 7.8.** *Variance of Error in Estimating a Population Mean, with and without Item Sampling*

| Source of variance | Estimate of variance component | Design IV-A; $p:(s \times f)$, 10 forms, 30 schools, 4 pupils/form/school | | Design IV-A; $p:(s \times f)$, 10 forms, 10 schools, 12 pupils/form/school | | Design IV-A; $p:(s \times f)$, 10 forms, 60 schools, 2 pupils/form/school | | Design V; $(p:s) \times f$, 1 form, 30 schools, 40 pupils/form/school | |
|---|---|---|---|---|---|---|---|---|---|
| | | Frequency | Contribution to $\sigma^2(\Delta)$ | Frequency | Contribution to $\sigma^2(\Delta)$ | Frequency | Contribution to $\sigma^2(\Delta)$ | Frequency | Contribution to $\sigma^2(\Delta)$ |
| Schools $s$ | 0.165 | 30 | 0.006 | 10 | 0.017 | 60 | 0.003 | 30 | 0.006 |
| Forms $f$ | 0.241 | 10 | 0.024 | 10 | 0.024 | 10 | 0.024 | 1 | 0.241 |
| $sf$ | 0.018 | 300 | 0.000 | 100 | 0.000 | 600 | 0.000 | 30 | 0.000 |
| Residual | 1.372 | 1200 | 0.001 | 1200 | 0.001 | 1200 | 0.001 | 1200 | 0.001 |
| $\widehat{\sigma^2}(\Delta)$ | | | 0.031 | | 0.042 | | 0.028 | | 0.248 |
| $\hat{\sigma}(\Delta)$ | | | 0.18 | | 0.20 | | 0.17 | | 0.50 |

determining the generalizability of scores for individual pupils, but that is not the concern of this G study.

A study of this character calls for a radical change in our mode of interpretation, for we are not interested in the universe score for the pupil, nor in the universe score for the school. The D study is intended to estimate the mean over the population of schools and the universe of items.

We ignore school size. As we state the problem, each school is assigned equal weight in the mean. In effect, we estimate the error variance in a mean based on schools of the same size, assuming score components independent of school size. The argument becomes much more complicated if we deny that assumption or allow schools to enter the universe mean with weights proportional to enrollments.

The problem, then, is to evaluate the likely magnitude of the error $\Delta$ under any proposed D-study design, where $X_{SP_sF}$ is the observed sample mean, $\mu$ is the mean over the population and universe, and $\Delta_{SP_sF}$ is the difference. An argument could be made for examining an error $\Delta_{SP_sF} - (\mu_{..F} - \mu)$, because the form component is irrelevant to the comparison of textbooks if the same forms are applied to all textbook groups. But interactions between test items and textbooks are quite likely. In our analysis, the textbook is a hidden fixed facet and the $F$ component includes any effects arising from item–text interaction. If $\sigma^2(\Delta)$ is used to judge a design where textbooks are compared on the same forms, all variation in item difficulty, in effect, is being treated as if it were a consequence of the textbook interaction. Calculations on the opposite assumption that the interaction is zero appear in the exercises.

Table 7.8 permits us to examine the errors of estimate, first under the assumption that the D study has the same design as the G study (except that $n'_s = 30$). For illustrative purposes we consider two further alternative designs: giving the same 10 items to all subjects, and decreasing the number of schools while taking more pupils per school. There are many alternatives, but these will give a sufficient sense of the way variance components may be used. In each of the designs of Table 7.8, 1200 pupils each take 5 items.

While the error variance for the first design is quite small numerically, a mean error square of $0.031^{1/2} = 0.18$ would rarely be acceptable in estimating a population mean on a 0–5 scale, for a survey of this kind. The other three designs do not give appreciably better results. As the table shows, using 10 schools gives a much poorer result than using 30; but a 60-school sample is not much better. If we are to evaluate the result on the basis of $\sigma(\Delta)$, it is clearly necessary to reduce $\sigma^2(F)$ by using more forms. If forty 5-item forms were used in a 30-school Design IV-A study, one form per pupil per school, $\hat{\sigma}(\Delta)$ would drop from 0.18 to 0.11. (Another line of extension is to consider making the test forms longer or shorter. But one cannot estimate $\sigma(\Delta)$ for the longer or shorter form unless Design V is used in the G study, for reasons that will become apparent.)

Reasoning such as this, applied to a pilot study, can be of considerable value in designing an efficient experiment or a study to estimate a population mean. Applied to a D study, it indicates the precision of the estimate obtained. The error of generalization over forms is considered in addition to the error from sampling of persons that usually enters the standard error of a mean.

### A study with Design V

It is not essential, and perhaps not desirable, to use an item-sampling design in the G study when one is evaluating item-sampling plans. Design V, $i \times (p:s)$, can estimate the components of interest. As an example, we employ data for a 15-item test administered to fifth grades in 19 schools. In any school where more than 18 pupils completed the test, we have reduced the number to 18 by random sampling. Item scores are analyzed, and the score scale has a 0–1 range. We need not give details of the estimation of components of variance; the results are as follows:

| | |
|---|---|
| Schools | 0.004 |
| Items | 0.019 |
| Schools × items | 0.002 |
| Within schools: | |
| $p, sp$ | 0.023 |
| $pi, spi, e$ | 0.117 |

The item-sampling D study with Design IV-A will confound the two within-school components shown above. Because components are assumed to be independent, the variance of a compound is simply the sum of the variances of the parts going into it. If each "form" in the item-sampling design consists of a single item, we estimate the within-$sf$ component of the item-sampling design as 0.140 ($= 0.023 + 0.117$). If a form is to be made up of five items, the $pi$, $spi$, $e$ component is sampled five times in the pupil's score and the $p$, $sp$ component only once. Using the average score over the five items, to keep the scale of components the same, the estimate of the within-$sf$ component is 0.023 + 0.023 or 0.046. The $f$ and $sf$ components would be reduced, respectively, to 0.004 and 0.0004 when five items are averaged. The effect of a change in $n'_i$ could not be determined in the Design IV-A G study above, where the $pi$ and $p$ effects were confounded. The reader can trace out how well alternative designs estimate a population mean, in the manner of the foregoing section.

In a D study that is chiefly concerned with understanding educational results, one might estimate an *item* mean $\mu$ for a fixed item. The observed mean is $X_{SPi}$, and $\Delta_{SPi}$ is the weighted sum of $\mu_s\sim$, $\mu_{si}\sim$, $\mu_{sp}\sim$, and $\mu_{spi}\sim$, $e$ components. (The $i$ component is not a source of error when the item is fixed.) One divides the four corresponding components of variance by the number of observations to be made *within* the item for that component. Suppose that 500 pupils, 50 in each of 10 schools, will be given the item. Then, from the estimates of variance components given above, we have:

| Component | Frequency | Contribution to $\sigma^2(\Delta)$ |
|-----------|-----------|-----------|
| $s$ | 10 | 0.0004 |
| $si$ | 10 | 0.0002 |
| $p, sp$ | 500 | 0.0001 |
| $pi, spi, e$ | 500 | 0.0002 |

Hence, $\widehat{\sigma^2(\Delta)} = 0.0009$. The value of $\hat{\sigma}(\Delta)$, 0.03, is probably adequately low for evaluative conclusions.

### E.  A Stratified Achievement Test

We have spoken of the possibility of regarding an achievement test as a sample from a stratified universe in which items are classified *a priori* according to content or task. "Stratified-parallel" tests formed by sampling within strata agree more closely than the analysis based on random sampling indicates.

#### Description of data

The NLSMA study provides illustrative data from a test having three sections, each containing six items. The test was given to 18 pupils in each of 24 schools. One would generalize over the universe of tests formed by repeatedly sampling 18 items, 6 from each stratum. We investigate generalizability of both the pupil score and the school mean. The design of the G study is (items:strata) × (pupils:schools)—$(i{:}j) \times (p{:}s)$, similar to that in analysis 4 of the Belgard study (see Table 7.5) except that $J^*$ is fixed. For each $j \in J^*$, $i$ is nested within $j$ in the universe. The components involving $i$ are defined differently from those in (2.19) and (2.20). The $i$ component, for example, has to be $\mu_{ij} - \mu_j$ $(j \in J^*)$. This makes the distinction, for example, between $\sigma^2(i \mid J^*)$ and $\sigma^2(ij \mid J^*)$ unnecessary. It is probably wise to think of pupils as nested within schools, in generalizing over pupils; this does not require the modification of any procedures, however, if the number of pupils per school can be taken to be very large.

#### Estimates of components and their interpretation

The four-way analysis of variance produces the mean squares in Table 7.9. We present the full set of "mixed-model" equations for expected mean squares, which may be compared with the random-model equations given in Table 7.5. The components would also be correctly estimated by applying

**TABLE 7.9.** *Estimates of Variance Components from a G Study with the Design* $(i{:}j) \times (p{:}s)$ *(Strata j Assumed To Be Fixed)*

| Source of variance | Analysis of variance as if crossed | | Mean square from analysis as nested | Expected mean square under the mixed model[a] | Estimate of variance component |
|---|---|---|---|---|---|
| | Sum of squares | Degrees of freedom | | | |
| Schools $(s\,\vert\,J^*)$ | 99.19723 | 23 | 4.31298 | $\sigma^2 + n_p\sigma^2(si,sji\,\vert\,J^*) + n_j n_i\sigma^2(p,sp\,\vert\,J^*)$ $+\, n_p n_j n_i\sigma^2(s\,\vert\,J^*)$ | 0.0114 |
| within $s$ | | | | | |
| Pupils $(p\,\vert\,J^*)$ | 10.29384 | $17\}$ | 0.39778 | $\sigma^2 + n_j n_i\sigma^2(p,sp\,\vert\,J^*)$ | 0.0122 |
| $sp\,\vert\,J^*$ | 151.99924 | $391\}$ | | | |
| Strata $(j\,\vert\,J^*)$ | 48.97333 | 2 | 24.48667 | $\sigma^2 + n_p\sigma^2(si,sji\,\vert\,J^*) + n_i\sigma^2(pj,spj)$ $+\, n_p n_i\sigma^2(sj\,\vert\,J^*) + n_s n_p\sigma^2(i,ji\,\vert\,J^*)$ $+\, n_s n_p n_i\sigma^2(j\,\vert\,J^*)$ | 0.0028 |

| | | | | |
|---|---|---|---|---|
| within $j \mid J^*$ | | | | |
| Items ($i$) | | | | |
| $ji$ | 163.77884 | 5$\rbrace$ | 16.97194 | $\sigma^2 + n_p\sigma^2(si,sji \mid J^*) + n_s n_p\sigma^2(i,ji \mid J^*)$ | 0.0386 |
| $sj \mid J^*$ | 90.80027 | 10$\rbrace$ | | $\sigma^2 + n_p\sigma^2(si,sji \mid J^*) + n_i\sigma^2(pj,spj \mid J^*)$ $+ n_p n_i\sigma^2(sj \mid J^*)$ | 0.0000 |
| | 14.05140 | 46 | 0.30547 | | |
| within $sj \mid J^*$ | | | | |
| $pj$ | 6.42477 | 34$\rbrace$ | | | |
| $spj$ | 157.10487 | 782$\rbrace$ | 0.20040 | $\sigma^2 + n_i\sigma^2(pj,spj \mid J^*)$ | 0.0036 |
| $si$ | 34.14399 | 115$\rbrace$ | | | |
| $sji$ | 61.87817 | 230$\rbrace$ | 0.27832 | $\sigma^2 + n_p\sigma^2(si,sji \mid J^*)$ | 0.0055 |
| $pi$ | 17.87616 | 85$\rbrace$ | | | |
| $pji$ | 32.10695 | 170$\rbrace$ | | | |
| $spi$ | 361.91664 | 1955$\rbrace$ | 0.17862 | $\sigma^2$ | 0.1786 |
| $spji, e$ | 681.29076 | 3910$\rbrace$ | | | |

[a] The value $\sigma^2$ is used here to represent $\sigma^2(pi,pji,spi,spji,e \mid J^*)$.

the equations of Table 7.5 except that the results would have to be further treated by evaluating the right side of:

$$\widehat{\sigma}^2(s \mid J^*) = \widehat{\sigma}^2(s) + \frac{1}{n_j} [\widehat{\sigma}^2(sj) + \widehat{\sigma}^2(pj,spj)]$$

and

$$\widehat{\sigma}^2(p,sp \mid J^*) = \widehat{\sigma}^2(p,sp) + \frac{1}{n_j} \widehat{\sigma}^2(pj,spj).$$

In interpreting components it should be remembered that with 0–1 scoring, no variance can exceed 0.25. The residual component, then, is large; it includes all miscellaneous effects from distraction and guessing, as well as from a pupil's ignorance regarding a particular item. The second largest effect is items within strata, a reflection of variation in item difficulty. The pupil (within school) and school effects are equivalent to standard deviations of about 0.1, on a 1-unit scale. It is remarkable that school means should contribute as much to variation in $X_{spji}$ as the pupil-within-school component does. Evidently, schools vary enormously in effectiveness, or in the quality of pupils they take in, or both.

The strata differ only slightly in average difficulty, according to the $j$ component of 0.0028. The extremely small $sj$ component implies that the order of difficulty of the strata is the same from school to school. The component for $p$ within $sj$ is only 0.0036, implying that factors specific to the strata are weak.[4] The component for $p$ within $s$ indicates that the strata are moderately intercorrelated. The factor common to all strata (which generates the $p$ component) accounts for four times as much of the within-stratum variance of pupil scores on single items as does the stratum-specific $pj$ factor (0.0121/0.0036). Over all pupils taken together, the general factor is an even more potent source of variance. The questions likely to be of greatest interest are these:

1. How great is $X_{spJI} - \mu_{spJ}$? This compares the pupil to his own universe score, recognizing that his assignment to a school is fixed. And how great is $(X_{spJI} - \mu_{spJ}) - (\mu_{sJI} - \mu_{sJ})$? This is pertinent where comparative standings of pupils within the school are of interest. (The number of pupils in the school will be treated as very large.)

2. How great is $X_{sP_sJI} - \mu_{sJ} = \Delta_{sP_sJ}$? This is the relevant question about the absolute school mean.

[4] The suggestion of Rabinowitz and Eikeland that the null hypothesis be accepted if $MS(p:sj)/MS(pi:sj)$ does not yield a significant $F$ is not followed. If items are classifiable on some logical ground, the hypothesis $\sigma^2(p:sj) = 0$ is not very reasonable.

**TABLE 7.10    Estimated Error of Generalization from a Stratified Test**

| Source of variance | Estimate of variance component | Frequency within $p$ | Contribution to variance of $X_{spJI} - \mu_{spJ}$ | Frequency within $s$ | Contribution to variance of $X_{sP_sJI} - \mu_{sJ}$ |
|---|---|---|---|---|---|
| Schools $s$ | 0.0114 | 1 | | 1 | |
| $p$ within $s$ | 0.0122 | 1 | | 18 | 0.001 |
| Strata $j$ | 0.0028 | 3 | | 3 | |
| $i$ within $j$ | 0.0386 | 18 | 0.002[a] | 18 | 0.002 |
| $sj$ | 0.0000 | 3 | | 3 | |
| within $sj$: | | | | | |
| $\quad pj, spj$ | 0.0036 | 3 | | 54 | 0.000 |
| $\quad si, sij$ | 0.0055 | 18 | 0.000[a] | 18 | 0.000 |
| $\quad$ residual | 0.1786 | 18 | 0.010 | 324 | 0.001 |
| $\hat{\sigma}^2(\Delta)$ | | | 0.012 | | 0.004 |

[a] Does not contribute to $\mathscr{E}\sigma^2(\delta)$.

The analysis for pupils in Table 7.10 indicates $\sigma(\Delta_{spJI})$ to be in the neighborhood of 0.1. The 18-item test does not locate the individual very exactly within the 1-point range. The variance of universe scores $\mu_{spJ}$ for pupils in the same school is estimated as 0.0122. The variance of $\Delta_{spJI} = (X_{spJI} - \mu_{spJ})$ is 0.012. To estimate observed-score variance the $i$ and $si$ components are ignored $(0.012 + 0.012 = 0.024)$. The within-school coefficient of generalizability is 0.51.

Assuming the 18 pupils to be a random sample of the within-school population, $\sigma(\Delta_{sP_sJI})$ for the school mean is estimated to be 0.06; this precision is not very adequate. If we convert the 0–1 scale into percentages (0–100), the standard error of the school mean is about 6 percentage points.

## EXERCISES

**E.1.** In the Medley–Mitzel study, what modification of the experimental plan would change the design to $o:(t \times r)$?

Suppose the G study had been carried out in that way and had yielded the "as it crossed" sum of squares and degrees of freedom shown in Table 7.1. Recombine these to get the mean squares and estimated variance components for the $o:(t \times r)$ design.

**E.2.** Write formulas for estimating variance components in the study diagrammed in Figure 7.1 and show how to calculate the estimates given in Table 7.2 (p. 196) from the rescaled mean squares.

**E.3.** Show how $\hat{\sigma}^2(r)$ is computed in Table 7.3 (p. 200). Prepare a diagram similar to those in Figure 7.3.

**E.4.** The older literature on the reliability of measures often stated rules of thumb regarding the level a coefficient should reach if a measure is to be used for individual measurement or group measurement. Thus, it has been suggested that the coefficient derived from parallel forms given on different occasions should reach 0.50 if a group is being appraised (rather than individuals). Does generalizability theory confirm the reasonableness of this rule, where the investigator's intent is to measure the attitude of one ethnic group toward another?

**E.5.** Pilliner (1965, p. 91) reports a study of a test with three kinds of arithmetic items. The analysis of variance yielded the following tabulation:

| Source | Sum of squares | Degrees of freedom | Mean square |
|---|---|---|---|
| Children (28) | 100.1471 | 27 | 3.7092 |
| Strata (3) | 4.2137 | 2 | 2.1066 |
| Items:strata (25 per stratum) | 97.5585 | 72 | 1.3550 |
| cs | 12.5374 | 54 | 0.2322 |
| c × i:strata | 305.6811 | 1944 | 0.1572 |

a. Estimate $\mathscr{E}\rho^2$ for the family of stratified–parallel tests. (Treat strata as fixed, items as random.)

b. Calculate $\widehat{\mathscr{E}\rho^2}$ for a 75-item test, generalizing over strata and items.

c. Collapse the analysis so as to ignore the stratification, into that for a persons × items design. Estimate $\mathscr{E}\rho^2$ for the family of random–parallel tests this test represents.

d. Compare the above results, explaining similarities or differences.

**E.6.** Use the data for Exercise 5 to compute $\hat{\sigma}(\Delta)$ for a school mean. Assume that there are 100 children per school in the relevant grade, that all of these are tested, that all interaction components involving school are zero, and, further:

a. that the test has 25 items in each of the three fixed strata, crossed with children.

b. that an item-sampling design is used, with 5 items from each of the three fixed strata (15 items in all) given to each subset of 20 children.

**E.7.** For each design in Table 7.8, calculate the error variance in the population mean that would be important for textbook *comparisons*. Assume that forms are applied in a similar manner to all textbooks, and that there is no item–textbook interaction.

**E.8.** Using the findings of the G study of Design V (p. 218), calculate $\widehat{\sigma^2}(\Delta)$ per form for each of these designs, on a 0–1 scale.

a. Design of 30 schools, 1 form of 10 items, 40 pupils per school per form.

b. Design of 30 schools, 10 forms of 10 items, 4 pupils per school per form.

c. Design of 30 schools, 20 forms of 5 items, 2 pupils per school per form.

### Answers

**A.1.**   If every rater visited every teacher, but no two raters visited a teacher on the same occasion, the design would be $o:(t \times r)$. Technically, one might defend adding the specification that only one teacher is observed on any given date. However, occasion variance seems far more likely to arise from events within the single classroom than from the date itself.

|          | Sum of squares | Degrees of freedom | Mean square | Component of variance |
|----------|----------------|--------------------|-------------|-----------------------|
| $t$      | 203            | 23                 | 8.83        | 0.67                  |
| $r$      | 13             | 1                  | 13.00       | 0.09                  |
| $tr$     | 17             | 23                 | 0.74        | (0)                   |
| residual | 403            | 192                | 2.10        | 2.10                  |

**A.2.**   $\sigma^2(i,pi,di,pdi,e) = $ EMS residual; hence $\widehat{\sigma^2}(i, \ldots) = $ MS residual; here, 0.21.

$$n_i \sigma^2(d,pd) + \sigma^2(i, \ldots) = \text{EMS within } p;$$

hence,

$$\widehat{\sigma^2}(d,pd) = \frac{1}{n_i}(\text{MS within } p - \text{MS residual})$$

Here

$$\widehat{\sigma^2}(d,pd) = 0.08$$

Similarly,

$$\widehat{\sigma^2}(p) = \frac{1}{n_i n_d}(\text{EMS } p - \text{EMS within } p)$$

Here, $\sigma^2(p) = 0.045$.

**A.3.**
The diagram indicates that EMS $r = A + B + C + G$.
And from Figure 7.3, EMS $pr = A + B + C$.

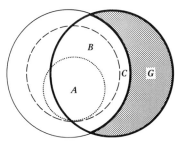

Numerically, MS $pr$ is greater than MS $r$. Therefore, the estimate of $G$ is negative and we take zero as $\widehat{\sigma^2}(r)$.

**A.4.** The result in such a study will be expressed as a sample mean (or median, etc.) for the responses of ethnic group $A$ when asked about ethnic group $B$. The question about precision of measurement has to do with the accuracy of this sample mean on the form used (assuming a crossed design) as an estimate of the population mean over a universe of forms. Consequently, one is concerned, not with the coefficient of generalizability, but with $\sigma(X_{Pi} - \mu)$. This depends not only on the instrument, but on the proposed sample size.

It does not report, as the coefficient does, the magnitude of the error relative to individual differences within the population. That does not bear on this investigator's inquiry. If one population of respondents has a completely uniform attitude $[\sigma(\mu_p) \equiv 0]$, one could still measure that attitude very accurately, even though $\mathscr{E}\rho^2 \doteq 0$.

It should be noted that $\sigma(\Delta)$ may well vary from one population of respondents to another, and that the standard deviation of responses among subjects of group $A$ referring to ethnic group $B$ may differ from their responses regarding group $C$.

**A.5.** The questions have to do with individual differences in a crossed D study. The components of variance required are:

| $c$ | $cs$ | $ci{:}s$ |
|---|---|---|
| 0.0463 | 0.0030 | 0.1572 |

This result is carried to more decimal places than is justified, for the sake of the following discussion:

   a. Treating strata as fixed,

$$\widehat{\mathscr{E}\rho^2} = \frac{0.0463 + \frac{1}{3}(0.0030)}{0.0463 + \frac{1}{3}(0.0030) + \frac{1}{75}(0.1572)}$$

$$= \frac{0.0473}{0.0494} = 0.959$$

   b. Treating the kinds of items in this test as randomly sampled from a universe of kinds of items (i.e., treating strata as random),

$$\widehat{\mathscr{E}\rho^2} = \frac{0.0463}{0.0493} = 0.939$$

   c. Carrying out the calculation without regard to stratification, one forms MS $ci + 318.24/1998 = 0.1593 = \widehat{\sigma^2}(ci)$. And $\widehat{\sigma^2}(c)$ now equals 0.0473. Then,

$$\widehat{\mathscr{E}\rho^2} = \frac{0.0473}{0.0473 + \frac{1}{75}(0.159)} = \frac{0.0473}{0.0494} = 0.957$$

d. For the domain of arithmetic tested, the kinds of items are not distinct; the small *cs* component of variance implies that items within a stratum are little more homogeneous than items from different strata.

The coefficient in *b* is lower than that in *a* because the proposed generalization is to a broader universe. The coefficient in *c* might be considered to be an estimate of the accuracy of generalization from the test to the universe score based on all items within the collection defined by fixed strata. This is another estimate of the value given in *a*, and the two agree closely. But the value in *c* mistakenly includes the contribution of the *cs* component in the error, and is smaller than it should be. If the *cs* component were large, or the number of items within strata small, the disparity could be important.

The result in *c* might also be considered an estimate of accuracy of generalization over all items of all kinds (i.e., to the same universe score as in *b*). In this interpretation, the result in *c* is an overestimate because it ignores the fact that items were cluster-sampled, and consequently, reports that the universe is more homogeneous than it actually is.

**A.6.** In this problem children are treated as fixed (i.e., the sample exhausts the population within the school) and strata are fixed. The components involving items are the sources of variance that contribute to $\sigma(\Delta)$. For items within strata, the component is estimated to be 0.0142, and for *ci* within strata it is 0.1572.

a. There are 75 items, and the *ci* interaction is sampled 7500 times.

$$\widehat{\sigma^2}(\Delta_{Pt}) = \tfrac{1}{75}\,(0.0142) + \tfrac{1}{7500}\,(0.1572) = 0.0002$$

$$\hat{\sigma}(\Delta) = 0.014$$

b. With the less complete test, the sample still exhausts the universe of the components for *c*, *s*, and *cs*. There are still 75 separate items, but only 1500 samples of the *ci* interaction.

$$\widehat{\sigma^2}(\Delta_{Pt}) = \tfrac{1}{75}\,(0.0142) + \tfrac{1}{1500}\,(0.1572) = 0.0003$$

$$\hat{\sigma}(\Delta) = 0.017$$

These two designs gave nearly the same value of $\hat{\sigma}(\Delta)$, but would not agree as well with smaller samples or a larger *ci* interaction.

**A.7.** In each case, remove the contribution of the *f* component from $\widehat{\sigma^2}(\Delta)$ as given in Table 7.8. The results are, in order, 0.007, 0.018, 0.004, 0.007. Where the emphasis is on comparison, the single-form design (no item sampling) serves better than it did for absolute measurement. The first or third design would be preferred, however, if textbook–form interactions were at all likely.

**A.8.**    Estimate

| Source | Estimate of variance Component | Design a[a] | | Design b[b] | | Design c[c] | |
|---|---|---|---|---|---|---|---|
| Schools | 0.004 | 30 | 0.00013 | 30 | 0.00013 | 30 | 0.00013 |
| Items | 0.019 | 10 | 0.00190 | 100 | 0.00019 | 100 | 0.00019 |
| *si* | 0.002 | 300 | 0.00001 | 3000 | 0.00000 | 3000 | 0.00000 |
| within *s* | | | | | | | |
| *p, sp* | 0.023 | 1200 | 0.00002 | 1200 | 0.00002 | 1200 | 0.00002 |
| *pi, spi, e* | 0.117 | 12,000 | 0.00001 | 12,000 | 0.00001 | 6000 | 0.00002 |

[a] 30 schools, 1 form of 10 items, 40 pupils per school per form.
[b] 30 schools, 10 forms of 10 items, 4 pupils per school per form.
[c] 30 schools, 20 forms of 5 items, 2 pupils per school per form.

# CHAPTER 8
# Multifacet Correlational Analysis

## A. Comparison of Correlational Analysis with Variance Analysis

Early discussions of multiple sources of error in test data (Gulliksen, 1936; Thorndike, 1947; Cronbach, 1947) were cast in correlational terms. It is useful to relate correlational analysis to score components and their variances. This will help the reader to grasp generalizability theory and will aid in judging what previously published correlational analyses of significant tests have said about the generalizability of their scores.

The number of published studies applying correlational analysis to data organized with respect to two or more facets is very limited. An occasional study can be found where two forms or half-tests on two occasions were given in a $i \times j \times p$ design. This design generates four scores to be correlated, representing the paired conditions $ij$, $i'j$, $ij'$, and $i'j'$. There are six pairs of scores, which yield six possible interclass correlations. Classical theory treats these correlations as if equal in the population. It is more likely, however, that scores having one condition in common (e.g., $ij$ and $i'j$) will correlate to a greater degree than will scores obtained under totally unlike conditions (e.g., $ij$ with $i'j'$). The multifacet model permits interpretation of the differences among correlations such as these.

### Comparability of results when conditions are equivalent

While the classical model represented in such works as Gulliksen's *Theory of Mental Tests* does not separate facets, its assumptions can be put into multifacet language. The sample of persons for the G study is treated as indefinitely large. With two facets, scores within conditions are assumed to

**231**

have uniform means $\mu_{ij}$, the variance over $p$ of observed scores $X_{pij}$ is assumed to be the same for every $ij$ combination, and the covariance over $p$ for two sets of observed scores is assumed to be uniform for all possible combinations of conditions

$$[\text{i.e.,}\ \sigma(X_{pij}, X_{pi'j'}) = \sigma(X_{pij'}, X_{pi'j}) = \sigma(X_{pij}, X_{pi'j'}) = \cdots].$$

The assumptions imply that the following score components equal zero: $\mu_i{\sim}$, $\mu_j{\sim}$, $\mu_{ij}{\sim}$, $\mu_{pi}{\sim}$, $\mu_{pj}{\sim}$, $\mu_{pij}{\sim}$. The corresponding components of variance vanish, leaving only $\sigma^2(p)$ and $\sigma^2(e)$. Under these assumptions, the covariance between scores observed under two conditions equals $\sigma^2(p)$, and the within-condition observed-score variance equals $\sigma^2(p) + \sigma^2(e)$. The ratio of these is the intraclass correlation between measures. The correlational analysis and the analysis of variance of G data yield identical results when the strong classical assumptions hold.

It makes very little difference whether data to be analyzed are arrayed with respect to two facets, or are arrayed in a one-facet layout—if the strict equivalence assumptions hold. A two-way analysis of variance components will yield a near-zero value for the variance components for conditions; in a three-way analysis the $i$, $j$, $pi$, $pj$, and $ij$ components will be near zero. Both analyses produce intraclass correlations. Alternatively, the investigator may compute interclass correlations for pairs of conditions and average them. All procedures will lead to the same result, save for fluctuations arising from sampling error.

If conditions are not fully equivalent, standardizing scores does not necessarily make them so. Standard scores have equal means and variances, but covariances need not be uniform.

Much of what follows deals with covariances. All statements made about covariances apply to correlations also, because a correlation is a covariance of standard scores. In this chapter all attention is on covariances over persons. (Covariances over conditions will appear in Chapter 9.)

*Variance components entering into covariances when conditions are not equivalent*

When conditions in the universe of admissible observations are organized with respect to two facets, three types of correlation or covariance are found. The distinction between types was made in Gulliksen's 1936 study where two essays per person were each graded twice. First, there is a co-variance for unlike conditions (e.g., between scores on two essays) each judged by a different grader. Second, there is an $i$-common covariance (essay common, grader different). Third, there is the analogous $j$-common covariance. Two covariances of each kind can be formed where $n_i = n_j = 2$, as is the case in the Gulliksen study. With $n_i = 2$ and $n_j = 3$, there are 15

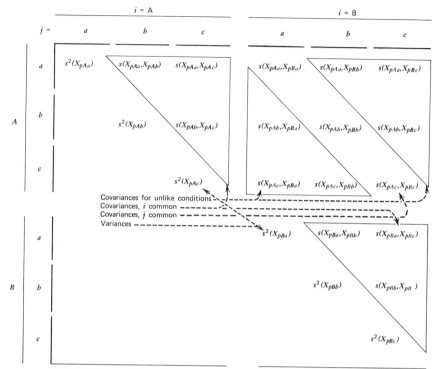

**FIGURE 8.1.** *Three Types of Covariance in a Study with Design VII.*

covariances, with 6, 6, and 3 of the respective types, as shown in Figure 8.1. It is reasonable to expect that the average values of covariances of different types will differ. Although Figure 8.1 deals with just six observations per person, one can imagine extending the table to represent all conditions in the universe.

Readers may recognize the resemblance of Figure 8.1 to the multitrait–multimethod matrix of correlations (Campbell & Fiske, 1959; see also Norman, 1967). However, the interpretation of a multitrait study where there is no intention to generalize over traits is somewhat different. Attention may also be drawn to the Stanley–Wiley paper (1962) on covariances in multifacet designs.

Consider the average of each type of covariance over all pairs of conditions in the universe. The expected value of $i$-common covariances, for example, is $\mathscr{E}\,\mathscr{E}\,\sigma(X_{pij}, X_{pij'})$.

$i\ j, j'$

The following equation regarding the expected variance of $X_{pij} - \mu_{ij}$

for Design VII is taken from Table 3.7 (for additional discussion, see Appendix).

(8.1) $\quad \underset{i,j}{\mathscr{E}}\, \sigma^2(X_{pij} \mid ij) = \sigma^2(p) + \sigma^2(pi) + \sigma^2(pj) + \sigma^2(pij,e)$

These additional equations hold for the three types of covariances:

$$\underset{i\ \ j\neq j'}{\mathscr{E}\ \mathscr{E}}\, \sigma(X_{pij},X_{pij'}) = \sigma^2(p) + \sigma^2(pi) \qquad (i\text{-common})$$

(8.2) $\qquad \underset{i\neq i'\ \ j}{\mathscr{E}\ \mathscr{E}}\, \sigma(X_{pij},X_{pi'j}) = \sigma^2(p) + \sigma^2(pj) \qquad (j\text{-common})$

$$\underset{i\neq i'\ \ j\neq j'}{\mathscr{E}\ \mathscr{E}}\, \sigma(X_{pij},X_{pi'j'}) = \sigma^2(p) \qquad \begin{array}{l}(\text{no common}\\ \text{condition})\end{array}$$

The expansion into components given in (8.2) applies also to the expected covariance of average scores. Accordingly, for $i$-common covariances, the upper line of (8.2) applies to the expectation of $\sigma(X_{piJ},X_{piJ'})$, provided that sets $J$ and $J'$ are drawn at random. Where many $i$ are involved, as in

$$\sigma(X_{pIJ},X_{pIJ'}),$$

$\sigma^2(pi)$ in (8.2) must be divided by $n_i$. In the $j$-common covariance, $\sigma^2(pj)$ must be divided by $n_j$.

Assuming random sampling, any $s(X_{pij},X_{pij'})$ from a G study is an unbiased estimate of the expected $i$-common covariance; etc. Equations (8.1) and (8.2) are entered with means of observed covariances of each type from the G study; components of variance are then estimated by solving the equations. If covariances between scores on whole tests are calculated (e.g., if $i$ represents test forms and $j$ represents graders), the equations yield components for whole test scores. Similar analyses can also be carried out for items and part-tests.

Covariances obtained from partially nested designs can also be used to estimate variance components. Some components will be confounded, just as in the analysis of variance for the same design.

### B. Theory for a Test Organized into Subtests

The Wechsler Verbal Scale will be taken as an example for multifacet correlational analysis. Its organization into subtests introduces especially interesting questions. We shall assume that in all D studies the same test form is to be given to all subjects. We also assume that no two persons will be tested on the same day, in order to avoid complications in the discussion.

Four facets are to be distinguished: subtests $j$, item-sets $i$ within subtests, days $d$, and trials $t$. The smallest item-set considered will be the half-subtest,

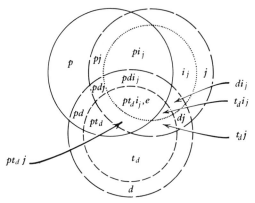

**FIGURE 8.2.** *Components in Scores on Item-Sets of the Wechsler Verbal Scale.*

for two reasons: the number of half-subtests is two in any subtest, consequently, $n_i = 2$ in every study (though the number of items varies); second, we have split-half data for subsequent illustrations of G studies, not item data. For any subtest such as Vocabulary, we assume that the universe contains an indefinitely large number of items that could be assembled into forms or half-length forms.

Any trial on any day with any item-set from any subtest falls within the universe of admissible observations. The universe of conditions has the structure $(t:d) \times (i:j)$, and there are observations on all conditions for all persons. Trials are nested within days in the universe and item-sets within subtests, as indicated in Figure 8.2. The nesting shown is present in the universe of admissible observations. Instead of writing such a pair of components as $t$ and $td$ separately, with $t:d$ we use the label $t_d$ for the trial-within-day effect (see pp. 63ff.). This, and a similar simplification for $i$-within-$j$, leaves us with 18 components of the observed score. Several of these components will be ignored in correlational analysis of G data from a crossed design.

*Universes of generalization*

What universe of generalization interests the Wechsler interpreter? More broadly, what are the conceivable universes of generalization? The interpreter will surely generalize over item-sets and over the moments when any item-set might be presented within a day. The interpreter usually generalizes over days also, because Wechsler scores are taken to describe the person at a given stage of development. After about age 8, a person's score has a "useful life" of a year or more under most circumstances.

Whether the Wechsler interpreter normally generalizes over subtests (from the specific Verbal tests to the Verbal domain) is arguable. The tasks may be regarded as a sample from a broad domain of eligible tasks, or as the entire set of relevant tasks. The former appears to be more usual. Just as the vocabulary items are recognized as no more than a sample of suitable words, Wechsler's formulation-of-definitions task is just one selection from a domain of usable verbal tasks. Other tasks include opposites, synonyms, verbal analogies, and picture vocabulary. The view that tasks are sampled is not at all strange to users of the Stanford–Binet. Form L and Form M are composed of largely different tasks representing a broad domain. Task-to-task variation was counted as a source of error in the basic Terman–Merrill reliability study where L and M were given on different days and the scores were correlated. Willingness to generalize over tasks was even more evident when Merrill later reduced the Stanford–Binet to the single form L–M and suggested that, if an L–M IQ needs to be checked, the Wechsler Scale for children serves to estimate the same thing. Interpreters of Wechsler Verbal (Ve) and Performance (Pe) IQs seem to regard them as representative of the respective domains of intellectual tasks.

There are six conceivable universes of generalization for a test score:

1. Generalization is over trials, that is, over administrations within a day. The person is tested repeatedly on a particular day, but the day typically differs from person to person. All observations in Universe 1 employ the same fixed set of items. This universe is included here only to square off a formal structure.

2. Generalization is over trials and item-sets. Each administration presents the same set of subtest tasks, with a new set of items, however. There is no attempt to generalize beyond the day on which the person is tested.

3. Generalization is over tasks, as well as over trials and item-sets. Again, the day is treated as fixed for the person. Each admissible test employs a new set of subtest tasks. That is, each testing uses a fresh set of subtests drawn from the general domain, not alternate forms of the original subtests. Subtest-specific effects thus become a source of error. A new subtest automatically brings in new items.

4. Generalization is over days, and consequently over trials as well. Observations are made on days within a certain time period, with subtest tasks and items fixed.

5. Generalization is over days (and trials) and over item-sets. Observations are made on days within a certain period, with subtests fixed but with items changing from day to day.

6. Generalization is over days (and trials) and tasks, and hence over items also. Observations are made on days within a certain period with subtests (and items) changing from day to day.

Score components $\mu_j\!\sim$ and $\mu_{i_j}\!\sim$ will be disregarded. One or both of these does enter the observed score and, for certain universes, enters the universe score, but they are of negligible importance because standardization greatly reduces them. Moreover, subtests are always crossed with persons; consequently variance components for $j$ and $i_j$ cannot enter the score variance. For purposes of the formal analysis the $d$ and $t_d$ components will be retained, because these are confounded with the person and enter the observed-score variance if not the universe-score variance.

Table 8.1 arranges the remaining components to show which ones contribute to the universe score for each of the six universes. All other components contribute to the observed score, but not to the universe score. In Universe 6, all components except that for persons contribute to the discrepancy between observed score and universe score, hence count as error. In Universe 5, subtests are fixed, so that the person × subtest component is part of the universe score, not of the error. And so on. The error is greatest in generalizing to the broad Universe 6, and least in generalizing to Universe 1. The set is partially ordered: $(6) \geqslant (3) \geqslant (2) \geqslant (1)$ and $(6) \geqslant (5) \geqslant (4) \geqslant (1)$.

The person–subtest interaction $\mu_{pj}\!\sim$ reflects the extent to which certain subtests are consistently easier or harder for one person than for others. The person–item interaction $\mu_{pi_j}\!\sim$ arises from a person's greater mastery of some items than of other items of the same character. We would expect reasonably large $pi_j$ interactions within Vocabulary, but interactions should be comparatively small within Digits Forward, as those items have few distinctive features.

The person–day component $\mu_{pd}\!\sim$ departs from zero if the individual does better on some days than on others. Practice effects or other trends common to all persons are not included in this interaction. The component $\mu_{pdj}\!\sim$ comes from day-to-day variation in ability to perform particular tasks. For example, some subtests may be sensitive to anxiety on the day of testing, while other subtests are insensitive.

### Six alternative experimental plans and their analyses

Table 8.2 catalogues G studies in which two scores for the person are obtained and correlated. The immediate retest (type I) is rarely used because of memory effects; it is included in the table for symmetry.

All studies of type IV are formally the same regardless of the interval between testings. The longer the interval, however, the larger the components for day and its interactions are likely to be. For example, the *pd* component of variance is likely to increase with greater lapse of time, and the *p* component to be reduced. The greater time interval often reflects choice of a broader universe of generalization. (The same is to be said, of course, about studies of type V, or VI.)

**TABLE 8.1.** *Components of Universe Score and Error with Various Universes of Generalization*

| | Universe of generalization | | | | | |
|---|---|---|---|---|---|---|
| | 1 | 2 | 3 | 4 | 5 | 6 |
| Day | Fixed[a] | Fixed[a] | Fixed[a] | Variable | Variable | Variable |
| Subtests entering test | Fixed | Fixed | Variable | Fixed | Fixed | Variable |
| Items entering test | Fixed | Variable | Variable | Fixed | Variable | Variable |
| **Score component** | | | | | | |
| person $\quad p$ | U[b] | U | U | U | U | U |
| person × subtest $\quad pj$ | U | U | U | U | U | |
| person × (item-set:subtest) $\quad pi_j$ | U | U | | U | | |
| person × day $\quad pd$ | U | U | U | | | |
| person × day × subtest $\quad pdj$ | U | U | U | | | |
| person × day × (item-set:subtest) $\quad pdi_j$ | U | U | | | | |
| day $\quad d$ | U | U | U | | | |
| day × subtest $\quad dj$ | U | U | U | | | |
| day × (item-set:subtest) $\quad di_j$ | U | U | | | | |
| (trial:day), person × (trial:day), and associated components $\quad t_d, pt_d, \ldots, e$ | | | | | | |

[a] Fixed within the person.

[b] U = component of universe score. Whenever U does not appear, the component ordinarily contributes to error of generalization.

**TABLE 8.2.** *Components of Variance Contributing to Test-Score Variance and Correlation in Different Types of G Study*

| | Experimental plan generating scores to be correlated | | | | | |
| --- | --- | --- | --- | --- | --- | --- |
| | I | II | III | IV | V | VI |
| Day (nested within the person) | One | One | One | Two | Two | Two |
| Procedure for obtaining scores | Retest | Two forms | Two forms | Retest | Two forms | Two forms |
| Handling of subtests | Item-sets combined | Same, with new items | New tasks | Item-sets combined | Same, with new items | New tasks |
| **Source of variation** | | | | | | |
| person | +ᵃ | + | + | + | + | + |
| person × subtest | + | + | | + | + | |
| person × (item-set:subtest) | + | | | + | | |
| person × day | + | + | + | | | |
| person × day × subtest | + | + | | | | |
| person × day × (item-set:subtest) | + | | | | | |
| day | + | + | + | | | |
| day × subtest | + | + | | | | |
| day × (item-set:subtest) | + | | | | | |
| trial:day, person × (trial:day), and associated components | | | | | | |

ᵃ + identifies components of variance raising the expected correlation between test scores obtained under this plan.

Plan II and plan V call for administering two forms of the conventional sort, where the same subtests appear in each form, but the items are changed. This is to be contrasted with Plans III and VI, which call for a second form employing subtests different from those of the first.

It would be possible to employ split-half designs in place of II, III, V, and VI. If the half-tests were separated in the test administration, they would conform precisely to the description in the table (e.g., two forms given on the same day), except that the forms are shorter than the usual test. The fact that half-subtest scores are usually obtained by a split-half scoring after the entire subtest has been administered as a unit means that the assumption of independence is violated. The two halves of the subtest are likely to be more highly correlated than they would be with independent administrations. There are two ways to split a test such as the Wechsler Verbal Scale. One can separate odd and even items, as is conventional. Making two short forms of each subtest is comparable to plans II and V. One can also divide the subtests into two groups; half-tests with different subtests resemble plans III and VI. In the argument that follows we shall continue to speak in terms of whole test forms, but in the numerical example we shall employ split-half data.

There is a one-to-one correspondence of experiments with universes. In a study of type IV, just the components of the universe-score variance for Universe 4 raise the correlation between tests. Correspondingly, for type V and Universe 5, type VI and Universe 6, etc. However, one can process data more complexly. As will be seen in the next sections, one experimental plan can give data on generalizability to several universes. Rather than examine this further theory, the reader may prefer to move at once to illustrative Wechsler data (p. 246).

*Multifacet analysis of variance.* In any one of the studies described in Table 8.2, the half-subtest scores could be processed by the analysis of variance components as in Chapter 2. (Analysis of whole subtest scores gives less complete information.) Alternatively, one could calculate a complete matrix of intercorrelations for all pairs of half-subtest scores. There are close correspondences between the two types of analysis. The following section describes the possible results from an analysis of variance under each of the six designs. It will then be shown how the findings from correlational analysis correspond to those results.

It is assumed that halves within a subtest are experimentally independent, and that having previously responded to an item or subtest task does not systematically affect a person's second performance on it. These are assumptions traditional in reliability studies. Figure 8.3 indicates what components of variance can be estimated from half-subtest scores in a retest study.

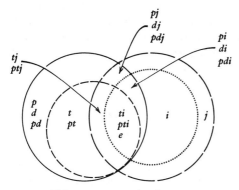

Eight components of variance

## (a)  Retest on same day. (I)

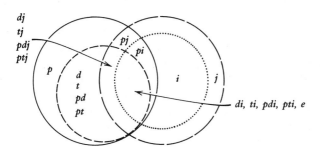

Eight components of variance

## (b)  Retest on two days. (IV)

***FIGURE 8.3.*** *Components of Variance Obtainable from Analysis of Variance of Half-Subtest Scores in Retest Studies.*

The first possibility to be considered is the same-day retest (plan I of Table 8.2). The upper half of Figure 8.3 represents this design. Person and day are confounded since persons are tested on different days. We treat trials as nested within $p,d$, though we could have classified first and second observations as distinct and so have treated $t$ as crossed, as we did with PICA data (p. 176). It should be noted that where $i$ appears in this and the following chart, this is an abbreviated notation for $i_j$ ($i$-within-$j$; see p. 235).

Correspondingly, any $t$ stands for $t$-within-$d$. It would be possible to describe the design for the same-day retest as $(i_j:j) \times (t:[p,d])$. The notational conventions are not very satisfactory, however, when we deal with such a complex design. This coded statement, for example, leaves somewhat ambiguous whether $i$ is crossed with $t$. (It is, in a retest.) We shall abandon the code for crossed and nested designs entirely in the remainder of this section, though the diagrams represent crossing and nesting faithfully.

The diagram for plan I indicates that analysis of variance estimates eight components, most of these being compounds. Only the components that fall within the $p$ circle affect a correlation, as the use of deviation scores eliminates other sources of variance from consideration. The six components identified within the $p$ circle for plan I can also be inferred from variances and covariances of half-subtest scores in a same-day retest study.

Plan IV is the same in all respects except that the retest takes place on a second day, and diagram (b) is correspondingly similar to (a). Trial $t$ is now joint with day, and both are nested within the person. Again, eight components of variance can be estimated, but the confounding is different.

The study may employ parallel forms with the same subtests and new items. Figure 8.4 includes representations for plans II and v (diagrams a and c, respectively); the two differ only in that $d$ is confounded with $p$ in one diagram, with $t$ in the other. The diagrams have a feature not previously encountered: an arc cutting off the lower portion, and an incomplete circle representing the residual term. For the reader who is interested in these matters we digress to spell out our basis for using these odd conventions.

It is envisioned that a different test form $f$ will be given on each trial. Because every subtest appears in both forms, we have $j$ crossed with $f$, and $i_j$ nested within $jf$. The large arc is a portion of the "$f$ circle," in a sketch representing $j \times f \times p$. By our usual convention, the outer part of that circle would represent an "$f$ component." But the $f$ score component proves to be nothing more than the average of the $\mu_j - \mu$; this average is zero, and is not reflected in the diagram. Where the $f$ and $j$ circles intersect we might display a $jf$ component. However, this "interaction" arises only from differences in the items chosen for the two forms, and therefore is indistinguishable from the $I_j$ component. We arbitrarily chose to erase a part of the $i$ circle rather than to erase part of the $j$ circle. A similar incomplete circle for forms appears in the diagrams for plans III and VI; the outer part of it is absorbed into the $j$ component. In all these diagrams, form is confounded with trial.

The reader will see that the upper and lower diagrams in Figure 8.4 differ only with respect to the way in which $d$ is confounded. The chief difference from left to right is that on the left there is a distinction between *pj-dj-pdj* and *tj-ptj* because these can be separated, whereas on the right they are confounded. The progression from eight to seven to six components, as we

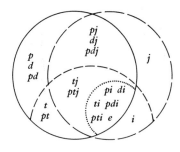

Seven components of variance

(a)   Parallel tests with same
tasks on same day. (II)

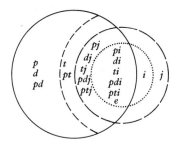

Six components of variance

(b)   Parallel tests with new
tasks on same day. (III)

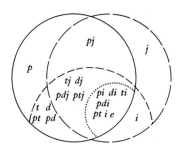

Seven components of variance

(c)   Parallel tests with same
tasks on two days. (V)

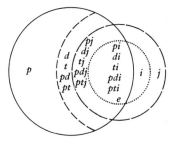

Six components of variance

(d)   Parallel tests with new
tasks on two days. ·(VI)

*FIGURE 8.4.   Components of Variance Obtainable from Analysis of Variance of
Half-Subtest Scores in Studies with Parallel Forms.*

move from plan I to plan III, or from IV to VI, is noteworthy. What would appear at first glance to be the more complex and more informative design estimates fewer components then the simple retest design. There are two countervailing considerations that may tend to make III or VI preferable to I or IV. First, the assumption of independence is less plausible in a retest design. Second, the parallel-form designs sample some components more thoroughly than a retest G study of the same overall size, and so will give better estimates of some of the variance components that appear in both studies. Accordingly, if $i$ is a half-subtest, and $n_j = 5$, plan I involves 10 distinct $i_j$, while II and III involve 20.

*Interpretation of the test-score correlation.* The traditional investigator has computed just one correlation in any correlational study, employing the total test score from each testing. He presents this result as a "reliability coefficient." The diagrams just examined clarify what causes the six coefficients for any given instrument to differ.

Any correlation is a ratio of a covariance to the product of two standard deviations. When the two scores correlated are observations from the same universe, the numerator can be regarded as an estimate of the expected covariance between pairs of scores obtained under the design used. The denominator can be regarded as an estimate of the expected observed-score variance. The expected observed-score variance under any design is made up of the components of variance falling within the $p$ circle of the corresponding diagram. The expected covariance and, therefore, the correlation is raised by the person component and also by whatever components within the $p$ circle arise from conditions that are the same for both measurements.

To be specific, consider plan I, the same day, test-retest design. All components within the $p$ circle in the upper part of Figure 8.3 enter the expected observed-score variance. Among those components, the ones lying outside the $t$ circle (i.e., $p,d,pd;$ $pj,dj,pdj;$ $pi_j,$ $di_j,$ . . . ,$pdi_j$) raise the retest correlation. Hence the retest correlation reports on the magnitude of this conglomerate of components, relative to the entire set of components involving or confounded with $p$. A component enters the conglomerate with a certain weight; for instance, $pj$ is sampled $n_j$ times in each score, so enters with weight $1/n_j$. The diagram is consistent with Table 8.2; the covariance does not include the components for trials, the person-trial interaction, etc. All these fall within both the $p$ and $t$ circles.

To make a similar interpretation of the other five types of covariance and correlation, working from the other diagrams, is left as an exercise.

*Interpretation of subtest and half-subtest covariances.* Far more information about components is obtained if one examines covariances for half-subtests. These can be organized so as to supply the same information about components involving $p$ as the analysis of variance offers.

Consider the same-day retest study. As diagrammed in (a) of Figure 8.3, the $p$ circle that contains the components of the observed-score variance has six segments. This divided $p$ circle is reproduced in diagram (a) of Figure 8.5. Common groups of elements within areas are keyed to the legend by arabic numbers. Each of the diagrams is identical; shading indicates which components of variance contribute to the quantity named in the label for the diagram. Diagram (b) represents the covariance for trials of the same subtest ($i$-$j$-common), averaged over subtests. This is seen to be the weighted sum of components of variance for $p,$ $d,$ $pd;$ $pj,$ $dj,$ $pdj;$ and $pi_j,$ $di_j,$ $pdi_j$.

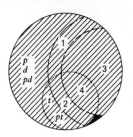

(a) Subtest
variance

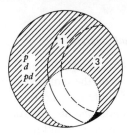

(b) Covariance
of like subtests
trials on two
( *i,j* common)

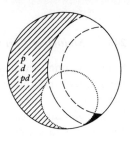

(c) Covariance of
unlike subtests
on two trials

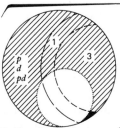

(b1) Covariance of
like halves of same
subtest on two trials
( *i,j* common)

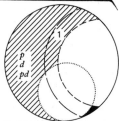

(b2) Covariance of
unlike halves of same
subtest on two trials
( *j* common)

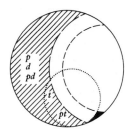

(d) Covariance of
unlike subtests
on same trial
( *t* common)

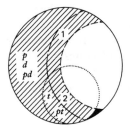

(e) Covariance of
unlike halves of same
subtest on same trial
( *t,j* common)

FIGURE 8.5. *Components Entering Average Variance and Average Covariances Obtainable from Subtest Scores and Half-Subtest Scores in a Same-Day Retest Study (Type*

$$1 = pj,\ dj,\ pdj$$
$$2 = t_d j,\ pt_d j$$
$$3 = pi_j,\ di_j,\ pdi_j$$
$$4 = t_d i_i,\ pt_d i_j^-,\ e).$$

Similarly, one can diagram covariances for unlike subtests on different trials (diagram c) and unlike subtests on the same trial (diagram d). Once numerical values for the three kinds of subtest covariance and the variance are calculated, one can form four simultaneous equations. These cannot be fully solved, of course, because there are six unknowns.

A full solution is achieved by considering the covariance for halves of the same subtest, within *and across* trials. For any subtest there are two values of *t-j*-common covariances: within trial from halves of the subtest (diagram e). There are two values of *i-j*-common covariances: between trials, from applications of the same half-subtest (diagram b1). Finally, there are two values of *j*-common covariances: between trials, from unlike halves of the subtest (diagram b2). Multiplying each of these mean half-subtest covariances by four yields estimates of mean subtest covariances of the three types. Averaging the estimates of *i-j*-common and *j*-common half-subtest covariances gives the *i-j*-common, between-trial covariance (diagram b) for the subtest.

When one equation is written to correspond to each circle of Figure 8.5, except the redundant diagram (b), there are six equations referring to observable quantities (five average covariances and one average variance). Therefore, the components can all be estimated. A complete analysis of the covariances for the retest study using split-half techniques gives the same information about the components of variance that contribute to individual differences as does the analysis of variance sketched in Figure 8.3. The analysis of variance is much more straightforward.

A similar procedure can be followed with every other experimental plan. The diagrams in Figures 8.3 or 8.4 can be used as a basis for a mapping of components of variance into covariances in the manner of Figure 8.5, and these give rise to the equations from which components of variance are estimated.

### C. Numerical Example: Interpretation of WPPSI Correlations

Previous correlational studies of tests have compared scores arising under two conditions of a facet, or two observations that differ with respect to two facets. Such information must be patched together to arrive at a multifacet interpretation. Though such reasoning is tortuous, it will be required to take full advantage of past studies.

Two correlational studies are reported in the manual for WPPSI, the Wechsler test for preschool ages (Wechsler, 1967). A split-half study was conducted at each of several ages; we shall give attention only to that for children of age 5½. In the second study, the test was given twice. From the correlations we shall infer variance components and examine the generalizability of the Verbal score to various universes. (Similar analyses could of course be made for the Performance section of the test.) The Verbal data are based on six subtests, though five would normally be used in D studies.

## Data

Sample 1: A group of 200 children, age 5½, who constituted the standardizing sample for the test at that age. Each subtest was split in half. We start from the published correlations for the halves within each subtest, corrected by the Spearman–Brown formula, the correlations between subtests, and the standard deviations for subtests. The correlations will be converted to covariances in the course of the analysis, and the standard deviations will be converted to variances.

Sample 2: A group of 50 children, tested at about age 5½ and again some weeks later. We start from the covariances between scores for the same subtests on the two occasions, the covariances between unlike subtests within and across occasions, and the variances for subtests on each occasion. While the manual gives correlations and standard deviations, we have used covariances supplied by Dr. Jerome Doppelt of The Psychological Corporation; the data for unlike subtests have not been published before.

It would be more satisfactory to deal with a single sample, but the original record sheets for sample 2 are not available and we cannot analyze that study within subtests. In fitting the two sets of findings together, we encounter the troublesome fact that sample 2 has a limited range. The standard deviation of IQs was 13.9 on the first test and 14.7 on the retest, instead of 15. In sample 1, the standardizing sample, the standard deviation is 15 and the subtest scaled scores have standard deviations in the range 2.9–3.1. (The departure from the ideal 3.0 very likely comes from smoothing the conversion table across ages.) We take as a working assumption that all components of variance except $\sigma^2(p)$ are the same in the populations the two samples represent, and designate the components for persons in the two populations as $\sigma^2(p:1)$ and $\sigma^2(p:2)$. The former population is the one of general interest. We have subtest variances from Study 1 and from Study 2. The covariances available are:

| | |
|---|---|
| Like subtests on two trials | Study 2 |
| Unlike subtests on two trials | Study 2 |
| Unlike subtests on same trial | Study 1 and Study 2 |
| Unlike halves of same subtest on same trial | Study 1 |

To determine what components of variance are to be estimated from these, Venn diagrams might be constructed in the manner of Figure 8.5. But we shall further simplify by treating all the following components of variance as negligible in size: $d$, $t$, $ti$, $di$, $tj$, and $dj$. It seems most unlikely that there are important systematic effects associated with day or trial. For example,

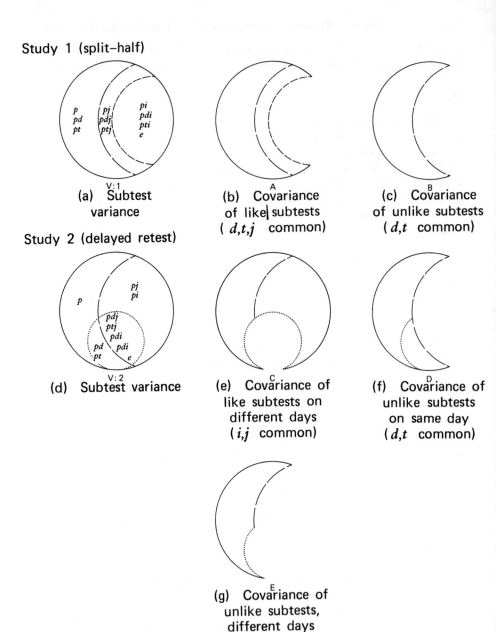

Study 1 (split–half)

(a) Subtest
variance
V:1

(b) Covariance
of like subtests
( $d,t,j$ common)
A

(c) Covariance
of unlike subtests
( $d,t$ common)
B

Study 2 (delayed retest)

(d) Subtest variance
V:2

(e) Covariance of
like subtests on
different days
( $i,j$ common)
C

(f) Covariance of
unlike subtests
on same day
( $d,t$ common)
D

(g) Covariance of
unlike subtests,
different days
E

FIGURE 8.6. *Composition of Variances and Covariance in WPPSI Data.*

Digit Span would not be particularly easy for all subjects on March 22. We continue to recognize all interactions involving persons. Figure 8.6 represents these data under the simplifying assumption; where $t$ and $i$ appear, read $t_d$ and $i_j$.

Study 1 is a split-half study with $n_i = 2$, $n_j = 6$, $n_p = 200$. Person, day, and trial are confounded; only a single day and trial are involved in the study and, therefore, there is considerable confounding of components. Figure 8.6a indicates what information enters the subtest variance in this study. Three components can be extracted from this variance (compare with Figure 8.5 or with the $p$ circle of Figure 8.3a, ignoring the $t$ circle):

$$\sigma^2(p,pd,pt_d)$$
$$\sigma^2(pj,pdj,pt_dj)$$
$$\sigma^2(pi_j,pdi_j,pt_di_j,e)$$

In Study 2 (retest), $n_j = 6$, $n_d = 2$, $n_p = 50$. Here, the variance contains four separable components, as seen in Figure 8.6d (compare with the $p$ circle of Figure 8.3b, ignoring the $i$ circle):

$$\sigma^2(p)$$
$$\sigma^2(pd,pt_d)$$
$$\sigma^2(pj,pi_j)$$
$$\sigma^2(pdj,pt_dj,pdi_j,pt_di_j,e)$$

These two breakdowns are less elaborate than those of previous figures, because in Study 1 there is only one trial, and in Study 2, half-subtest scores are not available.

Because our theory calls for decomposition of covariances, we transform the correlations reported for Study 1 to that form (Table 8.3). The first entries in Table 8.3 are the subtest reliability coefficients (split-half, corrected) and the subtest variances. Under the assumption that half-subtests have equal variances, the full-length coefficient multiplied by the subtest variance equals four times the half-subtest variance. The product formed here equals four times the covariance between halves of the same subtest. Under our assumptions, this is an unbiased estimate of the expected value of the covariance between forms of the same subtest administered at the same sitting.

The right-hand portion of Table 8.3 contains the subtest covariances. It is convenient to label each of the averages in Tables 8.3 and 8.4. In Table 8.3 the means are labelled v:1, A, and B.

v:1 is the mean of subtest variances.

A is the mean of covariances for like subtests (same task, same trial, same day, different items) estimated from half-subtest correlations.

*TABLE 8.3.   Data from WPPSI Split-Half Study*[a]

| Subtest | Corrected split-half correlation, variance, and product of these | | | Covariance matrix | | | | | |
|---|---|---|---|---|---|---|---|---|---|
| | | | | Inf | Voc | Ari | Sim | Com | Sen |
| Information | 0.81 | 9.0 | 7.3 | 5.85 | 6.00 | 5.30 | 5.58 | 5.22 | |
| Vocabulary | 0.85 | 9.0 | 7.6 | | 4.87 | 4.65 | 5.49 | 4.05 | |
| Arithmetic | 0.86 | 8.4 | 7.2 | | | 4.59 | 5.22 | 4.35 | |
| Similarities | 0.82 | 9.6 | 7.9 | | | | 4.56 | 4.37 | |
| Comprehension | 0.84 | 9.0 | 7.6 | | | | | 5.22 | |
| Sentences | 0.87 | 9.0 | 7.8 | | | | | | |
| Total | . | 54.0 | 45.4 | | | | | | 75.32 |
| Mean | | v = 9.00 | A = 7.57 | | | | | | B = 5.02 |

[a] Covariances calculated from data given in test manual, p. 29. Correlations and variances taken from test manual, p. 22.

B is the mean of covariances for unlike subtests on the same day and trial.

The data matrix for Study 2, given in Table 8.4, contains variances of subtests, *i-j*-common covariances (same subtests, different days), *d-t*-common covariances (unlike subtests, same day), and covariances for unlike subtests on different days. Averaging, we have:

v:2. Mean of subtest variances.
c. Mean of six *i-j*-common covariances.
D. Mean of 30 *d-t*-common covariances.
E. Mean of 30 unlike-subtest, different-day covariances.

### Estimation of components

We are now in a position to estimate the several components of variance of subtest scores that can be determined from correlations. Each diagram in Figure 8.6 can be read as an equation of the sort introduced in (8.1) or (8.2). From the split-half data we read

$$B = 5.02 = \widehat{\sigma^2}(p{:}1, pd, pt_d)$$

$$A = 7.57 = B + \widehat{\sigma^2}(pj, pdj, pt_dj)$$

$$2.55 = \widehat{\sigma^2}(pj, pdj, pt_dj)$$

$$v{:}1 = 9.00 = A + \widehat{\sigma^2}(pi_j, pdi_j, pt_d i_j, e)$$

$$1.43 = \widehat{\sigma^2}(pi_j, pdi_j, pt_d i_j, e)$$

**TABLE 8.4.** *Data from WPPSI Retest Study in Variance–Covariance Form*[a]

| | Day 1 | | | | | | Day 2 | | | | | |
|---|---|---|---|---|---|---|---|---|---|---|---|---|
| | Inf | Voc | Ari | Sim | Com | Sen | Inf | Voc | Ari | Sim | Com | Sen |
| **Day 1** | | | | | | | | | | | | |
| Information | **6.59** | | | | | | **5.86** | 4.87 | 3.38 | 2.96 | 4.11 | 4.72 |
| Vocabulary | 6.82 | **12.16** | | | | | 6.19 | **8.07** | 4.63 | 4.00 | 6.49 | 5.80 |
| Arithmetic | 3.72 | 4.87 | **4.55** | | | | 3.87 | 3.46 | **4.58** | 3.00 | 3.47 | 4.22 |
| Similarities | 3.58 | 5.87 | 2.89 | **8.13** | | | 4.28 | 4.50 | 2.91 | **4.62** | 3.73 | 3.19 |
| Comprehension | 5.26 | 7.67 | 3.44 | 3.75 | **8.57** | | 5.10 | 5.77 | 3.09 | 2.96 | **5.51** | 4.79 |
| Sentences | 4.40 | 5.76 | 4.23 | 2.61 | 4.56 | **8.82** | 5.38 | 4.14 | 4.24 | 2.48 | 3.87 | **6.00** |
| | Total off-diagonal, 69.43; mean, 4.63. | | | | | | Total off-diagonal, 64.28; mean, 4.28. | | | | | |
| | Total in diagonal, 48.82; mean, 8.13. | | | | | | Total in diagonal, 48.58; mean, 8.10. | | | | | |
| **Day 2** | | | | | | | | | | | | |
| Information | Total off-diagonal (same as upper right rectangle), 125.60; mean, 4.19 = E. | | | | | | **9.15** | 5.38 | 4.25 | 4.04 | 4.28 | 5.97 |
| Vocabulary | | | | | | | | **9.00** | 3.61 | 4.13 | 5.84 | 4.66 |
| Arithmetic | Total in diagonal, 34.64; mean, 5.77 = C. | | | | | | | | **7.36** | 3.41 | 3.32 | 4.00 |
| Similarities | | | | | | | | | | **7.33** | 3.65 | 3.48 |
| Comprehension | | | | | | | | | | | **7.41** | 4.26 |
| Sentences | | | | | | | | | | | | **8.33** |

Same day grand mean: off-diagonal, 4.46 = D; diagonal, 8.12 = v.

[a] Entries calculated from data supplied by The Psychological Corporation.

These components are on the scale of the total score for the whole subtest, not of the half-subtest. From the retest study:

$$E = 4.19 = \widehat{\sigma^2}(p:2)$$

$$D = 4.46 = E + \widehat{\sigma^2}(pd,pt_d)$$

$$0.27 = \widehat{\sigma^2}(pd,pt_d)$$

$$C = 5.77 = E + \widehat{\sigma^2}(pj,pi_j)$$

$$1.58 = \widehat{\sigma^2}(pj,pi_j)$$

$$v:2 = 8.12 = D + \widehat{\sigma^2}(pj,pi_j)$$

$$+ \widehat{\sigma^2}(pdj,pt_aj,pdi_j,pt_di_j,e)$$

$$8.12 = 4.46 + 1.58$$

$$+ \widehat{\sigma^2}(pdj,pt_aj,pdi_j,pt_di_j,e)$$

$$2.08 = \widehat{\sigma^2}(pdj, \ldots ,e)$$

To this point, we have simply interpreted two separate G studies, estimating aggregations of components. Further inferences can be drawn from the two sets of data together, under the working assumption that components other than $\sigma^2(p)$ are of similar magnitude for both groups of children. The confounded components overlap, as can be seen in Figure 8.5, and limited inferences are possible.

For sample 1, the component that confounds $p$ with $pd$ and $pt_d$ is 5.02. For sample 2 we have the separate values 4.19 and 0.27, implying a value of 4.46 for the compound. These two findings are consistent enough, because we know that the variation in sample 2 is less than in the standardizing sample.

Consider the remaining components. These have a different total in each study: in the first study, $2.55 + 1.43 = 3.98$, and in the second, $1.58 + 2.08 = 3.66$. These are not distressingly far apart, and perhaps the difference reflects sampling error. This suggestion is checked by noting that this set of components corresponds to the area $v:2$ less the area $D$ in Figure 8.6. In the retest study, each day provides data for an estimate of the quantities $v:2$ and $D$. On Day 1, $v:2 - D = 8.13 - 4.63 = 3.50$; on Day 2, the comparable figure is $8.10 - 4.28 = 3.82$. This strongly supports the notion that sampling variation is sufficient to account for the value of 3.98 in the first study.

It is of interest to try to isolate smaller bundles of components, and, in order to reach a more definite statement, we arbitrarily multiply the estimates entering $v:2 - D$ in the second study by 1.1 ($=$ ca. 3.98/3.66). The adjustment produces this information:

|  | $pj$ | $pi_j$ | Total from Study 2 adjusted by a factor of 1.1 |
|---|---|---|---|
|  |  |  | $1.7 = (C - E) \times 1.1$ |
|  | $pdj, pt_aj$ | $pdi_j, pt_ai_j, e$ | $2.3 = [v:2 - D - (C - E)]$ |
|  |  |  | $\times 1.1$ |
| Total from | 2.6 | 1.4 | $4.0 = (v:2 - D) \times 1.1$ |
| Study 1 | $A - B$ | $v:1 - A$ | $= v:1 - B$ |

From this, one can set upper and lower bounds for the components:

$$0.3 \leqslant \sigma^2(pj) \leqslant 1.7$$
$$0.9 \leqslant \sigma^2(pdj, pt_aj) \leqslant 2.3$$
$$0.0 \leqslant \sigma^2(pi_j) \leqslant 1.4$$
$$0.0 \leqslant \sigma^2(pdi_j, pt_ai_j, e) \leqslant 1.4$$

These bounds are too loose to be of great interest. The $pj$ components (specific factors in the subtests) are evidently much less influential in the subtest scores than is the general factor that runs through all verbal subtests. The general factor is reflected in the $p$ component of variance, which is only a little less than 5.0 in sample 1. The $pdj$, $pt_aj$ component may be as large as $pj$. This would imply a tendency for the shape of the profile of scores on the subtest tasks to change substantially from day to day even if the subtests were quite long. To pursue such matters properly requires a study in which all the components are estimated from the same, preferably large, sample. One might well divide the sample to have several substudies with different intervals between tests.

### Inferences about D data

The discussion that follows is no more than illustrative; any serious evaluation of WPPSI should be based on far more substantial data. Furthermore, some of the arguments to be developed in Chapters 9 and 10 should enter any attempt to draw conclusions about the generalizability of the profile of subtest scores or the Verbal composite.

**The Verbal score.** Consider the Verbal score to be the average of the scaled subtest scores obtained when the test is administered in the usual manner. This will later be rescaled to the standard deviation of 15 usually associated with IQ. The design of the D study is presumed to be $(i:j) \times (p,d,t)$, as in ordinary administration of the test. To be as definite as we might if the G data had all come from one sample, we arbitrarily assign the value 1.0 to $\sigma^2(pdi_j, pt_di_j, e)$. This is at the high end of the range 0.0–1.4, chosen because experience shows that there are substantial momentary fluctuations in performance. Assigning this value enables us to calculate the other previously indeterminate components, and the resulting values serve well enough in the ensuing illustration of technique:

| | |
|---|---|
| person (Study-1 population) | 4.7 |
| person × subtest | 1.3 |
| person × (item-set:subtest) | 0.4 |
| person × day, person × (trial:day) | 0.3 |
| person × day × subtest, person × (trial:day) × subtest | 1.3 |
| person × day × (item set:subtest), etc. | 1.0 |

Considering $n'_j = 5$,[1] the components entering the variance of the Verbal score are as follows:

| $p$ | $pj$ | $pi_j$ | $pd, pt_d$ | $pdj, pt_aj$ | $pdi_j, \ldots$ |
|---|---|---|---|---|---|
| 4.7 | 0.3 | 0.1 | 0.3 | 0.3 | 0.2 |

The expected observed-score variance is 4.7 + 0.3 (one-fifth of 1.3) + 0.1 + 0.3 + 0.3 + 0.2 = 5.9 for the Verbal score. (If a total score instead of an average had been used, this variance would be 25 times as large, i.e., 148. This is consistent with the value of 149 directly calculated from scaled scores for the standardization sample.)

The variance of universe scores depends on the universe. Making use of Table 8.1 we can construct Table 8.5. The table gives the universe-score variance for each universe, and the coefficient of generalizability (ratio of universe-score variance to observed-score variance). When the observed-score variance is rescaled to 225, that is, to the IQ scale, we have the universe-score and error-score ($\delta$) variances given in the last two lines of the table. In the table, $pd$ is called for and not $pd, pt_d$, because only the former enters the universe score. Inequalities have to be used for components involving $d$, since our estimates above include trial effects. Probably the trial effect is small relative to the day effect, save in the last component.

If investigators do indeed wish to generalize to Universe 6, the coefficient of generalizability is 0.80. The published coefficients calculated directly from retest scores (0.86) and split-half scores (0.94) give much too favorable an impression of the generalizability of Wechsler Verbal IQs. Coefficients of generalizability to Universe 6 are not inescapably low, because tests can be

---

[1] There is no need to consider $n'_i$, as components are already scaled to recognize that two half-subtests are added together.

**TABLE 8.5.** *Numerical Estimation of Universe-Score Variance for WPPSI Verbal Scale*

| Component of variance | Value when $n'_j = 5$ | Contribution to universe-score variance for Universe | | | | | |
|---|---|---|---|---|---|---|---|
| | | 1 | 2 | 3 | 4 | 5 | 6 |
| person | 4.7 | 4.7 | 4.7 | 4.7 | 4.7 | 4.7 | 4.7 |
| person × subtest | 0.3 | 0.3 | 0.3 | | 0.3 | 0.3 | |
| person × (item-set:subtest) | 0.1 | 0.1 | | | 0.1 | | |
| person × day | ≤0.3 | ≤0.3 | ≤0.3 | ≤0.3 | | | |
| person × day × subtest | ≤0.3 | ≤0.3 | ≤0.3 | | | | |
| person × day × (item-set:subtest) | ≤0.2 | ≤0.2 | | | | | |
| Estimated universe-score variance | | ≤5.9 | ≤5.6 | ≤5.0 | 5.1 | 5.0 | 4.7 |
| Estimated observed-score variance | 5.9 | 5.9 | 5.9 | 5.9 | 5.9 | 5.9 | 5.9 |
| Coefficient of generalizability | | ≤1.00 | ≤0.95[a] | ≤0.85 | 0.86[b] | 0.85 | 0.80 |
| Variances rescaled by 225/5.9 = 38.1: | | | | | | | |
| Expected observed scores (IQ) | | 225 | 225 | 225 | 225 | 225 | 225 |
| Universe scores (IQ scale) | | ≤225 | ≤214 | ≤191 | 194 | 191 | 179 |
| Error $\mathscr{E}\sigma^2(\delta)$ | | ≥0 | ≥11 | ≥34 | 31 | 34 | 46 |

[a] The split-half study reported in the manual led to a coefficient of 0.94. Calculating directly from the sample-1 components:

$$5.02 + \frac{2.55}{5} = 5.53 \quad \text{and} \quad \frac{5.53}{5.53 + \dfrac{1.43}{5}} = 0.95$$

Therefore, the small discrepancy does not arise from our mixing of data from two samples in the above table.

[b] The retest study reported in the manual led to a coefficient of 0.86, when rescaled by the test publisher to the wider range of the standardization sample.

designed with such generalization in view. Terman and Merrill calculated a coefficient of 0.91 from Stanford–Binet scores obtained under a design of type VI, taking Universe 6 as the universe of generalization. There are several possible reasons for their larger coefficient. The universe from which they sampled covered a small time period; they used 7-year-old, not 5-year-old, subjects; and their $n'_j$ was large because of the Stanford–Binet format.

The considerable drop from the Universe-5 coefficient to the Universe-6 coefficient is the key point to be understood. Both are what can be called coefficients "of stability and equivalence." If performance is unstable between testings, or if too few items are used to give an adequate sample of abilities, either coefficient will be lowered. But in generalizing to Universe 5, one looks on the test as a sample of performances within a fixed universe of tasks defined by the subtests. In Universe 6, the subtests are seen as representative of a larger class of possible subtests. Apparently, WPPSI gives fairly good data about a pupil's ability on this particular set of verbal tasks, but on another set of verbal tasks of the same general nature ranks would change appreciably. The Stanford–Binet has a considerably greater diversity of tasks, and is better adapted for estimating the Universe-6 score. The price of this is that no subtest scores can be interpreted. Single tasks are not represented by enough items in the Stanford–Binet to justify a breakdown of the IQ.

Inspection of the magnitudes of the components leads to suggestions for improving the generalizability of the Verbal score. The components involving item-sets are comparatively small, and one could add subtests (keeping $n'_i n'_j$ fixed) without changing the net contributions of these components. Shortening subtests would improve generalization over Universe 6 for the Verbal score; but it would reduce generalizability to the subtest universe score. The components involving $pd$ interactions account for much of the error variance in generalizing to Universes 4, 5, and 6. To reduce this error, one could administer half the test on one day and half on another. Assuming that the "day" portion is much larger than the "trial" portion in the fourth and fifth components, splitting administration of the test over two days would cut the $pd$ contribution by 0.01 or 0.02, and would raise the Universe-5 coefficient from 0.86 to about 0.89.

The inequalities are troublesome. The only way to separate "day" from "trial" components is to administer the same items twice on one day. The immediate retest design is ordinarily considered questionable because memory effects raise the test–retest correlation. With our analysis that considers internal consistency simultaneously with retest information, any such spurious evidence of consistency would raise the "day" components at the expense of the "trial" components, but it would not alter the coefficients for generalization to Universes 4, 5, and 6.

The ideal design for a G study is difficult to specify because investigators will set different priorities. A design that disentangles components more completely is likely to sacrifice precision of some estimates. To give some sense of the range of possibilities, consider the following design that (like a conventional parallel-form study) obtains 12 subtest scores:

Day 1: Subtests 1, 2, 3, 4, 5, 6; 1 and 2 repeated
Day 2: Subtests 3, 4, 7, 8.

This generates covariances for scores reflecting no common facets, or having *d*, or *j*, or *d* and *t*, or *d* and *j* in common. Something is now learned about the distinctive contributions of *d* and *t*. A split-half scoring adds further information.

If the investigator is interested only in Universes 4, 5, and 6, an elaborate breakdown such as we have made is not essential. From the matrix of covariances between and within testings, the first three components in Table 8.3 can be estimated. The observed-score variance can be obtained directly from the data. This information is all that is needed to arrive at the coefficients for generalization to Universes 4, 5, and 6. However, much information useful in altering the design of the D study is not obtained by this analysis.

A question regarding standard-score scales is raised by Table 8.3. It is hard to justify a conversion to IQ that holds the variance of observed scores constant. The universe-score variance is a function of the universe of generalization and the D study. Any modification of the D-study design, such as splitting the test administration between two days, will alter the observed-score variance. In fields other than psychology and education, units of measure are defined in terms of the ideal. It seems likely that, where standard scores are wanted, tests should be rescaled to set the standard deviation of universe scores equal to some preferred constant such as 10 (considering the proper universe of generalization). However, this would be troublesome for a test that is interpreted with reference to more than one universe because each universe would require a different numerical scale.

*Subtest scores.* It remains to apply the information collected in the G study to the evaluation of subtest scores. It would be possible to investigate each subtest separately, carrying out a G study on the halves of that particular subtest on two trials. Components estimated for all subtests taken together give a gross impression but do not indicate which subtests have the highest degree of generalizability.

The expected value of observed-score variance on an unspecified subtest is estimated from the components of variance, with $n'_j = 1$ and, in this case, no adjustment for $n_i'$ because components are already scaled to refer to whole subtests. The estimated observed-score variance is $4.7 + 1.3 + 0.4 + 0.3 + 1.3 + 1.0$, or $9.0$. The components of the subtest universe score and its variance depend upon one's intent. One is unlikely to generalize from just one subtest to a universe of subtests, and therefore Universes 6 and 3 are ignored. Information is too limited to make a report for Universe 1. For practical purposes Universe 5 appears to be the most likely universe of generalization. The universe-score variance is $4.7 + 1.3 = 6.0$ and the coefficient equals $6.0/9.0 = 0.67$. For Universe 4, the values are 6.4 and 0.71; for Universe 2, 7.6 and 0.84.

***Differences between subtests.*** Profile interpretation assumes that observed difference scores correspond reasonably well to difference scores in the universe. The universe presumably is one in which the pair of subtests is fixed, the items are variable, and the days of testing are variable. The theory for this kind of analysis will be delayed until Chapters 9 and 10, as two distinct variables are under consideration. Nevertheless, it may be useful to make a sketchy analysis here, with machinery already at hand.

Consider subtests $a$ and $b$, with the difference based on scores from the same trial (as is typically the case). The universe score is $\mu_{pa} - \mu_{pb}$. The components that enter the universe-score variance for the subtest are those for person and person $\times$ subtest. However, because the person component enters scores for both subtests, it does not enter the universe-score variance for the difference. As both subtests contribute interactions, that variance is $2\sigma^2(pj)$. The person $\times$ day component is ignored because this enters both scores and cancels out of the observed difference. The remaining components enter the observed-score variance with weight 2, because each subtest samples items and moments independently. Consequently, these estimates are obtained:

1. Expected observed-score variance for a difference between subtests given on same day $2(1.3) + 2(0.4 + 1.3 + 1.0) = 8.0$.
2. Universe-5 score variance for a difference between subtests $2(1.3) = 2.6$.
3. Coefficient of generalizability for difference score 0.32.
4. The value $\hat{\sigma}(\delta)$ for subtest difference score $5.4^{1/2} = 2.3$.

This $\hat{\sigma}(\delta)$ is appreciably larger than the standard error of about 1.8 used by the test developers in telling the reader of the manual what differences are likely to be significant. Variation from day to day has been considered to be a source of error, where the test developers base their figure on the split-half coefficient that treats day as fixed.

## EXERCISES

**E.1.** Complete this table:

| Study (as in Table 8.2) | I | II | III | IV | V | VI |
|---|---|---|---|---|---|---|
| Diagram (as in Figures 8.3, 8.4) | 8.3a | | | | | |
| Components adding to correlation | $p, d, pd$;<br>$pj, dj, pdj$;<br>$pi_j, di_j$,<br>$pdi_j$ | | | | | |

**E.2.** State a general rule, in terms of $n_i$ and $n_j$, regarding the number of $i$-common, $j$-common, and other covariances (over all persons) in a study with Design VII.

**E.3.** A universe of admissible observations is defined by a listing of numerous topics for essays and numerous graders. This universe is the same for all persons. In principle, a person may write on a topic more than once and a grader may score a particular essay more than once.

Complete the table (in the general style of Table 8.1) identifying four universes of generalization, and the score components that might contribute to the universe score. For the sake of simplifying the response, do not list any component such as grader that does not involve the person.

| | Universe of generalization | | | |
|---|---|---|---|---|
| | 1 | 2 | 3 | 4 |
| Topic | Fixed | Fixed | Variable | Variable |
| Grader | Fixed | Variable | Fixed | Variable |
| person | | | | |
| person × topic | | | | |
| person × grader | | | | |
| person × topic × grader | | | | |
| trial:person × topic | | | | |
| [trial:(person × topic)] × grader | | | | |
| trial × rescoring:(person ×<br>    topic × grader) | | | | |

**E.4.** The average observed-score variance for an essay (one topic, scored once, all persons scored by the same grader) is 100. The correlation between essays for two particular topics scored by the same grader is 0.70. What does this tell about components of variance in the universe of admissible observations described in Exercise 3?

**E.5.** Two simple random-parallel achievement tests were administered twice (Westbrook & Jones, 1968). The correlations are as follows:

|     | A1 | B1 | A2 | B2 |
|-----|----|----|----|----|
| A1  |    | 0.63 | 0.82 | 0.71 |
| B1  |    |    | 0.55 | 0.71 |
| A2  |    |    |    | 0.62 |
| B2  |    |    |    |    |

What inferences can be made about variance components, taking 1.00 as the observed-score variance for each test ($i$ = A or B) on each occasion ($j$ = 1 or 2)? The sample used in determining the correlations was modest in size.

**E.6.** Suppose that the number of Wechsler subtests were doubled, and the length of each subtest cut in half. Using the components of variance from which Table 8.5 was derived, what effect would this change have on the error variance and the coefficient of generalizability for each of the six universes?

**E.7.** Explain the difference between the coefficients for generalization to Universes 2 and 3 in Table 8.5.

**E.8.** Suppose that a clinician attempts to generalize from a single Wechsler subtest score to Universe 6—the person's expected score over all admissible verbal subtests. What is the coefficient of generalizability?

**E.9.** From the average covariances between subtests given in Chapter 9, Exercise 7 (p. 304), which components of variance for Verbal and Performance subtests can be estimated?

### Answers

**A.1.**

| Study | II | III | IV | V | VI |
|-------|----|----|----|----|----|
| Diagram | 8.4a | 8.4b | 8.3b | 8.4c | 8.4d |
| Components adding to correlation | $p, d, pd; pj, dj, pdj$ | $p, d, pd; t_d, pt_d$ | $p, pj, pi_j$ | $p, pj$ | $p$ |

**A.2.** There are $n_i(n_j)(n_j - 1)/2$ different values of $i$-common covariances, and $n_i n_j(n_i - 1)/2$ different values of $j$-common covariances. There are $(n_i)(n_i - 1) \times (n_j)(n_j - 1)/2$ different values of covariances with both $i$ and $j$ different. There are no $i$-$j$-common covariances.

**A.3.**

| | Universe | | | |
|---|---|---|---|---|
| | 1 | 2 | 3 | 4 |
| person | U | U | U | U |
| person × topic | U | U | | |
| person × grader | U | | U | |
| person × topic × grader | U | | | |

All other cells shown in statement of exercise are empty.

**A.4.** The sum of the following components is approximately 70: person, person × grader.

The sum of the following components is approximately 30: person × topic, person × topic × grader, trial:(person × topic), trial × grader, and the residual (trial × rescoring, etc.).

**A.5.**

| | | Components adding to correlation |
|---|---|---|
| Average of $i$-common correlations | 0.765 | $p, pi$ |
| Average of $j$-common correlations | 0.625 | $p, pj$ |
| Average covariance, $i, j$ different | 0.630 | $p$ |

**A.6.** Components were originally determined for whole subtests. A subtest of half length will have double the value originally determined, for any component involving $i_j$. However, those components are sampled 10 times in the test, hence their contribution to the Verbal score is the same as before. The components involving $j$ and not $i_j$ are now sampled 10 times rather than 5. We have these values:

| | | | | |
|---|---|---|---|---|
| person | 4.7 | | 4.7 | 4.7 |
| person × subtest | 0.13 | rounded to 0.1 | | 0.1 |
| person × $i_j$ | 0.08 | | 0.1 | 0.1 |
| $pd, pt_d$ | 0.3 | | 0.3 | ignoring $t_d$ portion ⩽0.3 |
| $pdj, pt_aj$ | 0.13 | | 0.1 | ⩽0.1 |
| $pdi_j$, etc. | 0.2 | | 0.2 | ⩽0.2 |
| Observed-score variance | | | | 5.5 |

| | 1 | 2 | 3 | 4 | 5 | 6 |
|---|---|---|---|---|---|---|
| Universe-score variance | ⩽5.5 | ⩽5.2 | ⩽5.0 | 4.9 | 4.8 | 4.7 |
| Coefficient | ⩽1.00 | ⩽0.95 | ⩽0.91 | 0.89 | 0.87 | 0.85 |

Theory requires the improvement in the coefficient for Universes 3 and 6. The improvement observed for Universes 4 and 5, and the lack of improvement for Universes 1 and 2, are functions of the particular numerical values used in the example.

**A.7.**   In each case, one is generalizing over scores obtained on a single day, but in Universe 2, admissible observations are confined to a single set of five fixed subtests. Generalization over subtest tasks (to Universe 3) is not very accurate when the observed score is based on only five tasks.

**A.8.**   The components given on p. 254 indicate that the observed-score variance is 9.0. The universe-score variance derives entirely from the person, hence the coefficient is 4.7/9.0, or 0.52.

**A.9.**   A three-way analysis of variance is reported on p. 282. The covariances allow one to estimate components of variance as follows:

| | Result for Verbal | Result for Performance |
|---|---|---|
| Persons. Take average covariance of unlike subtests in different forms. | 8.16 | 4.84 |
| Persons × subtests. Subtract person component from average covariance of like subtests in different forms. | $9.12 - 8.16$ $= 0.96$ | $6.34 - 4.84$ $= 1.50$ |
| Persons × forms. Subtract person component from average covariance of subtests in same form (both forms considered). | $(8.66 + 7.67)/2$ $- 8.16 = 0.01$ | $(5.59 + 4.17)/2$ $- 4.84 = 0.04$ |
| Sum of remaining components. Subtract above components from average variance. | 2.34 | 2.20 |

# CHAPTER 9
# Introduction to Multivariate Generalizability Theory

To this point, observations on a single variable have been considered. Examining the generalizability of multiscore instruments and composite scores requires an extension of the theory. To be sure, the theory so far developed is applicable to any composite score, because an observed composite score is a sample from a universe of composite scores formed according to the same combining rule. A G study in which two or more suitably independent values of the composite score are obtained for each individual can provide an estimate of the generalizability of a similar score in a D study. For example, the investigator may be interested in the difference score formed by subtracting the Wechsler Performance IQ from the Verbal IQ. If an abbreviated form of the notation to be developed below is used, the observed composite score can be labelled $_dX = _{Ve}X - _{Pe}X$. A conventional way to evaluate generalizability would be to test a sample of persons twice, preferably using two forms of the test, to obtain two values of $_dX_{pi}$ for each person. Either analysis of variance or a simple correlation would indicate how well observed differences agree with the universe scores $_d\mu_p$.

Instead of first computing the $_dX_{pi}$, a multivariate analysis of the Wechsler would treat the two values of $_{Ve}X$ and the two values of $_{Pe}X$ simultaneously. There would be G studies of the Ve and Pe scores, and an analysis of the correlations or covariances of Ve with Pe. This information can be recombined to reach the conclusions given by the analysis of the $_dX_{pi}$, but it also leads to additional conclusions. It permits one to ask about the optimal design of the D study. Thus, for example, the number of conditions for observing the Pe score perhaps should be greater than the number of conditions for the Ve score; this question can be investigated best by multivariate analysis.

A multivariate analysis of observed scores on the separate variables attends to information the univariate analysis discards, and takes into account complexities ignored in most test theory. This chapter develops a rationale for multivariate analysis and applies it to the study of correlations between variables. Chapter 10 applies the theory to linear composites (including difference scores) and to score profiles. Wherever possible, the present chapter will paraphrase relevant parts of Chapters 1 and 2, because the multivariate theory is a more general case of the theory already presented.

Multifacet analysis of multivariate data is not easy to discuss. To make the main features of multivariate G studies clear, we treat only the most basic cases and designs. Even so, there will be no shortage of novel ideas for the reader to ponder over.

The material presented in preceding chapters has been under development for many years, and most of it has been exposed to thorough criticism. Moreover, it has been put to use in a number of investigations. The multivariate extension of generalizability theory is relatively recent and less tested.

### A. *Formulation*

The basic concepts of generalizability theory, developed in Chapter 1 (pp. 14–29), apply, with slight rephrasing, to composites. That discussion need not be repeated here.

When there are two or more variables, the person has a universe score on each variable and relations of the components of one variable with those of another require examination. The universe-score variance, the expected observed-score variance, etc., for a composite variable depend on components of covariance as well as on the components of variance. Components of covariance, along with components of variance, determine the correlations between universe scores on any two variables and the expected correlations between observed scores obtained under various designs. Decomposition of covariance into components is a direct extension of the decomposition of variance. Our presentation is influenced by the work of Kenneth W. Travers (unpublished) that capitalized on Tukey's dyadic analysis of variance (1949) and Bock's discussion of multivariate analysis (1963).

*Observed scores and universe scores*

Only two variables need to be explicitly considered in developing this theory. Once the basic concepts are developed, the notation and equations extend in an obvious manner to any number of variables. To indicate that variable $v$ is under discussion, we employ a prescript, designating an observed score as $_vX_{pi}$, for example. In general expressions, a second variable is denoted by $v'$. Thus $\sigma(_vX_{,v'}X)$ is a general expression for a covariance; when $v' \equiv v$, this is a variance. For ease of reading we shall ordinarily refer to the specific variables $v_1$ and $v_2$, whose covariance is $\sigma(_1X_{,2}X)$.

To begin, assume that observations on any variable are classified with respect to just one facet. Let $_1X_{pi}$ represent a score on a variable $v_1$ arising from the observation of person $p$ under a single condition of facet $i$. Observations on $v_1$ can be made under an indefinitely large number of conditions $N_i$ of facet $i$. Let $_2X_{pg}$ represent a score on $v_2$; the conditions under which $v_2$ may be observed are said to be classified with respect to facet $g$. Assume also that $N_g$ is indefinitely large. We use the symbol $g$ to make the notation general. One specialization is to let $g$ represent the same facet as $i$, but to suppose that sampling of conditions for variable $v_1$ is independent of sampling for $v_2$. A further specialization is to assume that the same particular conditions $I$ are used to observe both variables.

A universe of admissible observations is defined, such that for person $p$ there is a score $_1X_{pi}$ for every condition of facet $i$, and a $_2X_{pg}$ for every condition of $g$. Sampling one $i$ and one $g$ yields a pair of scores $_1X_{pi}, _2X_{pg}$ for person $p$. Each possible pairing of a condition $i$ with a condition $g$ defines a pair of scores that could be observed.[1] As an example, consider as variables mechanical interest and mechanical ability. There is no reason to think that in the universe a certain form of the interest test is paired with any particular form of the ability test. Similarly, if the two variables are a test score and a rating, there is no reason to think of any test form as associated particularly with any one rater in the universe.

The decision to measure $v_1$ and the condition under which it is observed are assumed not to affect the scores on $v_2$.

The investigator most often wishes to generalize to the profile of universe scores $_1\mu_p, _2\mu_p$. Alternatively, the investigator may wish to generalize to a composite of the form $w_1 {}_1\mu_p + w_2 {}_2\mu_p$. The usual difference score is such a composite, having $w_1 = 1$ and $w_2 = -1$. To write generally, an observed composite is denoted by $\sum_v w_v {}_vX_{pi}$, where $v$ now stands for any member of the set of variables and $i$ for the appropriate member of the set of facets $i, g, \ldots$.

---

[1] Formally, the universe may be thought of as having the form $(i{:}v) \times p$, with variables fixed. There is a score $X_{pvi_v}$ for each condition. The placement of $v$ as a prescript seems to be a clearer notation. (If the set of $i$ is identical to the set of $g$, the universe is fully crossed: $i \times v \times p$.)

*Score components and components of covariance in the universe of admissible observations*

Along with $_v\mu_p$, the universe score for $p$ on variable $v$, we define the means $_v\mu_i$ and $_v\mu$, comparable to the $\mu_i$ and $\mu$ of earlier chapters. For each observation corresponding to a particular $v$, $p$, and $i$,

$$
\begin{aligned}
(9.1) \qquad _vX_{pi} = {}&_v\mu &&\text{(general mean for } v) \\
&+ _v\mu_p - _v\mu &&\text{(person effect)} \\
&+ _v\mu_i - _v\mu &&\text{(condition effect)} \\
&+ _v\text{res}
\end{aligned}
$$

This merely adds the symbol $v$ to the univariate model of (1.1).

Suppose that every condition of $i$ is taken in combination with every condition of $g$. Each combination yields a pair of scores. We shall refer to such a pair (and, more generally, to a series of scores on $n_v$ variables) as a vector of scores. The set of vectors that results from forming all possible pairs is termed the *universe of admissible vectors* of observations. (A restricted universe of vectors will be defined later.)

Any observed-score pair can be expanded as in (9.1):

$$
(9.2) \qquad
\begin{aligned}
_1X_{pi} &= {}_1\mu + ({}_1\mu_p - {}_1\mu) + ({}_1\mu_i - {}_1\mu) + ({}_1\text{res}) \\
_2X_{pg} &= {}_2\mu + ({}_2\mu_p - {}_2\mu) + ({}_2\mu_g - {}_2\mu) + ({}_2\text{res})
\end{aligned}
$$

Considering all persons and/or all conditions, each score component has a multivariate distribution. Where the univariate G study examined components of variance, in the multivariate study attention is given to both variances and covariances.

Testers are familiar with the within-condition variances of observed scores that we denote by $\sigma^2(X_{pi} \mid i)$. We have also directed attention to $\sigma^2(X_{pi})$, the variance of observed scores over all persons and conditions. With regard to the covariance of observed scores, a similar pair of definitions is required. The familiar covariance for two sets of scores (i.e., for two designated conditions) is $\sigma(_vX_{pi},_{v'}X_{pg} \mid i,g) = \mathscr{E}_p(_vX_{pi} - {}_v\mu_i)(_{v'}X_{pg} - {}_{v'}\mu_g)$. The covariance analogous to $\sigma^2(X_{pi})$ is $\sigma(_vX_{pi},_{v'}X_{pg}) = \mathscr{E}_i\mathscr{E}_g\mathscr{E}_p(_vX_{pi} - {}_v\mu)(_{v'}X_{pg} - {}_{v'}\mu)$. This expectation takes into account all admissible pairs of conditions of facets $i$ and $g$. These and other covariances to be defined are listed in Table 9.1.

There is a variance–covariance matrix for observed scores whose general element is $\sigma(_vX_{pi},_{v'}X_{pg})$. For two variables it can be written as

$$
(9.3) \qquad
\begin{vmatrix}
\sigma^2(_1X_{pi}) & \sigma(_1X_{pi},_2X_{pg}) \\
\sigma(_1X_{pi},_2X_{pg}) & \sigma^2(_2X_{pg})
\end{vmatrix}
$$

**TABLE 9.1.** *Definitions of Components of Variance and Covariance in the Multivariate Model*

| Univariate model | Multivariate model | | | |
|---|---|---|---|---|
| Variance | Variance | Covariance | Matrix[a] | Remarks[b] |
| $\sigma^2(X_{pi}\mid i)$ | $\sigma^2(_vX_{pi}\mid i)$ | $\sigma(_vX_{pi},_{v'}X_{pg}\mid i,g)$ | | Defined within conditions as $\mathscr{E}_p(_vX_{pi} - _v\mu_i)(_{v'}X_{pg} - _{v'}\mu_g)$ |
| $\sigma^2(X_{pi})$ | $\sigma^2(_vX_{pi})$ | $\sigma(_vX_{pi},_{v'}X_{pg})$ | $\sum_\mathbf{v}X_{pi}$ | Defined over all admissible pairs of conditions as $\mathscr{E}_i\mathscr{E}_g\mathscr{E}_p(_vX_{pi} - _v\mu)(_{v'}X_{pg} - _v\mu)$ |
| $\sigma^2(\mu_p)$ | $\sigma^2(_v\mu_p)$ | $\sigma(_v\mu_p,_{v'}\mu_p)$ | $\sum_\mathbf{v}p$ | |
| $\sigma^2(\mu_i)$ | $\sigma^2(_v\mu_i)$ | $\sigma(_v\mu_i,_{v'}\mu_g)$ | $\sum_\mathbf{v}i$ | Defined over all admissible pairs of conditions as $\mathscr{E}_i\mathscr{E}_g(_v\mu_i - _v\mu)(_{v'}\mu_g - _{v'}\mu)$ |
| $\sigma^2(\mu_{pi},e)$ | $\sigma^2(_v\mu_{pi},e)$ | $\sigma(_v\mu_{pi},e;_{v'}\mu_{pg},e)$ | $\sum_\mathbf{v}pi, e$ | Defined over all admissible pairs of conditions as $\mathscr{E}_i\mathscr{E}_g\mathscr{E}_p(_vX_{pi} - _v\mu_p - _v\mu_i + _v\mu)$ $\times (_{v'}X_{pg} - _{v'}\mu_p - _{v'}\mu_g + _{v'}\mu)$ |

[a] The notation in this column is explained on p. 272.
[b] These definitions apply to the universe of admissible observations. The analogous definitions for the universe of linked observations (see p. 271) have $\mathscr{E}_{i\cdot g}$ in place of $\mathscr{E}_i\mathscr{E}_g$.

The component for persons has a variance-covariance matrix such as this:

$$(9.4) \quad \begin{vmatrix} \sigma^2(_1\mu_p - _1\mu) & \sigma(_1\mu_p - _1\mu, _2\mu_p - _2\mu) \\ \sigma(_1\mu_p - _1\mu, _2\mu_p - _2\mu) & \sigma^2(_2\mu_p - _2\mu) \end{vmatrix}$$

The general term of such a matrix is $\sigma(_v\mu_p,_{v'}\mu_p)$; when $v \equiv v'$, one has a diagonal element, a variance. $\sigma^2(_1\mu_p - _1\mu)$ simplifies to $\sigma^2(_1\mu_p)$, etc. Previously, the variance of the person component has been written as $\sigma^2(p)$; a multivariate study requires notational distinctions such as these:

$\sigma^2(_vp)$—abbreviated form of $\sigma^2(_v\mu_p)$, the variance component for persons on variable $v$;

$\sigma^2(_1p)$—same specialized to variable 1.

For the condition component, there is a matrix whose general term is $\sigma(_v\mu_i,_{v'}\mu_g)$ that can be abbreviated to $\sigma(_vi,_{v'}g)$. And for the residual there is a similar matrix of $\sigma(_v\text{res},_v\text{res})$ or, in a better notation, $\sigma(_vpi,e;_vpg,e)$. Only the matrix for the person component is of basic interest. The components defined above, for scores obtained with all possible pairings of $i$ and $g$, are required in developing the theory, but the covariance components for conditions and residual to be estimated from a G study are defined differently. This will be made clear shortly.

The variance $\sigma^2({}_vX_{pi})$, analogous to the "total sum of squares" in analysis of variance, is partitioned as follows:

$$(9.5) \qquad \sigma^2({}_vX_{pi}) = \sigma^2({}_vp) + \sigma^2({}_vi) + \sigma^2({}_vpi,e)$$

This is the result—originally (1.2)—on which our univariate theory was based. It specializes to

$$(9.6) \qquad \begin{aligned} \sigma^2({}_1X_{pi}) &= \sigma^2({}_1p) + \sigma^2({}_1i) + \sigma^2({}_1pi,e) \\ \sigma^2({}_2X_{pg}) &= \sigma^2({}_2p) + \sigma^2({}_2g) + \sigma^2({}_2pg,e) \end{aligned}$$

The first of these equations has, on the right, the sum of the upper-left elements of the matrices for persons (9.4), conditions, and residual. A similar statement can be made about the off-diagonal elements:

$$(9.7) \qquad \sigma({}_1X_{pi},{}_2X_{pg}) = \sigma({}_1p,{}_2p) + \sigma({}_1i,{}_2g) + \sigma({}_1pi,e;{}_2pg,e)$$

Therefore, we may say that the matrix (9.3) is the sum of the variance–covariance matrices for person, condition, and residual components. Our concern is limited to the covariance matrix for persons. Because we are considering all possible pairs (every condition of $i$ with every condition of $g$), $\sigma({}_1i,{}_2g)$ is zero.[2] Correspondingly, $\sigma({}_1pi,e;{}_2pg,e) = 0$.

### Joint sampling in multivariate studies

Though all possible pairings of conditions of $i$ with conditions of $g$ usually define meaningful vectors, the tester may be interested in only a restricted set of pairs. The most obvious example arises where observations on $v_1$ and $v_2$ are classified with respect to the same facet. In this case ${}_1X_{pi*}$ is directly related to ${}_2X_{pi*}$, and less related to any other ${}_2X_{pi}$. The investigator may know that he will always, in his D study, pair each ${}_1X_{pi}$ with the corresponding ${}_2X_{pi}$ and not with any ${}_2X_{pi'}$. Thus, an observed composite score could be ${}_1X_{pA} + {}_2X_{pA}$ or ${}_1X_{pB} + {}_2X_{pB}$ but not ${}_1X_{pA} + {}_2X_{pB}$. To deal with this formally, it is necessary to use the concept of joint sampling, introduced originally in considering univariate designs such as $(i,j) \times p$ and $(i,j){:}p$ (see p. 37). Joint sampling is to be contrasted with independent sampling.

The multivariate G study collects observations for both variables on the

---

[2] As seen earlier (p. 27), certain products vanish when expectancies are taken. For example, $\underset{i}{\mathscr{E}}\,\underset{g}{\mathscr{E}}({}_1\mu_i - {}_1\mu)({}_2\mu_g - {}_2\mu)$ can be factored into the product of $\underset{i}{\mathscr{E}}({}_1\mu_i - {}_1\mu)$ and $\underset{g}{\mathscr{E}}({}_2\mu_g - {}_2\mu)$, and each of these is zero. If facets $i$ and $g$ are identical, we may make the same argument, because the number of conditions is indefinitely large, and every $i$ is paired with every other $i$.

same persons, with two or more conditions of facet $i$ and of $g$. Assume that, within variables, $n_i$ and $n_g$ are uniform for all $p$.

In independent sampling, the conditions of $g$ are selected without knowledge as to the $i$ selected, and vice versa. The G study samples $n_i$ conditions from the universe of $i$, and independently samples $n_g$ conditions from the universe of $g$. (E.g., the investigator might select at random a series of occasions on which to observe aggressiveness, and select independently occasions on which to observe constructiveness of play.) If independent sampling is continued indefinitely with replacement, one pair of observations being drawn at a time, the entire universe of admissible vectors is generated.

In joint sampling, pairs $i$, $g$ are to be sampled. One possibility is that facet $g$ is the same as facet $i$; then joint sampling ordinarily means that whatever condition of $i$ is drawn will be used to make observations on both $v_1$ and $v_2$. However, there can be joint sampling where $v_1$ and $v_2$ are observed under different conditions. The occasion for observing a child's attentiveness might be the class period immediately after a sample of blood is collected for analysis of blood-sugar level. The observation is more closely linked to this blood sample than to a blood sample taken prior to some other observation. By slight license, we speak of having taken observation and blood sample on the same occasion. Joint sampling can even occur where facet $i$ and facet $g$ are different in kind. For example, $i$ might represent informants whose ratings constitute $v_1$, and $g$ might represent situations in which the person is systematically observed by a trained team to get scores $v_2$. The two sets will be in correspondence when informant $i^*$ has formed his impressions of the person in situation $g^*$ (e.g., the sports field), and not in other situations where the team might also directly observe the person. There is joint sampling of informants and situations if we systematically choose informants who are ecologically paired with the observation situations we have chosen. There is independent sampling if the informants are chosen without regard to the sample of situations where observations are being made. With independent sampling, the French teacher can show up as an informant even if the trained team makes no observations in the French classroom.

A joint-sampling plan is, formally, a plan for drawing $i$, $g$ *pairs* according to some rule, instead of drawing $i$ and $g$ separately. Strictly speaking, to have a rule for drawing pairs one must specify that conditions are paired *a priori* within the universe of $i$ and $g$. The mathematical results to be stated regarding joint-sampled data hold even when pairs are sampled from a universe of $i$ and $g$ within which the pairing is entirely arbitrary. Independent sampling can therefore be seen as a degenerate case of joint sampling where the pairing makes no difference.

The theory can also accommodate the case where drawing a particular $i$ alters the probability that the various conditions of $g$ will be drawn, yet does not determine that a certain $g$ will be selected with probability 1.00. Subsequent statements are not worded to cover this possibility.

When joint sampling was considered in univariate studies, the number of conditions of $j$ was always equal to the number of conditions of $i$. Where there was a many-to-one correspondence of $j$ to $i$, we spoke of the design as $j:i$. In sampling $i$ and $g$, it is usual to have $n_i = n_g$ or $n_i' = n_g'$, but many-to-one correspondences are possible.

We refer to observations $_1X_{pi}$ and $_2X_{pg}$ as "linked" when $i$ and $g$ are jointly sampled. The "bullet" symbol will be used to indicate linkage; for example, in describing an experimental design we may include the expression $i \bullet g$. An $i$ can be linked with more than one $g$. Where we wish to emphasize the fact that $i$ and $g$ are independently sampled, we use the symbol $\circ$ ("hole") as in $i \circ g$. As we assume that the $i$ are sampled randomly, it follows that when $i \bullet g$ and $i' \bullet g'$, $i \circ g'$ and $i' \circ g$.

Where there are two facets, the model extends readily to the drawing of $i,g$ pairs and the drawing of $j,h$ pairs. Complex combinations might exist in the universe; for example, $i$ might be linked to $g$ and also to $h$, with $j$ independent. We shall ignore such possibilities in our discussion.

*The universe of linked observations.* When a very large number of $i$ and $g$ are drawn under a joint-sampling scheme, the observations on $v_1$ ultimately cover the universe of $N_i$ observations on $v_1$; similarly for $v_2$. Because of the pairing rule reflected in the sampling scheme, however, not all of the admissible score vectors will be obtained. If the $i$ and $g$ are put in one-to-one correspondence (implying $N_i = N_g$), only $N_i$ of the $N_i^2$ admissible pairs of scores may be considered. These pairs consistent with the sampling rule constitute *the universe of linked observations*. This is a subset of the universe of admissible vectors. The components of linked observations are the same as those listed earlier, and (9.2) applies. The vector of universe scores $_1\mu_p$, $_2\mu_p$ is the same for the universe of linked observations as for the universe of admissible vectors, and (9.4) applies. But only a subset of the original component vectors for conditions and residual enter into joint-sampled observed scores, hence the components of covariance are different.

Covariance components are mean products of score components, averaged over pairs of observations that could be formed by applying a certain sampling rule repeatedly to the universes of conditions $i$ and $g$. The sampling rule is an aspect of the experimental design. In univariate studies, the components to be estimated in a G study are components of the universe; the design is irrelevant. In a multivariate study, design still has no influence on variance components, but the design does define the universe of linked observations and hence the components of covariance.

Different joint-sampling rules ordinarily produce different covariances.

As one example, suppose observations are classified with respect to occasions, and the occasions for observing $v_1$ are selected jointly (i.e., in a one-to-one pairing) with occasions for observing $v_2$. The interval between the members of the pair could be a few minutes or it could be a few days. Same-day observations probably covary to a greater degree than same-week observations, and to a lesser degree than same-minute observations.

*Covariance components for the universe of linked observations.* The variances and covariances in the universe of linked observations are as follows:

|  | Variances | Covariance |
|---|---|---|
| Condition component | $\sigma^2(_1i); \sigma^2(_2g)$ | $^{\bullet}\sigma(_1i,_2g)$ |
| Residual component | $\sigma^2(_1pi,e); \sigma^2(_2pg,e)$ | $^{\bullet}\sigma(_1pi,e;_2pg,e)$ |
| All observations | $\sigma^2(_1X_{pi}); \sigma^2(_2X_{pg})$ | $^{\bullet}\sigma(_1X_{pi},_2X_{pg})$ |

The bullet in each case is a reminder that we are considering only the pairs that can be formed according to the joint-sampling rule. Therefore, the covariance component for $i$ is defined as

$$(9.8) \qquad {}^{\bullet}\sigma(_1i,_2g) = \underset{i\cdot g}{\mathscr{E}} \, (_1\mu_i - _1\mu)(_2\mu_g - _2\mu)$$

where the index $i \cdot g$ indicates that products only for linked $i$ and $g$ enter the expectation. The variances are the same as in independent sampling and in univariate studies. The covariance components for linked observations will not, in general, be zero.[3]

While linkage has to be considered in defining components of covariance for conditions and residual in the universe of linked observations, the definition of $\sigma(_1p,_2p)$ is not altered. In joint sampling, the analog of (9.7) is

$$(9.9) \qquad {}^{\bullet}\sigma(_1X_{pi},_2X_{pg}) = \sigma(_1p,_2p) + {}^{\bullet}\sigma(_1i,_2g) + {}^{\bullet}\sigma(_1pi,e;_2pg,e)$$

With each change in the sampling rule that links $i$ and $g$, a new universe of linked observations is defined and the linked covariance components change.

The reader may be puzzled by the inclusion of a covariance component for $e$, because errors are generally thought of as uncorrelated. It is possible, however, that the component $_1e_{pi}$ will correlate with $_2e_{pg}$. As a simple example, consider the case where $i \equiv g$, and variables 1 and 2 are two scorings of the same performance. Thus $v_1$ might be speed in arithmetic and

---

[3] The linked covariances are derived from a subset of the products of components that enter $\sigma(_1i,_2g)$—discussed on p. 268. That subset has a nonzero mean. The products in the subset for non-linked conditions have an expected value of zero. Thus, a set whose expected value is zero has been decomposed into subsets, one having a zero mean and one having a nonzero mean. This is possible because the proportion of the total set that consists of linked observations becomes vanishingly small as more and more observations are considered.

$v_2$ might be accuracy. On any problem sheet $i$, both scores are recorded. Any tendency for the subject to hurry faster than usual on a particular problem sheet is likely to raise $_1X_{pi}$ and to lower $_2X_{pi}$. We interpret this as a reflection of the $e$ component rather than of the $pi$ component, because there is no reason to think the person would work especially fast everytime he works on this particular problem sheet. The covariance for $e$ is likely to depart from zero whenever $_1X_{pi}$ and $_2X_{pg}$ are "not experimentally independent," as the traditional language has it.

Just as (9.3) could be seen as the sum of variance-covariance matrices for persons, conditions, and residual, we can consider the variance-covariance matrix whose general term is $^\bullet\sigma(_1X_{pi},_2X_{pg})$ to be the sum of three matrices: (9.4) for person components, a matrix for condition components that contains the linked covariance, and a matrix for the residual components containing their linked covariance. (9.6) gives the diagonal elements and (9.9) the off-diagonal elements.

### A general notation and statement of identities

To discuss a large number of variables, matrix notation is useful. We shall rarely employ this notation in subsequent discussion, but in the later evolution of multivariate generalizability theory such a flexible notation will probably prove useful. No significant use of matrix algebra is required in the statements made here.

Where previously we spoke of specific variables $v_1$ and $v_2$ we may write $\mathbf{v}$ for the series of $n_v$ variables $v_1, v_2, \ldots$. We let $i$ stand, not merely for the facet associated with $v_1$, but for whatever facet is associated with a particular variable, as in the general expression for a vector of observed scores $_\mathbf{v}X_{pi}$. This expression serves for the string of scores $_1X_{pi}, _2X_{pg}, \ldots$. We write $_\mathbf{v}\mu_p$ to represent $_1\mu_p, _2\mu_p, \ldots$; similarly for other components.

The general form of (9.1) expresses in vector notation a system of equations

$$(9.10) \qquad _\mathbf{v}X_{pi} = {}_\mathbf{v}\mu + ({}_\mathbf{v}\mu_p - {}_\mathbf{v}\mu) + ({}_\mathbf{v}\mu_i - {}_\mathbf{v}\mu) + ({}_\mathbf{v}\text{res}).$$

The observed vector of scores is thus described as a sum of vectors of score components. The vector of universe scores is $_\mathbf{v}\mu_p$.

With $n_v$ variables, the matrix of variances and covariances for observed scores (9.3) is extended to $n_v$ rows and columns; each diagonal entry is a variance, and each off-diagonal entry a covariance. We denote this matrix by $\sum {}_\mathbf{v}X_{pi}$; its general element is, as before, $\sigma(_vX_{pi},_{v'}X_{pg})$, but now $v = 1, \ldots, n_v$ and $v' = 1, \ldots, n_v$. There is a similar general symbol for each other matrix, as indicated in Table 9.1.

The various sums referred to in the two-variable case can be expressed in general equations. The statements made in (9.6) and (9.7) combine, and extend over all variables, in

(9.11)
$$\sum{}_{\mathbf{v}}X_{pi} = \sum{}_{\mathbf{v}}p + \sum{}_{\mathbf{v}}i + \sum{}_{\mathbf{v}}pi,e$$

It will be recalled that this statement applies to the universe of admissible observations, and that the off-diagonal elements in the second and third matrices on the right vanish.

For linked observations there is a comparable statement:

(9.12)
$$\cdot\sum{}_{\mathbf{v}}X_{pi} = \sum{}_{\mathbf{v}}p + \cdot\sum{}_{\mathbf{v}}i + \cdot\sum{}_{\mathbf{v}}pi,e$$

For each expression in $i$ there is a completely analogous expression in $I$, where $I$ stands here as a general symbol for $I, G \ldots$. We are familiar with the fact that $\sigma^2(I) = \sigma^2(i)/n_i$. Similarly, if $n_i = n_g$, $\cdot\sigma(_1I,_2G) = \cdot\sigma(_1i,_2g)/n_i$, and,

(9.13)
$$\cdot\sum{}_{\mathbf{v}}X_{pI} = \sum{}_{\mathbf{v}}p + \frac{1}{n_i}\cdot\sum{}_{\mathbf{v}}i + \frac{1}{n_i}\cdot\sum{}_{\mathbf{v}}pi,e$$

### B. *Varieties of Experimental Design*

For variable $v_1$, any of the designs employed in univariate studies may be used; similarly for $v_2$. We identify designs by means of a notation like that used earlier, with a small addition. Formerly, with one facet, the alternative designs were $i \times p$ and $i:p$. With two variables the following are possible:

$$_1(i \times p), {}_2(g \times p)$$
$$_1(i:p), {}_2(g:p)$$
$$_1(i:p), {}_2(g \times p)$$
$$_1(i \times p), {}_2(g:p)$$

We append $[i \bullet g]$ to the notation where sampling of $i$ and $g$ is joint. Where sampling is independent, we may write $[i \circ g]$, but ordinarily we merely omit the linkage notation. A design such as $_1(i:p), {}_2(g \times p), [i \bullet g]$ is little more than a hypothetical possibility; selecting a particular $g$ would have to imply selecting a particular $i$ or set of $i$ for each person, different from person to person. We shall ignore such designs.

With two facets and independent sampling, one may have any combination of the designs listed in Figure 2.4. Joint sampling may call for $i \bullet g$ or $j \bullet h$, or both. The design considered for the D study determines which G-study designs are suitable. If there is to be joint sampling of $i$ and $g$ in the D study, the G study must employ the same joint-sampling rule (or a complex rule that embodies the rule for $i$ and $g$) if it is to provide useful information. Suppose, for instance, that the D-study design is: $_1[j:(i \times p)], {}_2[h:(g \times p)]$, $[i \bullet g, j \bullet h]$; i.e., $_1$IV-A with $_2$IV-A, with joint sampling. Then the following

G-study designs, among others, will estimate the needed components of covariance:

$$_1\text{VII with } _2\text{VII: } _1(i \times j \times p), _2(g \times h \times p), [i \bullet g, j \bullet h]$$

$$_1\text{V-A with } _2\text{V-A: } _1[(j{:}i) \times p], _2[(h{:}g) \times p], [i \bullet g, j \bullet h]$$

$$_1\text{IV-A with } _2\text{IV-A: } _1[j{:}(i \times p)], _2[h{:}(g \times p)], [i \bullet g, j \bullet h]$$

The G study should embody any joint sampling and any crossing that will appear in the D study.

Values of $n_i$ and $n_g$ need not be equal, nor will $n_i'$ and $n_g'$ generally be equal in the D study. The latter is the more important to consider. We have seen earlier that the G study suggests how many conditions of a facet should be drawn in the D study to attain the level of precision desired for a particular decision. There is no reason to think that in any given study the same number will be recommended for $v_1$ and $v_2$, because the corresponding components of error variance will not in general be of the same size. Suppose $v_1$ is a measure of conduct in the English classroom, and $v_2$ a measure of conduct during the physical education period. Time-samples are taken of each and, because there is interest in day-to-day variability, observations are made for both variables on the same day. The G study may suggest that it is desirable to take two samples per pupil during the physical education period, and only one during the English period, if there is greater within-period variability for the former.

We have spoken as if a single G study on $v_1$ and $v_2$ is carried out. Instead, three separate studies may be used to estimate the variance and covariance components. (Something like this has often been done in correcting correlations for attenuation.) Estimates of variance components can be obtained from a G study observing $v_1$ several times in one sample of persons, and from a similar G study on $v_2$ using a second sample. The covariance component(s) can be estimated in a third study that makes just one or two observations on $v_1$ and $v_2$, both on the same persons.

A multiple-sample procedure is advantageous when one is considering several variables; this makes it difficult to collect extensive G data for every pair of variables on a single sample. The method is reasonable as long as the several samples of persons are large and represent the same population. To be sure, the conclusions reached by combining estimates are likely to be less accurate than those from an integrated G study that collects the same total number of scores for a single sample. Statistical theory for combining data from separate samples in a problem like this is still crude.[4]

---

[4] For another use of three-sample designs see p. 102.

It is a bit anomalous to refer to the third part of a three-sample study as a "G study." Consider the case where the developer of the Stanford–Binet test for the blind calculates components of variance from a G study on it. Independently, the developer of a version of the Block Design test for the blind does likewise. A third investigator, applying both tests to a new sample of blind subjects, wants to estimate the correlation of the two universe scores. He need only administer each test once, keeping conditions independent, to obtain the estimate of $\sigma(_1p,_2p)$ that, along with the components of variance, gives the desired correlation. While his study, with one condition per variable, is not truly a G study, it fulfills a G-study function.

## C. *Analysis of G Studies*

### *One-facet G studies*

The analysis of a multivariate study estimates components of covariance as well as components of variance, and to do this it forms sums of products and mean products. The analysis of jointly sampled observations is more elaborate than that for independent observations, and we shall begin with it.

*Joint sampling,* $n_i = n_g$. Assume, for the moment, that the G study has the design $_1(i \times p)$, $_2(g \times p)$, $[i \cdot g]$, $n_i = n_g$. Then sums of products (SP) are defined by

$$^\bullet SP_{1,2}p = n_i \sum_p (_1X_{pI} - {}_1X_{PI})(_2X_{pG} - {}_2X_{PG})$$

(9.14)
$$^\bullet SP_{1,2}i = n_p \sum_{i \cdot g} (_1X_{Pi} - {}_1X_{PI})(_2X_{Pg} - {}_2X_{PG})$$

$$^\bullet SP_{1,2}Total = \sum_p \sum_{i \cdot g} (_1X_{pi} - {}_1X_{PI})(_2X_{pg} - {}_2X_{PG})$$

The symbol $\sum_{i \cdot g}$ implies that sums are taken over linked $i,g$ pairs. The product for any $i$ with any $g$ to which it is not linked is ignored, as the expected value of such products is zero. There is a set of equations in the form of (9.14) for each pair of variables. If we write, more generally,

(9.15)
$$^\bullet SP_{v,v'}p = n_i \sum_p (_vX_{pI} - {}_vX_{PI})(_{v'}X_{pG} - {}_{v'}X_{PG})$$

it is evident that for $v \equiv v'$ the definition reduces to that for the sum of squares for persons. The equation calls for forming an average score over conditions and converting it to a deviation score. The product of the person's deviation scores on the two variables is calculated, and the sum over persons is taken. (This is precisely what one would do in computing a covariance of average scores in ordinary correlational work.)

Considering two or more variables one has matrices for $^\bullet\mathrm{SP}\,_vp$, $^\bullet\mathrm{SP}\,_vi$, and $^\bullet\mathrm{SP}\,_v\mathrm{Total}$; subtraction gives $^\bullet\mathrm{SP}\,_v\mathrm{res}$. All have the sum of the squares in the diagonal. Degrees of freedom $n_p - 1$, $n_i - 1$ and $(n_p - 1)(n_i - 1)$ are used as divisors to form matrices of mean products $^\bullet\mathrm{MP}\,_vp$, $^\bullet\mathrm{MP}\,_vi$, and $^\bullet\mathrm{MP}\,_v\mathrm{res}$, respectively. Each of these has mean squares in the diagonal, and (2.4) is applied to the mean squares to estimate variance components. The comparable equations for components of covariance have the form:

$$\mathrm{E}^\bullet\mathrm{MP}_{1,2}p = {}^\bullet\sigma(_1pi,e;_2pg,e) + n_i\sigma(_1p,_2p)$$

(9.16) $$\mathrm{E}^\bullet\mathrm{MP}_{1,2}i = {}^\bullet\sigma(_1pi,e;_2pg,e) + n_p{}^\bullet\sigma(_1i,_2g)$$

$$\mathrm{E}^\bullet\mathrm{MP}_{1,2}\mathrm{res} = {}^\bullet\sigma(_1pi,e;_2pg,e)$$

The $^\bullet\mathrm{MP}$ of the G study are substituted for the $\mathrm{E}^\bullet\mathrm{MP}$ to estimate the components.

From the crossed G study one can also estimate the covariance component $^\bullet\sigma(_1i,pi,e;_2g,pg,e)$ that would be relevant to a nested G study. Adding the $^\bullet\mathrm{SP}$ for conditions and residual, adding the degrees of freedom similarly, and dividing, gives $^\bullet\mathrm{MP}$ within $p$, which estimates the within-person covariance component.

If one has a nested G study with $n_i = n_g$, one calculates $^\bullet\mathrm{SP}\,p$ and $^\bullet\mathrm{SP}$ Total as in (9.14), and subtracts to get $^\bullet\mathrm{SP}$ within $p$. There are $n_i(n_p - 1)$ degrees of freedom.

(9.17) $$\begin{aligned} \mathrm{E}^\bullet\mathrm{MP}_{1,2}p &= {}^\bullet\sigma(_1i,pi,e;_2g,pg,e) + n_i\sigma(_1p,_2p) \\ \mathrm{E}^\bullet\mathrm{MP}_{1,2}\mathrm{res} &= {}^\bullet\sigma(_1i,pi,e;_2g,pg,e) \end{aligned}$$

*Joint sampling*, $\mathbf{n}_i = \mathbf{k}n_g$. The definitions and formulas may be extended to cover many-to-one sampling. Suppose that, for each $g$, $k$ values of $i$ are drawn; this of course implies multiple pairing in the universe of linked observations. The data on $_1v$ can be described in terms of $n_g$ sets of size $k$, and the set of conditions associated with any $g$ can be referred to as $I_g$. The score

$$_1X_{pI_g} = \frac{1}{k}\sum_{i\in I_g} {}_1X_{pi}\,.$$

If the design is crossed, one may apply the formulas given above to the scores $_1X_{pI_g}$, $_2X_{pg}$, because $n_I = n_g$. One arrives at estimates of $\sigma(_1p,_2p)$, $^\bullet\sigma(_1I,_2g)$, and $^\bullet\sigma(_1pI,e;_2pg,e)$. As the covariance of an average equals the average of the covariances for the elements entering the average, $^\bullet\sigma(_1I,_2g) = {}^\bullet\sigma(_1i,_2g)$; similarly for the residual. [This case is different from the $^\bullet\sigma(I,G)$ considered earlier where there was one-to-one sampling and each $i$ was linked to only one $g \in G$.] If $n_i = kn_g$,

$$^\bullet\sigma(_1I,_2G) = \frac{1}{n_g}\,{}^\bullet\sigma(_1i,_2g).$$

***Independent sampling.*** Where the G study is crossed and sampling is independent, with $n_i = n_g$, we have:

$$(9.18) \qquad \text{SP}_{1,2}p = n_i \sum_p ({}_1X_{pI} - {}_1X_{PI})({}_2X_{pG} - {}_2X_{PG})$$

There are $n_p - 1$ degrees of freedom.

$$(9.19) \qquad \text{EMP}_{1,2}p = n_i\sigma({}_1p,{}_2p)$$

No equations are written for conditions and residual. We know that these components of covariance in the universe of admissible observations are zero, and need not estimate them.

The use of the multiplier $n_i$ in (9.18) has no effect, as it is offset by the multiplier in (9.19). However, it is retained to maintain the similarity to the equation for SS $p$ and to the $^\bullet$SP $p$ for joint sampling. If $n_i \neq n_g$, the equation may be used as stated, though there is no logical justification for the multiplier.

***Discussion.*** The universe-score covariance remains the same, whether the design calls for joint or independent sampling. This should be obvious, because the linkage of $i$ with $g$ does not enter into the definition of the universe score ${}_1\mu_p$, ${}_2\mu_p$. In view of this, it may be puzzling that (9.16) and (9.19) for EMP $p$ differ. The covariance for residuals enters (9.16) because, with joint sampling, its expected value is not zero. When independently sampled data are treated as if jointly sampled, after randomly pairing $i$ and $g$ the estimates from (9.16) differ from those of (9.19) only because the $^\bullet$MP res used in evaluating (9.16) is based on a random subset of $i,g$ pairs rather than all pairs.

When there is joint sampling, components of covariance for conditions and residual can reasonably be expected to have nonzero values in the population. Suppose that the design calls for teacher $i$ to rate pupils $p$ on both ability $v_1$ and motivation $v_2$. Some teachers give higher ratings on the average than other teachers do; the $i$ component represents this bias. The constant errors in $v_1$ ratings are likely to covary (over teachers) with the constant errors in $v_2$ ratings. The covariance $^\bullet\sigma({}_1i,{}_2i)$ then would be positive. In another study, the covariance component might be negative; for example, if $v_1$ is a rating on ability and $v_2$ is a rating on anxiety. A positive covariance component for $pi,e$ would reflect other consistent tendencies. For example, Miss Smith rates Johnny higher than other teachers do on both ability and motivation either because of her stable bias in perceiving Johnny, or because she feels positive about Johnny at the time of rating; this raises the covariance.

The problem of hidden facets (p. 122) returns in various troublesome ways. There may be a hidden condition $j^*$ that enters all G-study observations on $v_1$ and does not enter observations on $v_2$. The supposed estimator of $\sigma^2({}_1p)$ from such a G study really estimates $\sigma^2({}_1p \mid j^*)$, as discussed in Chapter

4. Similarly, the supposed estimator of $\sigma(_1p,_2p)$ estimates $\sigma(_1p \mid j^*,_2p)$. Over studies using different $j$, $\sigma(_1p \mid j,_2p)$ will have the *expected* value $\sigma(_1p,_2p)$, but a particular $j^*$ may generate $v_1$ scores that covary to a greater or less degree with $_2\mu_p$ than the average condition $j$ does. The problem is slightly different if the same $j^*$ enters all G-study observations on both $v_1$ and $v_2$, as the expected value of the bias (over all $j$) is probably not zero. There is nothing new to be said here regarding strategies for coping with the hidden-facet problem.

*Alternative methods of analysis.* We need mention only briefly that there are alternative ways to estimate components of covariance. One can derive the desired results from successive analyses of variance of observed scores and of such averages as $(_1X_{pI} + _2X_{pG})/2$. This is because the variance of an average combines variances and covariances. (See Tukey, 1949.)

Also, score covariances can yield most of the components. In a linked G study, for example,

$$\mathscr{E} \underset{i \circ g}{\,^\circ s}(_1X_{pi} - _1X_{Pi},\, _2X_{pg} - _2X_{Pg}) = \sigma(_1p,_2p)$$

(9.20)

$$\mathscr{E} \underset{i \bullet g}{\,^\bullet s}(_1X_{pi} - _1X_{Pi},\, _2X_{pg} - _2X_{Pg}) = \sigma(_1p,_2p) + \,^\bullet\sigma(_1pi,e,_2pg,e)$$

Suppose one has drawn the following $i,g$ pairs: $1, a$; $2, b$; $3, c$. Then one averages the $1a$, $2b$, and $3c$ covariances to estimate $\mathscr{E} \circ s$, and averages covariances $1b$, $1c$, $2a$, $2c$, $3a$, $3b$ to estimate $\mathscr{E} \circ s$. Subtracting the second average from the first estimates the covariance component for residual. The $\hat{\sigma}(_1p,_2p)$ from (9.20) is identical to that from (9.16) applied to the same data, provided that $n_p - 1$ is used in computing the covariances. Covariances over persons give no basis for estimating the covariance component for conditions, however.

*Two-facet G studies*

The components of covariance for a two-variable two-facet universe are analogous to the components of variance for a one-variable two-facet universe. Resembling the variance component $\sigma^2(pi)$, for example, there is a covariance component $\sigma(_1pi,_2pg)$, the expected value of $(_1\mu_{pi} - _1\mu_p - _1\mu_i + _1\mu)(_2\mu_{pg} - _2\mu_p - _2\mu_g + _2\mu)$. This component is zero if sampling is independent, but for jointly sampled $i$ and $g$, it is nonzero in general.

We elaborate only on the G study that uses Design VII to collect data on both $v_1$ and $v_2$. When the reader understands this development he can develop multivariate procedures for analyzing other combinations of designs by extending the formulas of Chapter 2.

**Design VII with independent sampling.** Assume that the G study has the design $_1(i \times j \times p)$, $_2(g \times h \times p)$, $[i \circ g, i \circ h, j \circ g, j \circ h]$. This couples a Design-VII study of $v_1$ with a Design-VII study of $v_2$. There is no reason to make $n_i$ or $n_j$ equal to $n_g$ or $n_h$. Components of variance are estimated as if there were two separate univariate studies. With this design, the only non-zero component of covariance is that for persons; it is most convenient to estimate it from scores averaged over conditions:

$$(9.21) \quad \hat{\sigma}(_1p,_2p) = \frac{1}{n_p - 1} \sum_p (_1X_{pIJ} - _1X_{PIJ})(_2X_{pGH} - _2X_{PGH})$$

[Cf. (9.18) and (9.19).]

**Design VII with joint sampling.** Assume that in the G study Design VII is used for each variable independently; $i \bullet g, j \bullet h$, but $i \circ h, j \circ g$; $n_i = n_g$ and $n_j = n_h$. Because $n_i = n_g$ and $n_j = n_h$, sums of products are defined by (9.22).

$$^\bullet SP_{1,2}p = n_i n_j \sum (_1X_{pIJ} - _1X_{PIJ})(_2X_{pGH} - _2X_{PGH})$$

$$^\bullet SP_{1,2}i = n_p n_j \sum_{i \bullet g} (_1X_{PiJ} - _1X_{PIJ})(_2X_{PgH} - _2X_{PGH})$$

$$^\bullet SP_{1,2}j = n_p n_i \sum_{j \bullet h} (_1X_{PIj} - _1X_{PIJ})(_2X_{PGh} - _2X_{PGH})$$

$$^\bullet SP_{1,2}pi = n_j \sum_p \sum_{i \bullet g} (_1X_{piJ} - _1X_{pIJ} - _1X_{PiJ} + _1X_{PIJ})$$
$$(9.22) \qquad \times (_2X_{pgH} - _2X_{pGH} - _2X_{PgH} + _2X_{PGH})$$

$$^\bullet SP_{1,2}pj = n_i \sum_p \sum_{j \bullet h} (_1X_{pIj} - _1X_{pIJ} - _1X_{PIj} + _1X_{PIJ})$$
$$\times (_2X_{pGh} - _2X_{pGH} - _2X_{PGh} + _2X_{PGH})$$

$$^\bullet SP_{1,2}ij = n_p \sum_{i \bullet g} \sum_{j \bullet h} (_1X_{Pij} - _1X_{PiJ} - _1X_{PIj} + _1X_{PIJ})$$
$$\times (_2X_{Pgh} - _2X_{PgH} - _2X_{PGh} + _2X_{PGH})$$

$$^\bullet SP_{1,2}pij,e = \sum_p \sum_{i \bullet g} \sum_{j \bullet h} (_1X_{pij} - _1X_{piJ} - _1X_{pIj} - _1X_{Pij} + _1X_{pIJ}$$
$$+ _1X_{PiJ} + _1X_{PIj} - _1X_{PIJ})(_2X_{pgh} - _2X_{pgH} - _2X_{pGh}$$
$$- _2X_{Pgh} + _2X_{pGH} + _2X_{PgH} + _2X_{PGh} - _2X_{PGH})$$

[Cf. (9.14).]

While some of these sums of products will perhaps have negligible expected values, this need not be the case (see p. 271, regarding the $e$ component, for example). In general,

$$\text{E}^\bullet\text{MP}_{1,2}p = {}^\bullet\sigma(_1pij,e;_2pgh,e) + n_j{}^\bullet\sigma(_1pi,_2pg)$$
$$+ n_i{}^\bullet\sigma(_1pj,_2ph) + n_in_j\sigma(_1p,_2p)$$

$$\text{E}^\bullet\text{MP}_{1,2}i = {}^\bullet\sigma(_1pij,e;_2pgh,e) + n_j{}^\bullet\sigma(_1pi,_2pg) + n_p{}^\bullet\sigma(_1ij,_2gh)$$
$$+ n_pn_j{}^\bullet\sigma(_1i,_2g)$$

$$\text{E}^\bullet\text{MP}_{1,2}j = {}^\bullet\sigma(_1pij,e;_2pgh,e) + n_i{}^\bullet\sigma(_1pj,_2ph) + n_p{}^\bullet\sigma(_1ij,_2gh)$$
$$+ n_pn_i{}^\bullet\sigma(_1j,_2h)$$

(9.23)

$$\text{E}^\bullet\text{MP}_{1,2}pi = {}^\bullet\sigma(_1pij,e;_2pgh,e) + n_j{}^\bullet\sigma(_1pi,_2pg)$$

$$\text{E}^\bullet\text{MP}_{1,2}pj = {}^\bullet\sigma(_1pij,e;_2pgh,e) + n_i{}^\bullet\sigma(_1pj,_2ph)$$

$$\text{E}^\bullet\text{MP}_{1,2}ij = {}^\bullet\sigma(_1pij,e;_2pgh,e) + n_p{}^\bullet\sigma(_1ij,_2gh)$$

$$\text{E}^\bullet\text{MP}_{1,2}pij,e = {}^\bullet\sigma(_1pij,e;_2pgh,e)$$

Substituting the calculated mean products, one estimates all components of covariance. Each equation can be written in matrix notation, e.g.:

$$(9.24) \quad \text{E}^\bullet\text{MP}_v p = {}^\bullet\sum\nolimits_v pij,e + n_j{}^\bullet\sum\nolimits_v pi + n_i{}^\bullet\sum\nolimits_v pj + n_in_j\sum\nolimits_v p$$

Equations and degrees of freedom can be modified to allow $n_i = kn_g$ or the opposite.

If one has $i \bullet g$, but $j \circ g, j \circ h$, the expected mean products for $pij, e, pj, ij$, and $j$ are zero, and the corresponding sums of products would not be calculated.

*Partially nested designs.* A similar analysis applied to G studies in which the design is nested to some degree. The equations for sums of products given previously are applied. Whatever components of variance and covariance are confounded can be identified with the aid of Figure 2.4; the sums of products calculated for those components are pooled before calculating mean products. Expected mean product equations for confounded components can be written by analogy with equations for the expected mean square.

### Numerical Example: WISC and WAIS

To provide an example of components of covariance, we examine a further set of Wechsler data. These data were collected by Ross and Morledge (1967), and they were supplied to us through the courtesy of Dr. Ross. There are scores on subtests of WISC for a sample of thirty 16-year-olds, together with their scores on WAIS administered four weeks later. We shall work exclusively with scaled scores, ignoring the IQ conversion, and take the average over five subtests as the Ve or Pe scale score. (We ignore the Digit Span subtest so as to have five subtests in each scale. Ross and Morledge counted six Verbal subtests, prorating to get the Ve score for WISC.)

Formally this study is like the WPPSI retest study of Chapter 8, but here we organize the data differently. Scores are classified with respect to persons, a facet of test forms, and a facet of subtests nested within scales. Day and trial are confounded with form. We regard Ve and Pe as fixed variables, We consider WISC and WAIS subtests to be in one-to-one correspondence ignoring slight differences in form of task such as the shift from Coding to Digit Symbol.

Let $k$ and $\ell$ refer to test form (along with the confounding day and trial effects), and let $j$ and $h$ refer to subtests. This notation is chosen for consistency with Chapter 8. We presume that the investigator wishes to generalize over forms, days, trials, and subtests. The subtest scores have the form $_{Ve}X_{pjk}$ and $_{Pe}X_{ph\ell}$. Forms and subtests are crossed with persons and each other, within the variable. While the subtests of WISC and WAIS are in a sense linked, this is recognized by treating subtests as crossed with forms. The design, then, is $_{Ve}(j \times k \times p)$, $_{Pe}(h \times \ell \times p)$, $k \cdot \ell$. The linkage considers the fact that Ve and Pe are observed on the same form (WISC or WAIS) obtained on a single day and trial.

With this design there are seven components for each score. The nonzero covariance components are $\sigma(_{Ve}p,_{Pe}p)$, $^\bullet\sigma(_{Ve}k,_{Pe}\ell)$, and $^\bullet\sigma(_{Ve}pk,_{Pe}p\ell)$. The first has to do with scale universe scores, the second with an order effect or a standardization error common to the two scales of a form, and the third has to do with person-form-day effects common to the two scales (e.g., day-to-day variability of the person).

The variance components come from three-way analyses of variance of scaled subtest scores for Ve and Pe separately. This scale, it will be recalled, has a mean of 10 and standard deviation of 3 in the norm group. The summary listing in Table 9.2 includes the estimates for components of covariance, reached by procedures to be discussed below. The residual here includes the person × day × subtest effect and various interaction effects at the item level; these could be examined separately in the WPPSI data, where a split-half analysis was carried out (p. 250).

The Ve scale has a strikingly high person component of variance and a relatively small person × subtest component. For the Ross–Morledge 16-year-olds, there are evidently well-defined individual differences in verbal ability that are stable over days and subtests. The range of this sample appears to be somewhat greater than that in the standardization group.

**TABLE 9.2.** *Components of Variance and Covariance for Wechsler Subtest Scores*

|  | Estimate of variance components | | Estimate of covariance components |
|---|---|---|---|
|  | Ve | Pe | Ve × Pe |
| Persons | 8.16 | 4.84 | 5.54 |
| Forms, days, trials $(k, \ell)$ | 0 | 0 | ca. 0 |
| Subtests $(j,h)$ | 0.05 | 0.04 | 0[a] |
| Persons × forms | 0.01 | 0.04 | 0.02[b] |
| Persons × subtests | 0.96 | 1.50 | 0[a] |
| Forms × subtests | 0.05 | 0.15 | 0[a] |
| Residual | 2.33 | 2.20 | 0[a] |

[a] Component known to be zero.
[b] Assumed value. (See p. 283.)

Two components of covariance were estimated from the covariances of the scale scores. (Again, alternative procedures were available.) The covariances obtained from the sample (using 29 as the divisor of SP) are:

|  |  | Verbal | |
|---|---|---|---|
|  |  | WISC | WAIS |
| Performance | WISC | •6.13 | °6.15 |
|  | WAIS | °4.92 | •5.11 |

The two independent covariances, averaged, give $\hat{\sigma}(_{Ve}p,_{Pe}p) = 5.54$. We have entered this in Table 9.2.

The person component of covariance is large; evidently the universe scores for Ve and Pe correlate substantially, since $\sigma(_{Ve}p,_{Pe}p)$ is of the order of magnitude of the variance components for persons. The two linked covariances, which average 5.62, estimate $\sigma(_{Ve}p,_{Pe}p) + {}^\bullet\sigma(_{Ve}pk,_{Pe}p\ell)$; this implies that ${}^\bullet\sigma(_{Ve}pk,_{Pe}p\ell) = 0.08$. For reasons to be explained, we do not enter this in the table.

The attempt to estimate the covariance component for conditions encounters the difficulty that the estimate of the variance component for conditions is zero for each variable. If this is the case, the covariance component must also be zero. There are four means over persons, and one can calculate sums of products using the second equation in (9.22). Start with these values:

| | Mean scores | | Mean of deviations | |
|---|---|---|---|---|
| | Verbal | Performance | Verbal | Performance |
| WISC | 9.567 | 10.387 | +0.200 | +0.037 |
| WAIS | 9.167 | 10.313 | −0.200 | −0.037 |
| Mean | 9.367 | 10.350 | | |

The sum of products is $30 \times 5 \times [(0.200)(0.037) + (-0.200)(-0.037)] =$ 2.22. The mean product is also 2.22, because there is one degree of freedom. In the equation for $E^{\bullet}MS$, the covariances involving $j$ and $h$ are known to be zero, hence,

$$2.22 = 5^{\bullet}\sigma(_{1}pk,_{2}p\ell) + 150^{\bullet}\sigma(_{1}k,_{2}\ell)$$
$$2.22 = 5(0.08) + 150^{\bullet}\sigma(_{1}k,_{2}\ell)$$

and the covariance component is estimated to be 0.012. While this value is inconsistent with the variances, we can properly conclude that the covariance associated with forms is quite small. We have entered in the table that the estimate is about zero.

Similarly, the estimated component of covariance for person–form interaction (0.08) is larger than the estimated variance components (0.01, 0.04). The discrepancy is quite small, however. For the sake of some later examples we insert in the table a value of 0.02 in place of the calculated 0.08.

It may seem that nothing much was learned from the application of a fairly complex model to the Ve–Pe relation. There is, however, reason to take satisfaction in the apparent smallness of form and person–form components of covariance and variance. If, for example, there had been a sizable variance component for forms, this would imply either that scores tend to run higher for one of the two forms (because of poor standardization), or that there is a systematic order effect. And if variance components for persons × forms had been sizable, then the negligible size of the corresponding component of covariance would refute the interpretation that the person's variations show discrepancy between "good days" and "bad days." The finding of a small covariance along with large variances would have suggested that day-to-day (or form-to-form) variation is largely a within-scale phenomenon. In the present study, the findings are undramatic but not unimportant, as they rule out some counterhypotheses that, if true, would distress the interpreter.

### D. *Inferences Regarding Composite Scores*

The components of variance and covariance can be used to draw conclusions about the generalizability of any composite of the variables. Consider a composite with specified weights $w_v$. $\sum\limits_{v=1}^{n_v} w_v{}_v X_{pij}$ is a new variable; we

may denote it simply by $_w X_{pij}$ and its universe score (over all $i$ and $j$) by $_w \mu_p$.
Assume that single observations are made on the several variables. Then,
$$_w \Delta_{pij} = {}_w X_{pij} - {}_w \mu_p = \sum_v w_v ({}_v X_{pij} - {}_v \mu_p).$$

From the usual formula for the variance of a sum,

$$(9.25) \qquad \sigma^2({}_w\Delta) = \sum_1^{n_v} \sum_1^{n_v} w_v w_{v'} \sigma({}_v X_{pij} - {}_v \mu_p, {}_{v'} X_{pgh} - {}_{v'} \mu_p)$$

$$= \sum \sum w_v w_{v'} \sigma({}_v \Delta_{pij}, {}_{v'} \Delta_{pgh})$$

Now $_v\Delta$ is the sum of components. Whether these covary with the components of $_{v'}\Delta$ depends on the design of the D study.

The argument may be extended to D studies with $n_i'$, $n_j'$, etc. $> 1$. For example, if we make $n_i' n_j' = n_g' n_h' = \ldots$, all the foregoing statements hold, except for the necessary substitution of $I$ for $i$, etc. In (9.25), the covariance of $_v\Delta_{pIJ}$ with $_{v'}\Delta_{pGH}$ is required.

In the multivariate study, there are three possible regression procedures:

1. Estimate $_w\mu_p$ from $_w X_{pIJ}$ with the aid of the estimated $\mathscr{E}\rho^2(_w X, {}_w\mu_p)$.
2. For each $v$ in turn, estimate $_v\mu_p$ from the corresponding $_v X_{pIJ}$ with the aid of $\widehat{\mathscr{E}\rho^2}(_v X, {}_v\mu_p)$; then form $\sum\limits_v w_v {}_v\hat\mu_p$.
3. Make a multiple-regression estimate of $_w\mu_p$ from $_v X_{pIJ}$.

The third procedure is superior, and will be discussed in Chapter 10.

*Independent sampling in the D study*

Assume that $i \circ g$, $j \circ h$ in the D study. Then all covariance components except that for persons are zero, and

$$(9.26) \qquad \sigma^2({}_w\Delta) = \sum_v w_v^2 \sigma^2({}_v\Delta)$$

The value of $\widehat{\sigma^2({}_v\Delta)}$ will be calculated, as in any univariate study, by a formula fitting the design of the D study (see p. 84ff.). Nothing is new here. To minimize $\sigma^2({}_w\Delta)$ within some constraint on the total number of observations, one can adjust $n'$ separately for each variable and each facet. The universe-score variance is:

$$(9.27) \qquad \sigma^2({}_w\mu_p) = \sum_{v=1}^{n_v} \sum_{v'=1}^{n_v} w_v w_{v'} \sigma({}_v p, {}_{v'} p)$$

The expected observed-score variance for the composite consists of $\sigma^2({}_w\mu_p)$ plus the weighted variances of any other components that enter the expected observed-score variance of the $_v X_{pIJ}$. Thus, the only components of covariance entering $\mathscr{E}\sigma^2({}_wX_{pIJ})$ are those for persons, which enter through $\sigma^2({}_w\mu_p)$. What components of variance are to be considered depends on the design for observing each $v$; their contributions are calculated as in Table 3.7. As usual, the ratio of $\widehat{\sigma^2({}_w\mu_p)}$ to $\widehat{\mathscr{E}\sigma^2({}_wX)}$ is an estimate of $\mathscr{E}\rho^2({}_wX,{}_w\mu_p)$.

### Joint sampling in the D study

With the design $i \cdot g$, $n'_i = n'_g$, and $j \cdot h$, $n'_j = n'_h$ (but $i \circ h$ and $g \circ j$), $\sigma^2({}_w\Delta)$ includes terms of the form:

$$(9.28) \qquad {}^\bullet\sigma({}_1\Delta, {}_2\Delta) = \frac{1}{n'_i}\, {}^\bullet\sigma({}_1i, {}_2g) + \frac{1}{n'_j}\, {}^\bullet\sigma({}_1j, {}_2h)$$

$$+ \frac{1}{n'_i}\, {}^\bullet\sigma({}_1pi, {}_2pg) + \frac{1}{n'_j}\, {}^\bullet\sigma({}_1pj, {}_2ph)$$

$$+ \frac{1}{n'_in'_j}\, {}^\bullet\sigma({}_1ij, {}_2gh) + \frac{1}{n'_in'_j}\, {}^\bullet\sigma({}_1pij,e, {}_2pgh,e)$$

The G study has presumably estimated such covariance components. In $\sigma^2({}_w\Delta)$ they are to be weighted by $2w_1w_2$ and combined with variance components as in (9.25). With several variables, each pair for which there is linked sampling contributes a series of covariance terms.

Equation (9.27) again holds for the universe-score variance.

As shown in Table 9.3, the first step in determining the expected observed-score variance is to list the components of the deviation score for each variable. This is determined in the manner of Table 3.6. Table 9.3 considers three variables, each observed under a different design; the components of each deviation score appear across the top of the table. For each such component, a properly weighted variance term enters the observed-score variance. Each component of the deviation score that is jointly sampled gives rise to a covariance component that appears in the variance; there is also a covariance component for persons.

The first composite score assumes a crossed design for variables 1 and 2, with linkage of both facets. The expected observed-score variance is then the sum of eight components of variance and four components of covariance, as listed in the table.

In the second example, the crossed design is used to measure $v_1$ and a partially nested design for $v_3$. Moreover, there is an $i,g$ linkage only. The variance components for this composite are like those for the first composite, except that the nesting of $h$ within $p$ causes the variance components for $h$ and $gh$ to affect the observed score. There is no entry for the $j,h$ covariance, because with independent sampling, its expected value is zero. This also

**TABLE 9.3.** *Components Entering Expected Observed-Score Variance for Composite Scores*

| Suppositions regarding design | Components of deviation score under each design | | | | | |
|---|---|---|---|---|---|---|
| | $\mu_p\widetilde{\;}$ | $\mu_j\widetilde{\;}$ or $\mu_h\widetilde{\;}$ | $\mu_{pi}\widetilde{\;}$ or $\mu_{pg}\widetilde{\;}$ | $\mu_{pj}\widetilde{\;}$ or $\mu_{ph}\widetilde{\;}$ | $\mu_{ij}\widetilde{\;}$ or $\mu_{gh}\widetilde{\;}$ | $\mu_{pij}\widetilde{\;}, e$ or $\mu_{pgh}\widetilde{\;}, e$ |
| **Variables considered singly:** | | | | | | |
| $1[(i \times j) \times p]$ | $1^{\mu_p}\widetilde{\;}$ | | $1^{\mu_{pI}}\widetilde{\;}$ | $1^{\mu_{pJ}}\widetilde{\;}$ | $1^{\mu_{ij}}\widetilde{\;}$ | $1^{\mu_{pIJ}}\widetilde{\;}, e$ |
| $2[(g \times h) \times p]$ | $2^{\mu_p}\widetilde{\;}$ | | $2^{\mu_{pG}}\widetilde{\;}$ | $2^{\mu_{pH}}\widetilde{\;}$ | $2^{\mu_{gh}}\widetilde{\;}$ | $2^{\mu_{pGH}}\widetilde{\;}, e$ |
| $3[g \times (h:p)]$ | $3^{\mu_p}\widetilde{\;}$ | $3^{\mu_H}\widetilde{\;}$ | $3^{\mu_{pG}}\widetilde{\;}$ | $3^{\mu_{pGH}}\widetilde{\;}$ | $3^{\mu_{GH}}\widetilde{\;}$ | $3^{\mu_{pGH}}\widetilde{\;}, e$ |

**Components entering $\mathscr{E}\sigma^2(_wX)$**

Composite $w_{11}X + w_{22}X$, where:

$1^i \cdot {}_2g;\; 1^j \cdot {}_2h$

$n'_i = n'_g,\; n'_j = n'_h$

| | $\mu_p$ | $\mu_j / \mu_h$ | $\mu_{pi} / \mu_{pg}$ | $\mu_{pj} / \mu_{ph}$ | $\mu_{ij} / \mu_{gh}$ | $\mu_{pij}/\mu_{pgh}, e$ |
|---|---|---|---|---|---|---|
| | $\left.\begin{array}{l} w_1^2\sigma^2(_1p) \\ w_2^2\sigma^2(_2p) \\ 2w_1w_2\sigma(_1p,_2p) \end{array}\right\}$ | | $\dfrac{w_1^2}{n'_i}\sigma^2(_1pi)$ | $\dfrac{w_1^2}{n'_j}\sigma^2(_1pj)$ | | $\dfrac{w_1^2}{n'_in'_j}\sigma^2(_1pij,e)$ |
| | | | $\dfrac{w_2^2}{n'_g}\sigma^2(_2pg)$ | $\dfrac{w_2^2}{n'_h}\sigma^2(_2ph)$ | | $\dfrac{w_2^2}{n'_gn'_h}\sigma^2(_2pgh,e)$ |
| | | | $\dfrac{2w_1w_2}{n'_i}\sigma(_1pi,_2pg)$ | $\dfrac{2w_1w_2}{n'_j}\sigma(_1pj,_2ph)$ | | $\dfrac{2w_1w_2}{n'_in'_j}\cdot\sigma(_1pij,e;\,_2pgh,e)$ |

Composite $w_{11}X + w_{33}X$, where:

$1^i \cdot {}_3g;\; 1^j \circ {}_3h$

$n'_i = n'_g$

| | $\mu_p$ | $\mu_j / \mu_h$ | $\mu_{pi} / \mu_{pg}$ | $\mu_{pj} / \mu_{ph}$ | $\mu_{ij} / \mu_{gh}$ | $\mu_{pij}/\mu_{pgh}, e$ |
|---|---|---|---|---|---|---|
| | $\left.\begin{array}{l} w_1^2\sigma^2(_1p) \\ w_3^2\sigma^2(_3p) \\ 2w_1w_3\sigma(_1p,_3p) \end{array}\right\}$ | $\dfrac{w_3^2}{n'_h}\sigma^2(_3h)$ | $\dfrac{w_1^2}{n'_i}\sigma^2(_1pi)$ | $\dfrac{w_1^2}{n'_j}\sigma^2(_1pj)$ | $\dfrac{w_3^2}{n'_h}\sigma^2(_3gh)$ | $\dfrac{w_1^2}{n'_in'_j}\sigma^2(_1pij,e)$ |
| | | | $\dfrac{w_3^2}{n'_g}\sigma^2(_3pg)$ | $\dfrac{w_3^2}{n'_h}\sigma^2(_3ph)$ | | $\dfrac{w_3^2}{n'_gn'_h}\sigma^2(_3pgh,e)$ |
| | | | $\dfrac{2w_1w_3}{n'_i}\sigma(_1pi,_3pg)$ | | | |

accounts for the nonappearance of covariance terms for $pj$, $ph$ and $pij,e$; $pgh,e$.

Where a composite involves three or more variables, determination of the makeup of observed-score variance is more complex. The variance components may be listed for one variable at a time, as in Table 9.3. The covariance components may then be listed for each pair of variables in turn, taking the linkage for that pair into account. For example, consider the composite $w_{1\,1}X + w_{2\,2}X + w_{3\,3}X$, where the D study employs the design indicated in Table 9.3 and $_1i \bullet {}_2g \bullet {}_3g$ but $_1j \circ {}_2h \circ {}_3h$. There are 14 different variance components to consider, these being the ones listed in various lines of Table 9.3. Of the covariances listed, those involving $j$ and $h$ vanish. The following make up the expected observed-score variance: $2w_1w_2\sigma(_1p,_2p)$, $2w_1w_3\sigma(_1p,_3p)$, $2w_2w_3\sigma(_2p,_3p)$,

$$\frac{2w_1w_2}{n_i'} \bullet \sigma(_1pi,\,_2pg),\ \frac{2w_1w_3}{n_i'} \bullet \sigma(_1pi,\,_3pg),\ \text{and}\ \frac{2w_2w_3}{n_g'} \bullet \sigma(_2pg,\,_3pg).$$

(Note that denominators can be written with either $n_i'$ or $n_g'$.)

### E. *Relations Among Variables: the "Attenuation" Problem*

Reliability theory originated in Spearman's desire to interpret correlations between operationally distinct variables. He knew that two operationally realizable scores cannot correlate perfectly even when they reflect the same basic construct, because each is subject to random errors of observations that reduce correlation. He estimated the magnitude of these errors by making a reliability study of each variable, and then adjusted for the "attenuating" effect of errors from the original correlation by dividing by the square root of the reliabilities. In our notation his formula would be

$$(9.29) \qquad \hat{\rho}(_1\mu_p,_2\mu_p) = \frac{r(_1X_{pi},_2X_{pg})}{(\mathscr{E}_1\rho^2 \cdot \mathscr{E}_2\rho^2)^{1/2}}$$

The numerator on the right is ambiguous, because $i$ and $g$ may or may not be linked. The correlation for linked $i$ and $g$ is likely to differ from the correlation obtained when the conditions are not linked. Classical theory ignores the distinction, but generalizability theory forces us to take the probable difference seriously.

The problem is a broader one than that of estimating the correlation among universe scores. The G study observes a correlation for a score based on $n_i$ observations with another score based on $n_g$ observations, and can establish the equation for predicting one variable from the other. But one may be interested in obtaining the correlation or the regression equation in a

D study where there are $n_i'$ and $n_g'$ observations on the two variables, or where the design is modified in other respects.

## Universes classified with respect to one facet

Suppose that observations are classified only with respect to facet $i$ or $g$. The D study will generate observed scores $_1X_{pI}$ and $_2X_{pG}$ where $I$ is a set of $n_i'$ conditions and $G$ is a set of $n_g'$ conditions. Because our model does not assume strict equivalence, we have to estimate an expected correlation between pairs of tests used in various D studies with this design. The correlation is not presumed to be uniform for all pairs of tests so formed.

Where $i \circ g$, the covariance $^\circ\sigma(_1X_{pI} - {}_1X_{PI}, {}_2X_{pG} - {}_2X_{PG})$ has as its expected value $\sigma(_1p, {}_2p)$. This is true in a crossed or nested one-facet design, and for all values of $n_i'$ and $n_g'$. The estimate of the covariance component is available from the G study.

We estimate a correlation by dividing a covariance by the root mean square of the variances for the variables. The unbiased estimate of a variance does not produce an unbiased estimate of the standard deviation, but that is not a critical matter. Nor is it critical that we use a ratio of unbiased estimates, rather than an unbiased estimate of the ratio. Where $i \circ g$,

$$(9.30) \qquad \widehat{\mathscr{E}\,{}^\circ\rho}(_1X_{pI}, {}_2X_{pG}) = \frac{\hat{\sigma}(_1p, {}_2p)}{[\widehat{\mathscr{E}\sigma^2}(_1X_{pI}) \cdot \widehat{\mathscr{E}\sigma^2}(_2X_{pG})]^{1/2}}$$

The variance of observed scores takes on different values according to the design of the D study; Chapter 3 gave the necessary estimation formulas.

No matter what the design of the D study, as $n_i'$ and $n_g'$ increase, (9.30) approaches:

$$(9.31) \qquad \hat{\rho}(_1\mu_p, {}_2\mu_p) = \frac{\hat{\sigma}(_1p, {}_2p)}{[\widehat{\sigma^2}(_1p) \cdot \widehat{\sigma^2}(_2p)]^{1/2}}$$

This is Spearman's result, restated in terms of estimated variance and covariance components, and reached without assuming all test forms to have the same variance. Equations (9.30) and (9.31) can be evaluated whether $i \circ g$ or $i \bullet g$ in the G study. Spearman should be credited with considering linkage and proposing an adequate procedure for taking it into account in one-facet studies (1910, p. 277).

If $i \bullet g$ in the D study, the expected covariance of $_1X_{pI}$ with $_2X_{pG}$ is not equal to the covariance component for persons. The expected covariance in a crossed D study is

$$(9.32) \qquad \mathscr{E}^\bullet\sigma(_1X_{pI}, {}_2X_{pG}) = \sigma(_1p, {}_2p) + \frac{1}{n_i'}{}^\bullet\sigma(_1pi, e, {}_2pg, e)$$

In a nested D study, $\frac{1}{n_i'} \cdot \sigma(_1i,_2g)$ also enters the covariance. The denominator of the formula for estimating $\mathscr{E} \cdot \rho(_1X,_2X)$ is the same as in (9.30). With $i \cdot g$, in the limit as $n_i'$ and $n_g'$ increase we reach (9.31), whether the single facet of the design is crossed or nested.

Intuition may suggest that scores obtained with joint sampling will always correlate higher than those from independent sampling, whether the number of conditions is large or small. Ratings of leadership and intelligence given by the same rater do correlate higher than those given by different raters. But, when 10 raters are averaged, the individual biases (constant errors and biases favoring certain types of men) tend to cancel out. Consequently, as $n_i'$ increases, covariance arising from joint sampling adds less and less to the correlation.

The correlation between universe scores obviously cannot depend on the experimental design. A correlation between universe scores is a considerably more fundamental aspect of nature than an observed correlation, which is determined by the experimental designs. Therefore, conceptual interpretations of instruments should rest on estimated correlations between universe scores (Block, 1963).

Spearman's attenuation correction has a rather bad name because an occasional estimate exceeds 1.00. This may be a consequence of sampling errors or of hidden linkages; any underestimate of the universe-score variance or any overestimate of the covariance will inflate $\hat{\rho}(_1\mu_p,_2\mu_p)$. Despite these difficulties, the plain fact is that the correlation between universe scores answers the substantive question the typical correlational study asks. The uncorrected correlation manifestly gives only a tangential answer to that question. Such an answer may be quite misleading, notably when a low observed correlation is allowed to testify that constructs are essentially unrelated. The uncorrected correlation is as much subject to sampling errors of unknown direction and magnitude, and to effects of hidden facets, as is the corrected correlation. The remedy for sampling error is not to interpret uncorrected correlations, but to collect sufficient and appropriate data.

Some further discussion of the effect of hidden facets may be instructive. Suppose there is a three-sample study, and that the estimates of $\sigma^2(_1\mu_p)$ and $\sigma^2(_2\mu_p)$ are based on a proper random-model design, with no fixed, hidden condition. Then we have an unbiased estimate for the denominator of (9.31). But suppose that the study from which the covariance is obtained has a fixed $j^*$. Then the equations that ordinarily estimate $\sigma(_1\mu_p,_2\mu_p)$ will actually estimate $\sigma(_1\mu_p,_2\mu_p) + \sigma(_1pj^*,_2pj^*)$. The added term will most often have the same sign as the covariance of universe scores. Consequently, one tends to overestimate the absolute value of the numerator of (9.31) and the absolute value of $\rho(_1\mu_p,_2\mu_p)$. If a hidden facet $j*$ also influences the estimates for the denominator, the denominator will be overestimated by the addition of $pj*$ components of variance. The net effect when there are errors in both numerator and denominator is unforseeable.

In the two-facet study, to which we shall turn in a moment, there is also a risk of hidden facets. Those who employ the correction formula (9.30) or (9.31) are likely to think only of the facets taken explicitly into account. The interpreter should constantly bear in mind the question "What universe?," to recognize the explicit facets, the confounded hidden facets, and the fixed hidden facets. Fixed facets restrict the universe and the meaning of the correlation.

In the linear equation for predicting $_2X_{pG}$ from $_1X_{pI}$, the regression coefficient is $\sigma(_1X_{pI},_2X_{pG})/\sigma^2(_1X_{pI})$. Where we do not have data on the specific $G$ and $I$, we have to use the expected coefficient: $\mathscr{E}\widehat{\sigma}(_1X_{pI},_2X_{pG})/\mathscr{E}\widehat{\sigma}^2(_1X)$. This, however, is to be interpreted as $\hat{\sigma}(_1\mu_p,_2\mu_p)/\mathscr{E}\widehat{\sigma}^2(_1X)$ if $I \circ G$, and as $\bullet\hat{\sigma}(_1X_{pI},_2X_{pG})/\mathscr{E}\widehat{\sigma}^2(_1X)$ if $I \bullet G$. That is, different equations are required for the two cases.

Either coefficient can be evaluated for any $n_i'$ and $n_g'$. If $n_i'$ and $n_g'$ become indefinitely large, the regression slope becomes $\hat{\sigma}(_1\mu_p,_2\mu_p)/\sigma^2(_1\mu_p)$ for both linked and independent designs. The slope describes the expected universe score on one variable as a function of the other universe score; there are, of course, two such functions, the $_2\mu_p$-on-$_1\mu_p$ regression and the $_1\mu_p$-on-$_2\mu_p$ regression.

### Universes classified with respect to two facets

Scores $_1X_{pij}$ may be generalized to universe scores $_1\mu_p$, $_1\mu_{pi}$, or $_1\mu_{pj}$; for $v_2$, the possible universe scores are $_2\mu_p$, $_2\mu_{pg}$, or $_2\mu_{ph}$. Any of the first three scores, paired with any of the second three, defines a correlation. At first, it may seem pointless to consider any combination except $_1\mu_p$ with $_2\mu_p$, but that reaction is no more than an outcropping of the classical conception of the problem. Many a study in the past *has* estimated one of the other correlations and represented it as an estimate of $\rho(_1\mu_p,_2\mu_p)$. There is good scientific information to be found in almost any combination of the universe scores, when the correct interpretation is made.

Consider a study of English compositions. Let $v_1$ be a judgment on quality of writing and $v_2$ a judgment on quality of content. Let $i$ or $g$ be a topic, and $j$ or $h$ a judge. Now one might ask these questions:

1. Do writing ability and quality of ideas go together in general? This implies correlating $_1\mu_p$ with $_2\mu_p$.

2. Do writing quality and content quality on any one theme go together? Correlate $_1\mu_{pi*}$ with $_2\mu_{pi*}$.

3. Does the judge who likes a person's style also tend to like the quality of his ideas? Correlate $_1\mu_{pj*}$ with $_2\mu_{pj*}$.

(Meaning can also be given to the correlation of $_1\mu_p$ with $_2\mu_{pj*}$, $_1\mu_{pi*}$ with $_2\mu_p$, etc.).

The model defines one correlation of the first kind, $N_i$ of the second, and $N_j$ of the third. Because the $i$ represent a population of topics and the $j$ represent a population of raters, a general description of relations is given by a correlation of the first kind. A general answer to the second question is given by $\underset{i,g}{\mathscr{E}} {}^\bullet\rho(_1\mu_{pi},_2\mu_{pg})$. A similar expected value answers the third question in a general way. Correlations for separate topics and separate judges are valuable supplements. The three kinds of correlations are somewhat redundant, because their numerators always include $\sigma(_1p,_2p)$. Direct examination of the components of covariance sheds more light on the way in which judgments of content and quality covary than does comparison of correlations for universe scores.

To evaluate the correlation of $_1\mu_p$ with $_2\mu_p$, one divides $\hat\sigma(_1p,_2p)$ by the square root of the product of the two universe-score variances. That is, formula (9.31) serves. To evaluate $\mathscr{E} {}^\bullet\rho(_1\mu_{pi},_2\mu_{pg})$, one estimates the expected covariance of $_1\mu_{pi}$ with $_2\mu_{pg}$ as $\hat\sigma(_1p,_2p) + {}^\bullet\hat\sigma(_1pi,_2pg)$; this is the numerator. The variance estimates for the denominator are of the form $\widehat{\sigma^2}(p) + \widehat{\sigma^2}(pi)$. For the slightly less regular case $\rho(_1\mu_p,_2\mu_{pj})$, the numerator is simply $\hat\sigma(_1p,_2p)$.

Where the $i$ are topics for essays and $i \equiv g$ (the strongest possible linkage), the estimate of $\underset{i,g}{\mathscr{E}} {}^\bullet\rho(_1\mu_{pi},_2\mu_{pg})$ indicates the correlation expected on the average between ratings of content quality and writing quality, where each rating is the pooled judgment of all raters in the universe on the same essay. One may contrast this with $\mathscr{E} {}^\circ\rho(_1\mu_{pi},_2\mu_{pg})$, for ratings of different essays.

If a study is made with the design $_1(i \times j \times p)$, $_2(i \times h \times p)$, the same $I$ being used for observing $v_1$ and $v_2$, one may analyze data for any particular $i^* \in I$ to estimate ${}^\bullet\rho(_1\mu_{pi*},_2\mu_{pi*})$. Thus, if the $i$ are topics for essays, the $j$ may be judges who rate content while the $h$ are judges who rate writing. (The two sets of judges may be the same.) To learn that ratings of content and ratings of writing correlate more strongly for some topics than for others may be valuable, especially if one wishes to separate the two variables. Even if the same topics are not to be used again, it may be possible to describe how the topics that yield high correlations differ from those yielding low correlations. Such a conclusion would be helpful in selecting topics for future examinations.

The correlations for universe scores are actually limiting values of correlations between observed scores, as $n_i'$, $n_g'$, etc. increase. In principle, one can estimate the correlation for observed scores obtained with any number of conditions and any experimental design. Rather than trace out all the possibilities, we shall assume a crossed design for the measurement of both variables, and let $n_i' = n_g'$ and $n_j' = n_h'$. The covariance is then estimated (in general) by:

$$(9.33) \qquad \mathscr{E} \, {}^\bullet\sigma({}_1X_{pIJ}, {}_2X_{pGH}) = \sigma({}_1p, {}_2p) + \frac{1}{n_i'} \, {}^\bullet\sigma({}_1pi, {}_2pg)$$

$$+ \frac{1}{n_j'} \, {}^\bullet\sigma({}_1pj, {}_2ph) + \frac{1}{n_i'n_j'} \, {}^\bullet\sigma({}_1pij, e; {}_2pgh, e).$$

This applies directly when $i \bullet g$ and $j \bullet h$. If $i \bullet g$ and $j \circ h$, the last two covariance components vanish; if $i \circ g$ and $j \bullet h$, the second and fourth vanish. Estimating the expected observed-score variances as in Chapter 3 enables us to calculate the correlation.

### Relevance to causal inferences

Adjusting empirically observed relations for error of observation is required in all science. In the social sciences, particularly sociology and economics, there has recently been an increasing emphasis on attempts to reason causally from correlational findings (Blalock, 1968; Wittrock & Wiley, 1970, p. 351ff.). Path analysis and similar techniques set forth a model regarding the hypothetical flow of influence. An increase in the supply of money, for example, is thought conducive to a reduction in interest rates and an increase in new home construction. Basically, the model consists of a set of equations rather like regression equations, in which certain coefficients are specified *a priori* to be zero. Thus, in accounting for new home construction, the money supply and the rate of formation of new families would plausibly receive weight; one might argue for or against fitting a nonzero weight to the "influence" of sales figures for automobiles, depending on whether or not the automobile is regarded as competing with homes for the same dollars. The statistical procedures used attempt to estimate how much change in a dependent variable is associated with a certain change in one of the hypothesized causal indicators.

It is now recognized that errors of observation can distort the results of any such analysis. To take the simplest example, a model may postulate linear relations in which $v_1$ is the sole cause of $v_2$, and $v_2$ the sole cause of $v_3$. Then any "influence" of $v_1$ on $v_3$ is mediated by $v_2$ and the partial covariance of $v_1$ with $v_3$, $v_2$ constant, should be zero. This will be confirmed, however, only if the observed $v_2$ is a perfectly accurate measure of the corresponding construct in the model. Issues of validity as well as adequacy of sampling arise here; for present purposes, we need only connect the sampling concerns in causal inference with those of generalizability theory. It may be noted that Ross and Smith (1968, p. 348ff.) discuss generalizability (by that name) as a central matter in the interpretation of social experiments. They point not only to the hazards that arise from sparse data, but also to the hazard that, in selecting one indicator from a number of qualitatively different but equally admissible indicators, one will obtain inadequately generalizable data. In the same volume, Siegel and Hodge (1968, pp. 28–59) devote a chapter to the study of measurement error in the context of causal analysis. (See also Blalock, 1970, and Wiley & Wiley, 1970.)

Those who examine structural models have customarily mentioned plural sources of error, but have been less than explicit in identifying facets and estimating their contributions to variance. It is fairly common, for example, to collect social or economic data from a panel of informants in the community, or a panel of firms. It is obvious that these informants or firms are intended to represent a universe, and that a small or ill-chosen sample will tend to introduce an error into the indicator based on their reports. Variability is associated with the selection of a date of inquiry (assuming that the indicator is intended to be representative of a period rather than an instant in time). The particular wording of a question also alters results. While a growing number of studies attempt to assess the impact of error and to correct for it, we have found no study in this vein that decomposes the error variance, and on that basis recommends a better design of a measurement procedure. It is in this respect that multifacet theory is particularly superior to the older correction for attenuation.

The concept of correlated error (i.e., of linkage introduced by joint sampling) is highly pertinent to causal analysis and has been recognized in that research more often than in psychological measurement. The use of the same panel to supply data on two variables inevitably introduces linkage. Whether this linkage is "correlated error" or not depends on the causal model. To collect data on family formations, and at a later time on housing starts in the same community, introduces linked error if the community data are taken to represent national data on each variable. But if the model is concerned with community-specific effects (e.g., "In a community where the rate of family formations is large, one expects such-and-such a trend in local housing starts") the parameters to be estimated are community specific. Then the linkage is a part of the phenomenon to be described rather than an irrelevance. It seems likely that benefits will follow from a careful restatement of some of the problems of causal inference using concepts from generalizability theory.

One other phenomenon noted in the causal-inference context may be mentioned in passing. It is not a phenomenon that generalizability theory can now treat effectively, as our sampling model assumes that sampling condition $i$ for $p$ in observing $v_1$ does not affect $p$'s score on $v_2$. When variables are observed one after the other, there is a possibility of "propagated error," to use the term suggested by David Wiley (personal communication). Suppose, for example, that one is studying a pupil's participation in classwork as a function of his score on the preceding week's test, the study perhaps being motivated by some theory regarding the effect of increased or decreased competence and self-esteem. Suppose that the teacher gives the pupil a mark greater than the universe score he deserves—greater, let us say, than the mark this teacher would assign on several re-readings of the paper. There is error, in that the reported measure of competence departs from the universe score. The mark is an independent variable influencing the mediating variables of self-esteem and teacher esteem. Consequently, the "error" is as much a part of the treatment as is the accurate part of the score. So far as the teacher and pupil know, the pupil who received an $A$ "is an $A$ student" and the classroom interaction will proceed on that assumption. If esteem, rather than competence, determines the interaction, then the correlation between an interaction measure and the prior fallible mark will very likely be greater than the correlation of the interaction measure with the universe score on that prior test. This is one of many kinds of problem involving successive observations in a changing situation that the model based on sampling of conditions of observation does not describe adequately.

### Illustrative studies of universe-score relationships

*A TAT study.* The relatively simple example to be considered first is hypothetical, though suggested by a published paper. To study relations between motivational variables, four TAT pictures were presented to subjects at the same sitting and scored by a single rater. Scores on two variables will be considered. The investigator reports an observed correlation of 0.25, which implies a considerable degree of independence. The only reliability information offered is that a second scorer agreed with the first: $r = 0.90$ for $v_1$ and 0.80 for $v_2$. The implication is that error of measurement is not a source of concern, and that the observed correlation probably indicates how independent the constructs $v_1$ and $v_2$ are. No formal correction for attenuation was made.

Applying the concepts of our work, the first question is, what is the universe of generalization? The investigator has acted as if he were concerned with generalization over scorers and not over other facets. A score has the form $_vX_{pIjo}$, where $i$ is picture, $j$ is scorer, and $o$ is occasion; this investigator carried out a G study pertinent to universe scores $_v\mu_{pIo}$. The observed standard deviations are assumed to be unity so that we can write in terms of covariances. The value of 0.25 then represents the covariance of $_1X_{pIjo} - {}_1X_{PIjo}$ with $_2X_{pIjo} - {}_2X_{PIjo}$, which, on the average, equals $\sigma(_1p,_2p) + {}^\bullet\sigma(_1pi,_2pi)/4 + {}^\bullet\sigma(_1pj,_2pj) + {}^\bullet\sigma(_1po,_2po) + {}^\bullet\sigma(_1pij,_2pij)/4 + {}^\bullet\sigma(_1pio,_2pio)/4 + {}^\bullet\sigma(_1pjo,_2pjo) + {}^\bullet\sigma(_1pijo,e;_2pijo,e)/4$. The two-scorer study allows us to estimate $\sigma^2(_1\mu_{pIo})$, which equals $\sigma^2(_1p) + \sigma^2(_1pi)/4 + \sigma^2(_1po) + \sigma^2(_1pio)/4$, and $\sigma^2(_2\mu_{pIo})$, which is similar in form. Obviously, the investigator is in no position to evaluate $^\bullet\rho(_1\mu_{pIo},_2\mu_{pIo})$ even if that is his desire. A covariance would be required for these universe scores, and the only covariance available is affected at least to some extent by the linkages of $j$ as well as those of $i$ and $o$. His report permits no inference about any universe-score correlations.

If generalization is over judges, it is necessary for him to divide[5]

$$^\bullet\hat\sigma(_1\mu_{pIo},_2\mu_{pIo})$$

by $\hat\sigma(_1\mu_{pIo})$ and $\hat\sigma(_2\mu_{pIo})$. The divisors have been directly estimated but the covariance has not. The author should not have correlated scores given by the same judge; $[^\bullet s(_1X_{pIjo},_2X_{pIj'o}) + {}^\bullet s(_1X_{pIj'o},_2X_{pIjo})]/2$ is an unbiased estimate of $^\bullet\sigma(_1\mu_{pIo},_2\mu_{pIo})$.

In fact, with two scorings of a reasonable number of protocols, one could inquire into generalizability over pictures as well as over scorers. Consider the following covariances:

$$^\bullet\sigma(_1X_{pijo},_2X_{pijo}) \quad \text{estimates} \quad {}^\bullet\sigma(_1\mu_{po},_2\mu_{po}) + {}^\bullet\sigma(_1pi,pio;_2pi,pio)$$
$$+ {}^\bullet\sigma(_1pj,pjo;_2pj,pjo)$$
$$+ {}^\bullet\sigma(_1pij,pijo,e;_2pij,pijo,e)$$

$$^\bullet\sigma(_1X_{pijo},_2X_{pi'jo})^6 \quad \text{estimates} \quad {}^\bullet\sigma(_1\mu_{po},_2\mu_{po}) + {}^\bullet\sigma(_1pj,pjo;_2pj,pjo)$$

$$^\bullet\sigma(_1X_{pijo},_2X_{pij'o}) \quad \text{estimates} \quad {}^\bullet\sigma(_1\mu_{po},_2\mu_{po}) + {}^\bullet\sigma(_1pi,pio;_2pi,pio)$$

$$^\bullet\sigma(_1X_{pijo},_2X_{pi'j'o}) \quad \text{estimates} \quad {}^\bullet\sigma(_1\mu_{po},_2\mu_{po})$$

With two judges and four pictures, there are 8 covariances of the first kind that can be averaged, 12 of the second kind, etc. Subtraction of these average covariances from one another enables one to estimate each of the four covariance components. The comparable variance components can be derived from a three-way analysis of variance. [The covariance components could also be calculated from sums of products, as in (9.22) and (9.23).] From the variances and covariances, one can estimate $^\bullet\rho(_1\mu_{po},_2\mu_{po})$, $^\bullet\rho(_1\mu_{pIo},_2\mu_{pIo})$ for any $n_i'$, $^\bullet\rho(_1\mu_{pJo},_2\mu_{pJo})$ for any $n_j'$, and also $^\bullet\rho(_1X_{pIJo},_2X_{pIJo})$ for any $n_i'$, $n_j'$.

[5] $^\bullet\sigma(_1\mu_{pio},_2\mu_{pio})$ is a sum of $p$, $pi$, $po$, and $pio$ covariance components—linked, in that conditions $i$ and $o$ are the same for both variables. $^\bullet\sigma(_1\mu_{po},_2\mu_{po})$ is a sum of $p$ and $po$ covariance components.

[6] $^\bullet\sigma(_1X_{pi'jo},_2X_{pijo})$. In the next line there is a similar alternative.

Some investigators of motivational constructs would prefer not to treat occasions as fixed. Recent experiences affect TAT responses, implying that covariances of scores linked by sampling of occasions will be larger than those for independent scores. It follows that this G study would have been more informative if two pictures had been applied to each subject on each of two occasions, and each protocol had been rated twice. If this seems too expensive, various incomplete designs could be used. From such a design one could estimate not only the correlations above, but also $\rho(_1\mu_p,_2\mu_p)$, $\bullet\rho(_1\mu_{pI},_2\mu_{pI})$, $\bullet\rho(_1\mu_{pJ},_2\mu_{pJ})$, $\bullet\rho(_1\mu_{pO},_2\mu_{pO})$, $\bullet\rho(_1\mu_{pIJ},_2\mu_{pIJ})$, etc.

We have not developed this example numerically, but it should be obvious that these correlations need not be close to the correlation of 0.25 for observed scores. At the end of the chapter, Exercise 2 deals numerically with a somewhat simpler problem of this kind.

*Wechsler Verbal and Performance scores.*    The Ross data, for which components were determined earlier (p. 282), allow us to illustrate the multiplicity of correlations among universe scores. The covariance component for persons for the Ve and Pe scores was estimated to be 5.54, the component for forms $k$, $\ell$ was 0.00, and that for persons × forms was 0.02. Even though the estimates are based on small samples, they will be adequate to illustrate the kind of reasoning needed.

Suppose that one proposes to generalize over subtests and forms (and, because forms are given on different days, over days as well). The universe scores are $_{Ve}\mu_p$ and $_{Pe}\mu_p$, and their covariance is given by $\sigma(_{Ve}p,_{Pe}p)$, which equals 5.54. The variance of $_{Ve}\mu_p$ is given by the person component of variance for that score, estimated to be 8.16; the corresponding value for Pe is 4.84. From (9.30) we then have $\hat{\rho}(_{Ve}\mu_p,_{Pe}\mu_p) = 5.54/(8.16 \cdot 4.84)^{1/2} = 5.54/6.28 = 0.88$. This may be compared with the expected correlation of linked observed scores (i.e., scores obtained on the same form and day using the average of 5 subtests). This correlation is $5.56/(8.83 \cdot 5.62)^{1/2} = 0.79$.

If we generalize over forms (and days) but not over subtests, the universe scores are $_{Ve}\mu_{pJ\bullet}$ and $_{Pe}\mu_{pH\bullet}$. The covariance of these is again estimated to be 5.54, as the person–subtest component of covariance is zero. (There is no linkage of subtests.) The estimated variance of $_{Ve}\mu_{pJ\bullet}$ is now estimated by $\widehat{\sigma^2}(_{Ve}p) + \widehat{\sigma^2}(_{Ve}pj)/5$, according to the argument developed in Chapter 4; similarly for Pe. Hence,

$$\hat{\rho}(_{Ve}\mu_{pJ\bullet},_{Pe}\mu_{pH\bullet}) = \frac{5.54}{(8.16 + 0.19)^{1/2}(4.84 + 0.30)^{1/2}}$$

$$= \frac{5.54}{2.89 \cdot 2.27} = 0.85$$

A different question is to be raised about the agreement of Ve and Pe scores obtained by exhaustive testing on a single occasion. When one is interested in short-term fluctuations in mental efficiency, is the Ve–Pe distinction profitable? This inquires about the correlation of $_\mathrm{Ve}\mu_{pk*}$ with $_\mathrm{Pe}\mu_{pk*}$, generalization being over subtests. The covariance is the sum of the person and person–form components (unweighted, because $n'_k = 1$). If we take 0.02 as the person–form component of covariance (see p. 283) we have $5.54 + 0.02 = 5.56$. The universe-score variances are analogous; for Ve, $8.16 + 0.01 = 8.17$, and for Pe, 4.88. $5.56/(8.17 \cdot 4.88)^{1/2} = 0.88$. The correlation for exhaustive Ve and Pe measurements on a fixed day is therefore no higher than the correlation for scores based on an exhaustive series of measurements spread over many days, according to these data. The correlation of 0.88 does imply a very considerable degree of overlap between the constructs represented by Ve and Pe.

Still another variant inquires about generalization over items, leaving subtests and occasions fixed. To determine the correlation of $_\mathrm{Ve}\mu_{pJ*k*}$ with $_\mathrm{Pe}\mu_{pH*k*}$ requires evidence on components for items or half-tests, not available for the Ross–Morledge data. This would be answered by a split-half investigation. We suspect that most investigators trained in classical methods of analysis would think that dividing the Ve–Pe correlation (within a testing) by the square roots of reliability coefficients based on splits within subtests is *the* proper way to correct for error of measurement. Judging from the earlier study of WPPSI, it is likely that items-within-subtests account for about one-fourth of the variance arising from person–subtest interactions in the present study, because item and subtest effects are confounded here. Suppose, then, that we take the values as follows:

|  | Ve | Pe | Ve × Pe |
|---|---|---|---|
| Persons | 8.16 | 4.84 | 5.54 |
| Person × subtest (unconfounded) | 0.72 | 1.13 | — |
| Person × (items:within subtest) | 0.24 | 0.37 | — |
| Person × (form, occasion) | 0.01 | 0.04 | 0.02 |

Then the estimated universe-score variance with subtests, form, and occasion fixed will include the person, person × subtest, and person × form components above, the subtest component being multiplied by 1/5. The correlation of universe scores is estimated to be $5.56/(8.31 \cdot 5.10)^{1/2} = 0.85$.

Figure 9.1 provides a summary that enables one to trace the successive changes as generalization is broadened. The covariance changes little, and that change is due to the departure from within-day linkage. The variances decrease as fewer components are taken into the universe score. And, in this study, the correlation tends to increase as the universe of generalization broadens, but that will not inevitably be the case.

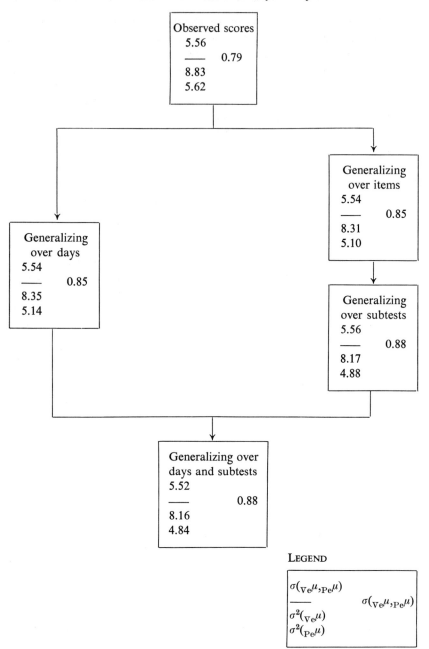

*FIGURE 9.1.    Changes in Covariance, Universe-Score Variances, and Correlation with Shifts in the Universe Definition (Data are for Wechsler Verbal and Performance Scores)*

*Validating a test of aptitude for science.* A more complex example of correlations among universe scores (correlations corrected for attenuation) is provided by research of Lesser, *et al.* (1962). The Hunter Science Aptitude Test ($v_1$) has two forms, individually administered. The test score is $_1X_{pijo}$, where $i$ = form, $j$ = examiner, and $o$ = occasion. A group of 58 children were given both forms, within a two-week period; for practical reasons, each form was given by a different examiner. The interclass correlation was 0.64.[7] The correlation may be considered to be an estimate of the squared correlation of test score with universe score $_1\mu_p$, the mean that would be approached if the subject were tested on many occasions during a period of two weeks or so by many examiners with many forms.

The criterion score $_2X_{pGHO}$ was a score over seven achievement tests, each given at the end of an instructional unit. Here $g$ = item and $h$ = unit. Its reliability was estimated to be 0.82 by a split-half analysis, based on splits within the subtests. The coefficient is to be interpreted as the squared correlation of observed score $_2X_{pGHO}$ with the score $_2\mu_{pHO}$ obtained from a universe of items representing the seven fixed units, each unit being given the same raw-score weight as in the total score used by Lesser. Occasion $O$ is treated by Lesser, *et al.* as if fixed; it is a set of seven occasions, each associated with one $h$.

The observed correlation of aptitude with achievement is 0.77 for Form A, 0.74 for Form B. By the usual Spearman formula, we have two estimates of the correlation of $_1\mu_p$ with $_2\mu_{pHO}$: 1.06 and 1.02. These values exceed unity, presumably because of inadequate sampling of persons and conditions. From these facts, one can only conclude as Lesser, *et al.* did: "The predictive validity is about as high as it can be, given the test reliability and the reliability of the criterion."

A further statement by the authors shows a misunderstanding. "These coefficients probably represent slight overcorrections because the parallel-forms reliability estimates of Forms A and B take into account error variance resulting from day-to-day variations in the children's performances as well as inter-examiner variability. The correction for attenuation assumes reliability estimates that exclude both these sources of error variance." This seems to mean that the correction should count person–day and person–examiner components as true variance. Our theory leads to a different view. One might consider several different universes of predictor information; let us concentrate on these three:

---

[7] The authors apply a correction formula developed by Angoff that introduces issues we must ignore. We shall extract from their discussion as if this special formula had not been used.

1. Universe 1: All forms, with a single examiner and on a single day. The universe score is $_1\mu_{pjo}$.
2. Universe 2: all forms, with a single examiner; many days within a certain period of time. The universe score is $_1\mu_{pj}$.
3. Universe 3: all forms, many examiners, many days within a certain period of time. The universe score is $_1\mu_p$.

Each one of these has its own correlation with $_2\mu_{pHO}$, and each of these corrected correlations can be estimated if the corresponding coefficient of generalizability is determined from the G study of $v_1$. The G study carried out by Lesser, *et al.* estimated the coefficient for Universe 3, and the attenuation formula therefore gave the correlation of the criterion universe score with $_1\mu_p$. Nothing in our theory (or in the classical theory, for that matter) "assumes" that the correlation with the universe score for Universe 1 should be estimated, as the quoted words of Lesser, *et al.* suggest. Indeed, if a genie were to offer to supply one of the three universe scores as a predictor, we would certainly ask him for the score on Universe 3, because person–examiner and person–day interactions must interfere with prediction of classroom learning.

The universe score $_2\mu_{pHO}$ is not the only possibility for the criterion. One might have planned to sample criterion measures for any unit on several occasions, as well as sampling several sets of items. The universe that $_2\mu_{pHO}$ describes consists of all items suitable for the given units, the test on each unit being administered on just one occasion; call this Universe 01. It may be contrasted with Universe 02: all items suitable for the given units, administered on several occasions within a certain period of time. The universe score is $_2\mu_{pH}$. The G study actually conducted estimated a coefficient of generalizability for Universe 01; this is larger than the coefficient for Universe 02. The latter is lowered by variance arising from person–occasion interaction. Consequently, we expect the corrected correlation for Universe 02 to be higher than the coefficient the authors reported.

Consider a different question. What if the criterion universe had been defined in terms of a collection of units ($N_h$ large) of which the given units are considered to be independent random samples? Does not a test intended for a talent search seek to predict achievement in primary science units generally, rather than achievement in the seven fixed units? Then the universe score $\mu_p$ (call this Universe 03) would be the mean score over all units (and over all items suitable for a given unit) over various occasions. Because generalization over the broader Universe 03 would lead to a smaller $\mathscr{E}\rho^2$, an even greater corrected coefficient is to be anticipated.

Admittedly, when the sample data produce an estimated correlation between universe scores of 1.00, it means little to talk of universes for which the correlation would be still larger. But the rationale we have set forth makes it even more obvious that chance inflated the value reached by Lesser, *et al.*

## EXERCISES

**E.1.** Children responded to tape-recorded dramas by describing the speaker's emotions, and the degree of insight or comprehension was scored. Successive tapes dealt with the themes of happiness, anger, anxiety, and sadness. The intercorrelation of scores on different tapes (different themes) was approximately 0.35. The author (Rothenberg, 1970) says: "While the correlations are not so high as to lead to the the conclusion that all four tapes are measuring exactly the same thing, they are high enough to justify pooling into one measure of social sensitivity."

Suppose that a study of similar subjects had shown that the correlation between two tapes on the *same* theme was approximately 0.50. How would this information bear on the advisability of pooling the scores? Suppose the correlation between tapes on the same theme was approximately 0.80, instead; how would this influence the decision to pool?

**E.2.** Children are observed in a preschool setting; observers $j$ are crossed with occasions $o$. Two scores are recorded on each observation: $v_1$, seeking help; $v_2$, task persistence. The intercorrelations are calculated for all degrees of linkage. The

**TABLE 9.E.1.** *Correlations for Observations of Children*

| $v$ | $j$ | $o$ | $v_1$: Seeking help | | | | $v_2$: Task persistence | | | |
|---|---|---|---|---|---|---|---|---|---|---|
| | | | $j=1$<br>$o=1$ | $2$<br>$1$ | $1$<br>$2$ | $2$<br>$2$ | $1$<br>$1$ | $2$<br>$1$ | $1$<br>$2$ | $2$<br>$2$ |
| 1 | 1 | 1 | 1.00 | 0.50 | 0.70 | 0.30 | $-0.20$ | 0.00 | $-0.20$ | $-0.15$ |
| 1 | 2 | 1 | | 1.00 | 0.30 | 0.70 | 0.00 | $-0.20$ | $-0.15$ | $-0.20$ |
| 1 | 1 | 2 | | | 1.00 | 0.50 | $-0.20$ | $-0.15$ | $-0.20$ | 0.00 |
| 1 | 2 | 2 | | | | 1.00 | $-0.15$ | $-0.20$ | 0.00 | $-0.20$ |
| 2 | 1 | 1 | | | | | 1.00 | 0.50 | 0.70 | 0.30 |
| 2 | 2 | 1 | | | | | | 1.00 | 0.30 | 0.70 |
| 2 | 1 | 2 | | | | | | | 1.00 | 0.50 |
| 2 | 2 | 2 | | | | | | | | 1.00 |

(artificial) data in Table 9.E.1 may be regarded as averages over many observations, hence as good estimates of the respective correlations. To minimize complications, observed scores are to be thought of as standardized.

a. Calculate the components of variance and covariance that can be determined from these correlations (covariances of standard scores).

b. Calculate $\rho(_1\mu_p, {_2}\mu_p)$.

c. Calculate $^{\bullet}\rho(_1\mu_{po}, {_2}\mu_{po})$.

d. Explain what the two correlations mean and why they differ.

**E.3.** Suppose that in Exercise 2 the variance and covariance components had been as follows:

|  | *p* | *pj* | *po* | *pjo, e* |
|---|---|---|---|---|
| Variance component (either variable) | 0.30 | 0.40 | 0.20 | 0.10 |
| Covariance component | −0.12 | −0.35 | −0.15 | −0.08 |

What do these components of covariance and variance tell us about observer behavior?

**E.4.** Levin, Rohwer, and Cleary (1971) administered four paired-associate lists, two per occasion. The form lists employed sets of pairs that were equally difficult to learn, other things being equal. On each occasion, one list paired verbal stimuli with verbal response terms, and the other list paired pictures. The following average correlation (over six different groups) summarizes the findings with school children.

Same kind of list, different occasions     $r = 0.50$ for verbal, 0.50 for pictures
Same occasion, different kind of list     $r = 0.27$
Different kind of list, different occasion     $r = 0.20$

Assume that the standard deviation for a verbal list is always 2, and that for a picture list is 3.

a. What can be learned about components of variance and covariance for the two variables, verbal learning and pictorial learning?

b. What more could be learned if a study were to employ two separate verbal lists and two separate picture lists on each occasion? (Eight lists in all.)

c. A V–P difference is formed, using scores on the same occasion to get the observed difference score. ($w_1 = 1$, $w_2 = -1$). What is the coefficient of generalizability from this score to the difference score in the universe (many lists of each kind, on many occasions)?

d. How would the answer in *c* be modified if the universe score were defined in terms of numerous verbal lists on one set of occasions, and numerous picture lists on a different series of occasions?

e. How would the answer in *c* be modified if the observed difference score were based on V and P lists on separate days?

**E.5.** Table 9.E.2 summarizes a combined variance–covariance analysis of scores and ratings on the Gross Forms Test (Gross & Marsh, 1970), a measure of creativity

**TABLE 9.E.2** *Variance–Covariance Analysis of Scores and Ratings on Gross Geometric Forms Test*

| Source of variance | Degrees of freedom | Score = $v_1$ | | Rating = $v_2$ | | Score × Rating | |
|---|---|---|---|---|---|---|---|
| | | Sum of squares | Mean square | Sum of squares | Mean square | Sum of products[a] | Mean product |
| $p$ | 29 | 127.987 | 4.413 | 1069.494 | 36.882 | 773.118 | 26.658 |
| $i{:}p$ | 270 | 171.362 | 0.635 | 541.470 | 2.005 | 301.074 | 1.115 |
| $r$ | 5 | | | 1.550 | 0.310 | | |
| $pr$ | 145 | | | 304.355 | 2.099 | | |
| $ri, pri, e$ | 1350 | | | 405.000 | 0.300 | | |

[a] The sum of ratings by six raters was used in this computation.

for young children. The child is to make "something," given a set of 48 different colored cardboard geometric forms and a feltboard. After each trial the examiner asks the child to tell about what he has made; the examiner takes notes on the product and the explanation. Later each product is scored on a six-point scale by formal rules.

As one step in the construct validation of the test, Gross investigated the appropriateness of the formal scoring system by asking six "creative" people in the community to rate (on a five-point scale) the products of 30 children. The correlations between the total rating over 10 products and the total formal score for the same products ranged from 0.69 to 0.83 for various raters (average, 0.75). No further interpretation was made by the authors.

The design of this G study was $_1[j^* \times (i{:}p)]$; $_2[r \times (i{:}p)]$; $n_j = 1$, $n_r = 6$, $n_i = 10$, $n_p = 30$. Variable 1 refers to the formal score with one scorer $j^*$ and $v_2$ refers to the ratings given by raters $r$. Because each child's products $i$ were unique to him, this facet is nested in persons. Facet $i$ is identical for the two variables.

a. Construct a table of estimated variance and covariance components similar to Table 9.2.

b. Calculate the coefficient of generalizability for the average formal score on 10 products, generalizing to a universe of products (scorer fixed). Calculate the coefficient of generalizability from an average rating given by 6 raters to 10 products, generalizing to a universe of raters and products.

c. Estimate the correlation between the formal score on the universe of products (using one scorer) and the rating for a universe of raters and products.

d. It is estimated (from other data) that the generalizability of scores on 10 products over a universe of scorers is 0.94. What, then, is the expected correlation between a universe of scorers and a universe of raters on a universe of products?

e. What conditions of testing have been held constant in this example that limit the universe of generalization and could affect interpretation of the result in *d*?

**E.6.** In the example of p. 290 what importance might the following correlations have?

a. $\rho(_{1}\mu_{pi*},{_{2}}\mu_{p})$

b. $\rho(_{1}\mu_{pj*},{_{2}}\mu_{p})$

**E.7.** One can process the Ross–Morledge data (p. 280) with four fixed variables: WISC–Ve, WISC–Pe, etc. Consider subtests as variables, nested within scales and

*TABLE 9.E.3    Mean Variances and Covariances for Ross–Morledge Data*

| | WISC Ve subtests | | WISC Pe subtests |
|---|---|---|---|
| Mean variance | 12.98  (5) | Mean variance | 9.79  (5) |
| Mean covariance | | Mean covariance | |
| within WISC Ve | 8.66  (10) | within WISC Pe | 5.59  (10) |
| with like Ve | | with like Pe | |
| subtests of WAIS | 9.12  (5) | subtests of WAIS | 6.34  (5) |
| with unlike Ve | | with unlike Pe | |
| subtests of WAIS | 8.16  (20) | subtests of WAIS | 4.84  (20) |
| with Pe subtests | | with Ve subtests | |
| of WISC | 6.13  (25) | of WISC | 6.13  (25) |
| with Pe subtests | | with Ve subtests | |
| of WAIS | 4.92  (25) | of WAIS | 6.15  (25) |

| | WAIS Ve subtests | | WAIS Pe subtests |
|---|---|---|---|
| Mean variance | 9.93  (5) | Mean variance | 7.37  (5) |
| Mean covariance | | Mean covariance | |
| within WAIS Ve | 7.67  (10) | within WAIS Pe | 4.17  (10) |
| with like Ve | | with like Pe | |
| subtests of WISC | 9.12  (5) | subtests of WISC | 6.34  (5) |
| with unlike Ve | | with unlike Pe | |
| subtests of WISC | 8.16  (20) | subtests of WISC | 4.84  (20) |
| with Pe subtests | | with Ve subtests | |
| of WAIS | 5.11  (25) | of WAIS | 5.11  (25) |
| with Pe subtests | | with Ve subtests | |
| of WISC | 6.15  (25) | of WISC | 4.92  (25) |

jointly sampled ($j \bullet h$) for like scales. (E.g., WISC Arithmetic joint with WAIS Arithmetic.)

The average variances and covariances in Table 9.E.3 were calculated. The numbers in parentheses are the numbers of values contributing to the various means.

a. Estimate the covariance component $\sigma(_{\text{WISC–Ve}}p,_{\text{WISC–Pe}}p)$

b. Estimate $\sigma(_{\text{WISC–Ve}}p,_{\text{WAIS–Ve}}p)$

c. Estimate $^\bullet\sigma(_{\text{WISC-Ve}}pj,_{\text{WAIS-Ve}}ph)$

d. Estimate $^\bullet\sigma(_{\text{WISC-Pe}}pj,_{\text{WAIS-Pe}}ph)$

e. Estimate $\sigma(_{\text{WISC-Ve}}p,_{\text{WAIS-Pe}}p)$

**E.8.** Using the components from p. 282 and p. 297, estimate the following correlations for a test that employs a single verbal and a single performance subtest.

a. The expected correlation of Ve and Pe observed scores on the same occasion.

b. The correlation of $_{\text{Ve}}\mu_{pk*}$ with $_{\text{Pe}}\mu_{pk*}$.

c. The correlation of $_{\text{Ve}}\mu_{pj*k*}$ with $_{\text{Pe}}\mu_{pj*k*}$.

**E.9.** Two moral attitudes are measured, each by a 13-item test requiring judgment. Each item is in the form suggested by Piaget, where two undesirable actions of children are described and the subject is to judge which action is worse and why. The correlation of the scales is low; the score reflecting tendency to judge in terms of the child's intentions (rather than the extent of damage he causes) correlated 0.19 with a score from other problems on knowing the meanings of rules (H. Harris, 1970).

The question arises whether these are two independent types of moral development or the low correlation reflects the fact that observed scores are poor measures of the respective universe scores.

a. What universe of generalization is relevant to this inquiry?

b. The investigator reports separate investigations of test-retest reliability and scorer agreement. What universe-score correlations can be inferred from these two findings?

c. What additional G-study information would be extremely useful?

## Answers

**A.1.** If 50% of the observed variance on any tape arises from person-tape variability, and 35% is common over themes, only 15% is left for person × theme variance. This argues strongly that little information on individual differences is lost by pooling themes. If, however, the within-theme correlation for two tapes is 0.80, about 45% of the variance arises from person × theme interaction. Then much might be learned by retaining separate scores for the several themes. Rothenberg is apparently interested in the universe score for response to all possible dramas on the several themes, and probably regards the themes as fixed. The small magnitude of the person–theme interaction argues that if themes were selected randomly rather than by the stratified pattern of one for each theme, the results would be almost equally suitable as a basis for generalization. It would of course be possible to generalize over the universe of tapes on each single theme, but this will be of little value unless several tapes on the theme are combined in a single "subtest" score.

The interpretation is somewhat different when the within-theme correlation for two tapes is 0.80, since about 45% of the variance arises from person–theme interaction. Now one can generalize to the universe score for any single theme with considerable accuracy with a very small number of tapes on that theme. And it is crucial to use a stratified design in which each theme is properly represented if one is to generalize from the total score, to the universe score for pooled themes.

A.2.  a.

| Types of correlation | Value for same variable | Value across variables | Components included |
|---|---|---|---|
| $j, o$ common | 1.00 | −0.20 | $p, pj, po, pjo, e$ |
| $j$ common | 0.70 | −0.20 | $p, pj$ |
| $o$ common | 0.50 | 0.00 | $p, po$ |
| independent | 0.30 | −0.15 | $p$ |

| | $p$ | $pj$ | $po$ | $pjo, e$ |
|---|---|---|---|---|
| Variance component (either variable) | 0.30 | 0.40 | 0.20 | 0.10 |
| Covariance component | −0.15 | −0.05 | +0.15 | −0.15 |

b. $\rho(_1\mu_p, _2\mu_p) = \dfrac{-0.15}{0.30^{1/2}0.30^{1/2}} = -0.50$

c. $^\bullet\rho(_1\mu_{po}, _2\mu_{po}) = \dfrac{-0.15 + 0.15}{0.50^{1/2}0.50^{1/2}} = 0.00$

d. The former correlation indicates the extent to which the trait of seeking help is correlated with the trait of persistence, considering all occasions during a certain period, observed by an indefinite number of observers. The second correlation is that within an occasion, and asks: Assuming thorough observation, to what extent does the child who seeks help on a certain occasion persist during *that particular* performance? The negative $p$ component of covariance (and the correlation where it is numerator) implies that persistent children are typically not those who seek help. The positive $po$ component implies that when a child seeks help he is likely also to persist longer on that occasion than he usually does. (Note that $^\bullet\rho(_1po\sim, _2po\sim) = 0.75$.) The $p$ and $po$ covariances offset each other; general persistence correlates negatively with general tendency to seek help, but on any one occasion there is no relationship between these aspects of behavior.

A.3.  The large $pj$ components of variance indicate that observers form substantially different impressions of the same child; a child one sees as consistently dependent, the other may see as much less dependent. There is a strong tendency for observers who rate a child as characteristically dependent also to rate him as characteristically lacking in persistence. The correlation of $_1\mu_{pj}\sim$ with $_2\mu_{pj}\sim$ is −0.87. There is a similarly strong correlation (−0.80) between components $_1\mu_{pjo}\sim, e$ and $_2\mu_{pjo}\sim, e$. When an observer reports a tendency to seek help that is not reported by other observers, he tends also to report an exceptional absence of persistence. The two results together strongly suggest a semantic linkage such that a perception of behavior—even a random error in perception—implies to the observer opposite things regarding the two variables.

**A.4.   a.**

| | Components of variance | | Components of covariance |
|---|---|---|---|
| | V | P | |
| Persons | $0.50 \times 4 = 2.0$ | 4.5 | $0.20 \times 6 = 1.2$ |
| Person $\times$ occasion<br>Residual } | 2.0 | 4.5 | $0.07 \times 6 = 0.4$ |

b. Examining the list-to-list covariance on the same occasion would enable one to separate person $\times$ occasion variance from the residual variance for each variable. In the design with only four lists, occasion and list (within variables) are confounded.

c. $^{\circ}\sigma(_VX,_PX) = 1.2$

  $^{\bullet}\sigma(_VX,_PX) = 1.6$

  $\widehat{\mathscr{E}}\rho^2(_dX,_d\mu) = \dfrac{2.0 + 4.5 - 2(1.2)}{4 + 9 - 2(1.6)} = \dfrac{4.1}{9.8} = 0.42$

d. No change. The universe-score variance is not altered.

e. $\widehat{\mathscr{E}}\rho^2 = \dfrac{4.1}{4 + 9 - 2(1.2)} = \dfrac{4.1}{10.6} = 0.39$

**A.5.   a.**

| Estimated variance components | | Estimated covariance components |
|---|---|---|
| Formal score $v_1$ | Rating $v_2$ | Score $\times$ Rating |
| Persons $\widehat{\sigma^2}(_1p \mid j^*)$ $= 0.378$ | $\widehat{\sigma^2}(_2p) = 0.551$ | $\hat{\sigma}(_1p \mid j^*,_2p) = 0.426$ |
| Products within persons $\widehat{\sigma^2}(_1i_p,e \mid j^*) = 0.635$ | $\widehat{\sigma^2}(_2i_p) = 0.284$ | $^{\bullet}\hat{\sigma}(_1i_p \mid j^*,_2i_p) = 0.186$ |
| Raters and related interactions | $\left\{ \begin{array}{l} \widehat{\sigma^2}(_2r) = (0) \\ \widehat{\sigma^2}(_2pr) = 0.180 \\ \widehat{\sigma^2}(_2ri_p,pri_p,e) = 0.300 \end{array} \right.$ | |

The notation $i_p$ follows the convention of Chapter 8, for a facet that is nested in the universe. The products are different for each individual. We have to interpret the $e$ component with some care. Because there is only one "trial" on a particular product, the product cannot be regarded as a member of a class of products representing the same condition of the facet $i$. There is variation of the scores assigned by person $j^*$ when he scores the drawings repeatedly, and this $e$ is confounded with the $i_p$ component of variance for $v_1$. Error of the same sort enters the ratings by any rater, but appears at another point in the variance analysis for $v_2$. The expected covariance for these two kinds of judge inconsistency is zero, because the raters are sampled independently of the scorer.

b. $\widehat{\rho^2}(_1X_{pIj^*},_1\mu_{pj^*}) = 0.856$; $\widehat{\rho^2}(_2X_{pIJ},_2\mu_p) = 0.897$

c. $\hat{\rho}(_1\mu_{pj^*},_2u_p) = \dfrac{0.426}{(0.378 \times 0.551)^{1/2}} = 0.934$

d. Approximately 1.00.

e. Examiner and occasion are both constant in this study. Both numerator and denominator of $d$ are probably inflated by these hidden facets. Consequently, the correlation of universe scores over scorers, raters, trials, examiners, and occasions cannot be estimated.

**A.6.** a. Does the person who is rated well for his style on topic $i^*$ generally win approval for his ideas? If this were higher than the correlation for other topics, it would suggest that $i^*$ is a topic where the person's ideas influence the rating of his style.

b. Does $j^*$ tend to give high or low style ratings to the person whose ideas are generally regarded as good? If this were higher than the correlation for other judges, it would suggest that $j^*$ tends more than others to take ideas into account in rating style.

**A.7.** a. $\hat{\sigma}(_{\text{WISC–Ve}}p,_{\text{WISC–Pe}}p) = 6.13$

b. $\hat{\sigma}(_{\text{WISC–Ve}}p,_{\text{WAIS–Ve}}p) = 8.16$. (Considering like subtests would bring in a $pj$ covariance.)

c. $^\bullet\hat{\sigma}(_{\text{WISC–Ve}}pj,_{\text{WAIS–Ve}}ph) = 9.12 - 8.16 = 0.96$

d. $6.34 - 4.84 = 1.50$

e. 4.92

**A.8.** a. $\dfrac{5.54 + 0.02}{(8.16 + 0.72 + 0.24 + 0.01)^{\frac{1}{2}}(4.84 + 1.13 + 0.37 + 0.04)^{\frac{1}{2}}}$

$$= \dfrac{5.56}{7.16} = 0.73$$

b. $\dfrac{5.56}{(8.17)^{\frac{1}{2}}(4.88)^{\frac{1}{2}}} = 0.88$

c. $\dfrac{5.56}{(8.89)^{\frac{1}{2}}(6.01)^{\frac{1}{2}}} = 0.76$

**A.9.** a. It appears that one would like to generalize over questions, occasions and scorers.

b. The information on scorer agreement permits us to evaluate the correlation between measures of the two kinds of attitude, very accurately scored, with fixed questions and occasions. The retest information indicates the correlation for scores over a universe of trials, with fixed questions and scorers. It should be possible to estimate the person–scorer and person–occasion components of variance by assuming the person–scorer–occasion interaction to be negligible. Then one could infer the correlation of two kinds of problems over a universe of scorers and occasion, still with questions fixed.

c. The retest data could have been analyzed by a two-facet design [questions × (occasion:person)] so as to evaluate the components associated with questions and question–occasion interaction. This is essential to estimate the correlation between scores in a universe where questions are a variable facet.

# CHAPTER 10

# Multivariate Estimation of Universe Scores: Profiles, Composites, and Difference Scores

Multivariate data have power considerably greater than previous psycho-metric treatments of profiles, composites, change scores, etc. have capitalized upon. This chapter combines generalizability theory with an extension of ideas that otherwise have been discussed only in the context of measurement of change (Lord, 1956, 1958; Harris, 1963).

Observing each person on several variables generates a score profile. The interpreter is usually interested in the profile of universe scores, and any procedure that enables him to reach a better estimate of that profile will be helpful. It is possible not only to obtain sounder information on the scores originally defined, but to reorganize and simplify the variate set by eliminating dimensions regarding which the information is highly fallible. After discussing profiles, we go further into the attenuation problem and then turn to composite scores, including difference scores and change scores.

### A. The Profile of Universe Scores

Universe scores are far more fundamental than observed scores, because the latter are affected by the design for data collection and the vagaries of each performance. The scientist is concerned with relations among constructs, which are always universe scores or functions of universe scores. To him, observed scores are no more than a basis for inference. Universe scores are especially to be emphasized when two or more variables are interpreted simultaneously, because differences among the person's observed scores on the variables may result solely from inadequate observation and describe nothing of significance. In counseling for example, it seems that one should examine the best available estimate of the person's profile of universe scores, rather than interpret his profile of observed scores as it stands. The observed profile shape may convey radically wrong information when the coefficient of generalizability varies considerably from one variable to the next, or when the mean profile for a subpopulation has a distinctive, irregular shape.

While the discrepancy between observed score and universe score is acknowledged by the confidence-band technique, that technique has serious limitations (p. 131 ff.). Its assumptions are rarely satisfied, and it ignores information. With norm-referenced profiles there is the further difficulty that the percentile or standard-score distribution for observed scores is not pertinent to universe scores (see p. 146). Consequently, the profile "shape" suggested by the display of score bands will not correspond to the shape of the profile of standardized universe scores. We shall explore the possibility of estimating each universe score in turn by regression methods. The profile of such estimates can be displayed on either absolute or norm-referenced scales.

Regressing may or may not make a great difference in the conclusions drawn about a person. Under some circumstances (if scores are extremely reliable; or if scores are equally reliable and equally intercorrelated, and have equal means in the pertinent subpopulation), the profile shape remains essentially the same. But for most variables employed in profiles, and in differences and other composites, these conditions are not satisfied. We suspect that substantive conclusions drawn from multivariate data, in either counseling or research, will often be altered if our proposals are followed.

Adoption of our proposals will pose serious difficulties for the test inter-preter, because he is inevitably guided by his experience. Clinical interpreters have had experience only with observed-score profiles, and they will have to build up a whole new body of experience to make sound use of estimated universe-score profiles. (The labor of calculating regression estimates of a series of scores might be a considerable deterrent, but computerized scoring can remove this difficulty.) It is fair to question whether an innovation that requires extensive retraining of investigators is worthwhile. But the fact that interpreters already possess a frame of reference for interpreting, for example, the observed difference between Wechsler Ve and Pe IQs can scarcely be a justification for rejecting a more reasonable estimate of the universe-score difference. On occasion, regression methods will indicate that a certain person whose observed-score difference is negative most probably has a positive universe-score difference (or vice versa). Any procedure that stands "obvious" conclusions on their heads in this manner is too important to brush aside for reasons of convenience or tradition.

## Assumptions for one-facet universes

In presenting univariate theory, it was pointed out that regression estimates from crossed data are untrustworthy when observed scores under various conditions do not have identical relations to the universe score. In this chapter, to reduce complications, *it is assumed that all conditions are equivalent.* Specifically, it is assumed that for any $v$, $v'$, $i$, $i'$, etc., and for any subpopulation for which a regression equation is formed,

$$\sigma^2(_vX_{pi}) = \sigma^2(_vX_{pi'})$$

$$\mathscr{E}_p(_vX_{pi}) = \mathscr{E}_p(_vX_{pi'})$$

(10.1)
$$\rho^2(_vX_{pi},_v\mu_p \mid i) = \rho^2(_vX_{pi'},_v\mu_p \mid i')$$

$$^{\circ}\sigma(_vX_{pi},_{v'}X_{pg}) = {}^{\circ}\sigma(_vX_{pi'},_{v'}X_{pg'})$$

$$^{\bullet}\sigma(_vX_{pi},_{v'}X_{pg}) = {}^{\bullet}\sigma(_vX_{pi'},_{v'}X_{pg'})$$

Essentially, these statements imply that the *factorial composition of scores arising from a condition i is the same as for scores arising from a condition i'.* We thus return to classical assumptions. Random sampling from a universe of conditions is retained, as is our concern with the design of the G and D studies and the concept of joint sampling. But we assume, formally, that conditions within the single-facet universe are equivalent.

The requirement of strict equivalence is unnecessarily severe. When $n'_i$ is large and sampling of $i$ is random, scores $_vX_{pI}$ and $_vX_{pI'}$ are likely to be nearly equivalent even if the $_vX_{pi}$ taken singly are not. When conditions are nested within persons in the D data, the means, covariances, etc. tend toward equality even without an equivalence assumption. (See also the discussion of assumptions on p. 100 ff.). If conditions are far from equivalent, however, one must use the procedures discussed in this chapter with great caution. The problems arise not from our model so much as from the fact that virtually any interpretaiion proposed involves generalization. Complex generalizations are difficult or impossible to make when conditions of observation are not roughly equivalent.

Scores being compared have to be expressed in "the same" metric. It is not necessary that the distribution of observed scores or universe scores for $v$ be the same as that for $v'$. What is necessary is a reciprocal relationship, such that when a score of 12 (say) on $v_1$ is said to "correspond to" a score of 27 on $v_2$ for the purpose of some interpretation, it is implied that a score of 27 on $v_2$ "corresponds" to 12 and only 12 on $v_1$. Further, the relation must be monotone; i.e., scores 12, 13, and 14 on $v_1$ may be mapped respectively into 27, 29, and 35 on $v_2$, or into 27, 23, and 20 (preserving the order) but not into 27, 35, and 29. Any correspondence is set up in a particular context, with reference to a certain kind of interpretation. How many ounces of butter corresponds to a cup of corn oil? There is one answer if we think in terms of cost, another if we think in terms of calories, and a third in terms of effect on the arteries. Psychologists should not rely exclusively on equipercentile correspondences.

To simplify further it is assumed that all variables in the G study enter the D study and vice versa.

*Statistical considerations as agenda for the future.* Classical test theory was developed in terms of population parameters. Until the Lord–Novick text of 1968 few papers and almost no textbook presentations took sampling of persons into account, or studied sampling distributions effectively. This statistical side of classical theory is now beginning to yield to attack, but even with strong assumptions about test equivalence, progress is slow. We have connected the weaker model of generalizability theory for single variables with statistical reasoning through the medium of the jackknife procedure.

As we turn to multivariate problems, we shall retreat as far as possible into that ideal world of indefinitely large samples where statistical problems do not arise. The majority of writers on classical theory, and even those applying analysis of variance components to psychometrics, have likewise stopped at the edge of the statistical cliff. They barely caution the reader that statistics are not parameters; after that, they avert their eyes from the edge and proceed as blithely as though their data stretched to the horizon; we shall do the same. In multivariate country the statistical abyss is deeper and more dizzying than the one with which we are, so to speak, at home. All the warnings ever uttered about shrinkage, collinearity, the need for cross-validation, and the problems of estimating factor scores for individuals apply with great force to what is discussed in this chapter.

By assuming very large samples, we are able to set out many valuable concepts and to identify what we would like to estimate. Once this is stated clearly, it is reasonable to hope that mathematical statisticians will come to our aid, and provide better solutions or identify the boundary conditions that set limits on the use of large-sample solutions.

### Univariate and multivariate estimation

Under the equivalence assumption with an $i \times p$ design, one has the usual equation for estimating $_v\hat\mu_p$ from the observed score on $v$:

$$(10.2) \qquad _v\hat\mu_p = [\rho^2(_vX_{pI}, _v\mu_p)](_vX_{pI} - _v\mu) + _v\mu$$

This is essentially a restatement of (3.2), employing parameters for the population to which $p$ belongs. We write $\rho^2$ in place of $\widehat{\mathscr{E}\rho^2}$, because with equivalent conditions and a large sample the several $\rho^2$ are equal and very accurately estimated. We shall customarily simplify in this manner in this chapter, writing as if the G study yielded parameters rather than estimates.

A reader of Kelley (see p. 103) might determine the raw score mean, standard deviation, and reliability for ninth-grade boys of the DAT Verbal Reasoning score, and regress the observed score of each ninth-grade boy by means of (10.2). Repeating this univariate analysis for each $v$ in turn forms a new profile, expressed in terms of the original units. One can convert this to a standard-score profile of estimated universe scores, using scales such that one unit on the $v$ scale equals $\hat{\sigma}(_v\mu_p)$. This procedure does not recognize that an observation on $v_1$ contributes information regarding the universe score on any variable that $_1\mu_p$ correlates with. We now propose a method of estimating universe scores that is truly multivariate.

The findings from a G study are to be brought to bear on D data to estimate the $n_v \times n'_p$ values of $_v\mu_p$. Suppose that the G study takes the form of $n_i$ administrations of a battery of $n_v$ measures. The D study consists, let us say, of a single administration of the battery. Let $_*\mu_p$ be the universe score for the variable $v^*$ whose universe score we are at the moment attempting to predict. Generalization is over administrations, and the regression equation for estimating the universe score $_*\mu_p$ from the D data takes the form

$$(10.3) \qquad _*\hat{\mu}_p = \sum_{v=1}^{n_v} \beta_{*v} \left(_vX_{pi} - _v\mu\right) + _*\mu$$

The symbol $\beta$ stands for a population regression coefficient, not necessarily standardized.

The anticipated design of the D study is taken into account in estimating variances and covariances. The variances of predictors are estimated as in Chapter 3 and the covariance as in Chapter 9. Where the predictors $_vX$ and $_{v'}X$ are scores within a single battery, $i \bullet g$; it is appropriate to use $^\bullet\sigma(_vX,_{v'}X)$, calculated within occasions, as a covariance between predictors. The covariance of any predictor with the target variable $_*\mu_p$ has the form $\sigma(_vX,_*\mu_p)$, which is equivalent to the covariance component $\sigma(_vp,_*p)$ for any predictor except $_*X_{pi}$; there, one uses $\sigma^2(_*p)$.

The analysis will produce a $\rho^2(_vX_{pi},_*\mu_p)$ for each $v^*$. This is a squared multiple correlation equal to $\rho^2(_*\hat{\mu}_p,_*\mu_p)$, hence a new sort of coefficient of generalizability. Such a coefficient will almost always be greater than the coefficient from univariate analysis for the same universe score (and never less). Using all the information in the profile will produce a superior estimate of the universe score whenever the $_vX_{pi}$ are correlated, though the difference may be trifling. A new, smaller $\sigma^2(_*\varepsilon)$ accompanies the estimate.

As far as we know, the determination of profiles for test interpretation by multiple regression has hitherto been proposed in only one context. In the middle 1960's the 16 PF Test of Cattell was provided with a computer-scoring service that estimates factor scores by entering the 16 observed scores into estimation equations (Cattell, Eber, & Tatsuoka, 1970, pp. 37–38). While the theory and technical characteristics of the analysis have not been published, Professor Cattell has provided us with a draft of an article on the subject (Eber, Cattell, & Delhees, in preparation). Estimating factor scores differs only in detail from the estimation of scores on universe-defined variables. According to the available reports, 16 PF scores that are predicted in a univariate fashion with $\rho^2 = 0.75$, for example, are predicted from all data with $\rho^2$ near 0.85 (corrected for shrinkage). To obtain this accuracy of generalization from the univariate estimate one would have to double the test length. The multivariate estimates of the 16 PF scores are more highly intercorrelated than the conventional estimates; this reflects, not a true reduction in differential information, but the elimination of some undependable differences the observed profile reports.

In general, the profile of estimated universe scores generated by a multiple-regression procedure (i.e., the vector of $_\text{v}\mu_p$) will differ from the observed profile $_\text{v}X_{pi}$ and from the profile formed by estimating each $_*\mu_p$ from the corresponding $_*X_{pi}$ alone. For most persons, the new profile will be flatter, as an extreme example will make obvious. Suppose $_vX$ and $_{v'}X$ are actually two equally good measures of the same variable, so that $_v\mu_p = _{v'}\mu_p$ for all $p$. Then the best estimator of $_v\mu_p$ and of $_{v'}\mu_p$ will be a function of $(_vX_{pi} + _{v'}X_{pi})/2$. No matter what difference in observed scores was produced by errors of measurement, the two estimated universe scores will be equal, as are the actual universe scores. While it will distress the person who seeks information from differences within a profile to be confronted with relatively flat profiles, the regression estimates are closer to the truth.

*Numerical example: the Wechsler Performance Scale.* As a simple example of multivariate estimation we shall use the Ross–Morledge data from p. 282. We use the mean of scaled scores on Verbal subtests as $_\text{Ve}X$ and the mean for Performance subtests as $_\text{Pe}X$. We treat the WISC and WAIS data as two samples of the same kind of performance; we propose to generalize from data on one form over a universe of Performance measurements. Symbols $h$ and $j$ represent subtests, and $k$ and $\ell$ represent forms. This universe has two facets, but the concepts discussed in this chapter in terms of one facet can be applied.

The multiple-regression technique is most advantageous when subtests are regarded as random. Generalization is over subtests, items within subtests, days, and trials within days. The data are presented in Table 10.1. For this example, the means are arbitrarily set at 10, but we employ the actual standard deviations for the Ross sample. (The variance of the universe score

**TABLE 10.1.**   *Data Used in Forming a Multiple-Regression Estimate of the Universe Score for the Wechsler Performance Scale*

| Variable | Mean | Standard deviation | Correlation with | | |
|---|---|---|---|---|---|
| | | | 1 | 2 | 3 |
| 1. Verbal Scale, observed score | 10.0 | 2.97 | 1.000 | 0.790 | 0.847 |
| 2. Performance Scale, observed score | 10.0 | 2.37 | | 1.000 | 0.928 |
| 3. Performance Scale, universe score (over subtests and days) | 10.0 | 2.20 | | | 1.000 |

is 4.84, as determined in the G study discussed on p. 282.) Each correlation is calculated from covariance and variance components. The value of 0.847, for example, is the covariance component $\sigma_{(\mathrm{Ve}p,\mathrm{Pe}p)}$ of 5.54 divided by 2.97 and 2.20.

The univariate equation for prediction of $_{\mathrm{Pe}}\mu_p$ from the observed Performance score is

$$_{\mathrm{Pe}}\hat{\mu}_p = 0.861_{\mathrm{Pe}}X_{pJ\ell} + 1.39.$$

The multiple regression equation is

$$_{\mathrm{Pe}}\hat{\mu}_p = 0.639_{\mathrm{Pe}}X_{pJ\ell} + 0.224_{\mathrm{Ve}}X_{pHk} + 1.36.$$

The squared multiple correlations are as follows:

Univariate    $\widehat{\mathscr{E}\rho}^2{}_{(\mathrm{Pe}X,\mathrm{Pe}\mu_p)} = \widehat{\mathscr{E}\rho}^2{}_{(\mathrm{Pe}\hat{\mu}_p,\mathrm{Pe}\mu_p)} = 0.861$

Multivariate    $\widehat{\mathscr{E}\rho}^2{}_{(\mathrm{Pe}\hat{\mu}_p,\mathrm{Pe}\mu_p)} = 0.896$

The increase of 0.035 is practically significant. It implies a 25% reduction in the error variance—a benefit equivalent to what one would gain by lengthening the Performance Scale by a quarter. (That is, the gain is greater than one would get by adding a subtest to the scale, since that would lengthen it by one fifth.)

The multiple-regression technique has little advantage when $_{\mathrm{Ve}}\mu_p$ is taken as universe score. The squared correlation rises from 0.925 to 0.934, which implies a 10% reduction in error variance—worth having, but unimpressive.

The components calculated from the Ross–Morledge data can also be used to form regression equations for estimating the universe scores with subtests fixed. For this limited generalization, the two-variable estimation equation reduces error variance by 1/9 for Pe and by 1/16 for Ve. In general, when the univariate coefficient of generalizability is large, the introduction of information on a second variable is unlikely to produce much improvement.

**Parameters determining improvement in prediction.** When a predictor is added, $\rho^2(_*\hat\mu_p,_*\mu_p)$ is increased in an amount equal to the squared part correlation of the added predictor with $_*\mu_p$. (This increment is labelled $a$ below.) The greater the part correlation, the greater is the resulting multiple correlation. It is instructive to ask what the correlation of the added information with the target variable must be to increase the coefficient of generalizability by any given amount. While the relationship could be expressed in many ways, we have chosen as parameter the correlation between the observed $_*X$ and the auxiliary predictor $_vX$. Let $_vX$ be any score or score composite that does not include $_*X$. Then the increase in the multiple correlation depends on $\rho(_vX,_*X)$. If $a$ is the squared part correlation $\rho^2[_*\mu_p,(_vX\cdot_*X)]$, and if $_vX$ is observed independently of $_*X$,

$$(10.4) \quad a\rho^2(_*X,_*\mu_p)[1 - \rho^2(_vX,_*X)] = \rho^2(_vX,_*X)[1 - \rho^2(_*X,_*\mu_p)]^2$$

The curves plotted in Figure 10.1 indicate the correlation of $_vX$ with $_*X$ required for several different increments. Large increments are of course not possible when $\rho^2(_*X,_*\mu_p)$ is already large.

Equation (10.4) is developed under the assumption that the conditions for observing auxiliary predictors are sampled independent of the conditions

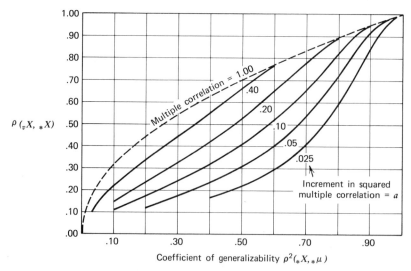

**FIGURE 10.1.** *Increment in Squared Multiple Correlation when a Supplementary Predictor $_vX$ is Used in Predicting the Universe Score $_*\mu$.*

for $_*X$. Where there is linkage, the equation takes on a far more complex form:

$$(10.5) \qquad a\rho^2(_*X,_*\mu_p)[1 - {}^\bullet\rho^2(_vX,_*X)]$$

$$= {}^\bullet\rho^2(_vX,_*X)[1 - \rho^2(_*X,_*\mu_p)]^2\left[1 - \frac{{}^\bullet\sigma(_vpI,_*pI)}{{}^\bullet\sigma(_vX,_*X)[1 - \rho^2(_*X,_*\mu_p)]}\right]^2$$

When the component ${}^\bullet\sigma(_vpI,_*pI)$ is zero, this reduces to (10.4). Where the component is positive, the value of $a$ calculated from (10.4), but using the correlation for the linked observed scores, will be too large; that is to say, the increment will be less than Figure 10.1 suggests. The component could be negative, making the increment greater than Figure 10.1 suggests, but this is most unlikely when ${}^\bullet\rho(_vX,_*X)$ is positive.

The figure can be used both to assess the general value of auxiliary predictors and to judge what can be expected from particular data. As a rule of thumb, we suggest that a gain of 0.05 is worthwhile if the coefficient of generalizability is above 0.40 and a gain of 0.025 is worthwhile if the coefficient is 0.85 or above.

We may apply this to specimen data without attempting to be very precise. Cronbach (1970, pp. 355, 358) has assembled data on the DAT and GATB batteries. For the former, several scores have generalizability coefficients (over forms and days) near 0.80. The chart suggests that an auxiliary predictor correlating 0.60 or better with the observed score will give a useful increment. For nearly every one of the DAT tests there are other tests in the battery for which intercorrelations reach this level; consequently, unless linkage produces a substantial covariance component, the regression technique would have practical value. The Verbal Reasoning test, for example, has a generalizability coefficient of 0.78. If one predicts the universe score from a combination of the observed Verbal score with an independent Grammar score (which has a correlation with Verbal of perhaps 0.70), the squared multiple correlation would be raised by perhaps 0.07, to 0.85.

For GATB some coefficients of generalizability are in the range 0.81–0.86, and these tests have no intercorrelation above 0.45. It follows that a multiple-regression estimate for their universe scores would add next-to-nothing. For Form Perception, however, the coefficient is 0.72, and another test (Clerical Perception) correlates 0.66 with it. The squared multiple correlation would then be around 0.80, a worthwhile gain. A similar, slightly smaller gain appears to be possible in the motor area, where coefficients for single tests are in the neighborhood of 0.65 to 0.75 and intercorrelations are around 0.50.

**Limits on the number of usable predictors.** In applying multiple-regression methods, one must be wary lest he capitalize on chance. The number of predictor weights that may properly be fitted will depend on the sample size.

It will often be advisable to reduce the number of predictors by combining them. Rarely will more than two supplementary composites produce a worthwhile increment in the accuracy with which the universe score is predicted. One procedure is to form *a priori* composites as we illustrate below.

*Numerical example: the DAT profile.* To expand on the possibilities of regression estimation of universe-score profiles and to draw attention to decisions the investigator must make, we apply methods suggested above to a set of data from the Differential Aptitude Tests. These data were supplied by Jerome Doppelt of The Psychological Corporation. The study is analyzed and reported in a deliberately sketchy fashion; a full presentation for all target variables would become so extensive that it would distract the reader from the main ideas of this chapter. Moreover, a good deal of trial and error will be required with various sets of data before any model of procedure can be set forth or any substantive conclusions drawn. This illustration is an early trial of the technique and the equations reached are emphatically not recommended for practical use.

The data come from the testing of boys in Grade 10 of two communities in Pennsylvania; in one school two forms (same tests, different items) were given a month apart, and in the other there was a six-month interval. There are 174 persons in all. (Some of these data are similarly pooled in the DAT manual, pp. 4–5. Means and standard deviations are available, but to simplify, we have treated the data as if all means were zero and all standard deviations unity. Consequently, we derive standardized regression weights. Because the standard deviations of tests did change from occasion to occasion, and sometimes departed from those in the standardizing sample, taking them into account would greatly complicate the discussion.)

The universe of generalization for each test presumably consists of various sets of items and various occasions within a period of some six months. This choice of universe appears to be consistent with the interpretations commonly made of DAT scores, although a longer time interval is sometimes implied in the interpretation.

There are two symmetric matrices of intercorrelations of observed scores within forms (within occasions), and an asymmetric matrix of correlations across forms (across occasions). We collapse the two within-form matrices into a single matrix of $^\bullet r(_vX, _{v'}X)$ by averaging. (For example, we average the two correlations of the Verbal and Numerical tests. Also, we make the between-forms correlation matrix symmetric by averaging across the diagonal; thus, the correlation of Form A Verbal with Form M Numerical is averaged with that for Form M Verbal and Form A Numerical. Call these correlations $^\circ r(_vX, _{v'}X)$. This method of introducing symmetry is consistent with the equivalence assumptions of Chapters 9 and 10.

We take $°r(_vX,_vX)$, the diagonal entry in the second matrix, as an estimate of $\rho^2(_vX,_v\mu)$. We estimate $\rho(_vX,_{v'}\mu)$ by $°r(_vX,_{v'}X)/[°r(_{v'}X,_{v'}X)]^{1/2}$, and $\rho(_v\mu,_{v'}\mu)$ by $°r(_vX,_{v'}X)/[°r(_vX,_vX)°r(_{v'}X,_{v'}X)]^{1/2}$. This corrects for attenuation to form two new matrices of correlations. These are assembled into a supermatrix:

$$\begin{bmatrix} \mathbf{A} & \mathbf{B} \\ \mathbf{B}^T & \mathbf{C} \end{bmatrix}$$

where the entries in **A** are the original $^\bullet r(_vX,_{v'}X)$, entries in **B** are the $\hat\rho(_vX,_{v'}\mu)$, and entries in **C** are the $\hat\rho(_v\mu,_{v'}\mu)$. The entries in **A** serve as predictor inter-correlations and the entries in **B** are predictor–criterion correlations. The matrix **C** is prepared only if the computer program used calls for a square matrix; **C** plays no role in the calculations. If any entry in **C** had exceeded 1.00, however, it would have been prudent to scale down the entries in the corresponding row and column of **C** and **B** by a sufficient amount to correct this anomaly (or else to reject the batch of data as too much perturbed by sampling error).

To give some impression of the data without introducing excessive detail, Table 10.2 presents the key correlations for one variable, Abstract Reasoning, and the coefficients of generalizability for all eight scores.

The added variable VERTO is a composite of Verbal, Spelling, and Grammar, equal weights being given to the three standardized scores; TOTAL is a similar composite of all eight scores. The composites were introduced as a simple means of capturing most of the predictor variance in two scores. TOTAL is a first centroid factor among the tests, and VERTO recognizes a group factor among the three verbal tests. This use of composites allows a kind of reduced-rank solution. While ours is not likely to be the most powerful solution, it is easily interpreted, and from the data it appears that use of principal components would give much the same final efficiency.

**TABLE 10.2.** *Illustrative Data Entering into Multivariate Estimation of DAT Abstract Universe Score*

| | Variable | | | | | | | | | |
|---|---|---|---|---|---|---|---|---|---|---|
| | Ver | Num | Abs | Spa | Mch | Clr | Spl | Gra | VERTO | TOTAL |
| Correlation with Abstract observed score on same form and occasion | 0.591 | 0.585 | 1.000 | 0.667 | 0.507 | 0.279 | 0.395 | 0.532 | 0.583 | 0.795 |
| Estimated correlation with Abstract universe score | 0.676 | 0.663 | 0.828 | 0.712 | 0.598 | 0.402 | 0.424 | 0.609 | 0.656 | 0.858 |
| Coefficient of generalizability | 0.779 | 0.793 | 0.686 | 0.823 | 0.679 | 0.483 | 0.828 | 0.741 | — | — |

**TABLE 10.3.** *Coefficients of Generalizability for DAT Scores Estimated by Multiple-Regression Methods*

| Universe score to be estimated | Squared multiple correlation where estimate is made from | | | Proportionate reduction in error variance with | |
|---|---|---|---|---|---|
| | Corresponding observed score | Three preselected predictors[a] | Eight observed scores[b] | Three predictors | Eight predictors |
| Verbal | 0.780[c] | 0.845 | 0.857 | 0.30 | 0.35 |
| Numerical | 0.794[c] | 0.835 | 0.839 | 0.20 | 0.22 |
| Abstract | 0.686 | 0.810 | 0.817 | 0.39 | 0.42 |
| Space | 0.823 | 0.836 | 0.844 | 0.07 | 0.12 |
| Mechanical | 0.679 | 0.725 | 0.764 | 0.14 | 0.26 |
| Clerical | 0.483 | 0.555 | 0.619 | 0.14 | 0.26 |
| Spelling | 0.828 | 0.851 | 0.864 | 0.13 | 0.21 |
| Grammar | 0.741 | 0.832 | 0.835 | 0.35 | 0.36 |

[a] The corresponding observed score, plus VERTO and TOTAL.
[b] Because stepwise procedure with redundant variables introduced discrepancies in the third decimal place, these values were calculated directly from the eight original scores.
[c] Values differ by 0.001 from values of original coefficients because of rounding.

Calculations were carried out by a stepwise regression program under these constraints: to predict any $_v\mu_p$, the corresponding $_vX$ was to be entered as the first predictor; at the second and third steps, VERTO and TOTAL were to be entered in whatever order gave the larger second-step multiple correlation; then six of the remaining observed scores were entered, in whatever order maximized the multiple correlation at each step.

As Table 10.3 shows, the increase in multiple correlation after the third step was small in most instances. We shall later consider in some detail the Mechanical and Clerical scores for which convergence was relatively slow. The benefit from making a multiple-regression estimate can be judged from the increase of $\rho^2(_v\hat{\mu}_p, _v\mu_p)$. Numerically small gains at the upper end of the scale must not be regarded as trivial. To maintain perspective, we have expressed the increment in the coefficient in terms of the change in error variance. It will be recalled that, in ordinary univariate estimation of a universe score, lengthening a test by half (multiplying $n_i'$ by 1.5) reduces the error variance by 33%. Roughly, this degree of improvement is achieved with three predictors, for Verbal, Abstract, and Grammar. While the gains for other universe scores are less impressive, they are attained at negligible cost.

What coefficient shall be accepted as the best representation of the gain from combining predictors? In predicting Abstract we might stress the coefficient after four steps rather than that after three steps, or the somewhat inflated coefficient for eight predictors.

The *F*-ratio for the increment from the fourth predictor, Spelling, is significant, and further predictors add no significant increment. A similar argument could be applied for other universe scores. Except for Mechanical and Clerical, this decision is quite unimportant; any later coefficient is very nearly the same as that for three predictors.

The next question is, which regression equation should be used? For any target variable there are eight sets of regression weights, each later equation using a small additional bit of information. In writing equations, we translated any weight calculated for VERTO by dividing any weight by $\sigma$(VERTO) and assigning the quotient as an added weight for Verbal, Spelling, and Grammar. We similarly distributed the weight for TOTAL uniformly over all the original scores. Applying this rule, we have the set of alternative equations for Abstract displayed in Table 10.4. The most important thing to note in the table is how little difference there is between weights at the various later stages. The choice of one equation rather than another, from the fourth through the eighth, will almost never alter the predicted universe score for an individual to a perceptible degree.

We have decided to examine the eight-predictor equation for each universe score, determined directly rather than stepwise. Although weights in this equation capitalize on chance to some slight degree, there is no reason to

**TABLE 10.4.**  *Regression Weights for Predicting Universe Score on Abstract after Each Stage of the Stepwise Analysis*

| Step number and predictor added | Standardized regression weight for | | | | | | | | |
|---|---|---|---|---|---|---|---|---|---|
| | Ver | Num | Abs | Spa | Mch | Clr | Spl | Gra | $\rho^2(_v\hat{\mu}_p, _v\mu_p)$ |
| 1. Abstract | | | 0.83 | | | | | | 0.686 |
| 2. TOTAL | 0.09 | 0.09 | 0.49 | 0.09 | 0.09 | 0.09 | 0.09 | 0.09 | 0.794 |
| 3. VERTO | 0.04 | 0.15 | 0.46 | 0.15 | 0.15 | 0.15 | 0.04 | 0.04 | 0.810 |
| 4. Spelling | 0.11 | 0.15 | 0.44 | 0.15 | 0.15 | 0.15 | −0.07 | 0.11 | 0.818[a] |
| 5. Mechanical | 0.11 | 0.15 | 0.44 | 0.15 | 0.13 | 0.15 | −0.07 | 0.11 | 0.818 |
| 6. Grammar | 0.10 | 0.15 | 0.44 | 0.15 | 0.13 | 0.15 | −0.07 | 0.09 | 0.818 |
| 7. Numerical | 0.10 | 0.14 | 0.44 | 0.16 | 0.13 | 0.16 | −0.07 | 0.09 | 0.818 |
| 8. Clerical | 0.12 | 0.14 | 0.43 | 0.17 | 0.12 | 0.15 | −0.07 | 0.09 | 0.818 |
| Weights by single-stage calculation | 0.12 | 0.14 | 0.44 | 0.16 | 0.13 | 0.15 | −0.07 | 0.09 | 0.817 |

[a] Last significant increment. This value and the $\rho^2$ at all later steps are slightly inflated because of collinearity and consequent exaggeration of rounding errors.

regard it as less meaningful than any equation with fewer variables. We have, however, stressed the more conservative multiple correlation based on three steps. (If sample size were much smaller, one would be far more hesitant to adopt the final weights.)

For Abstract, it appears that the prediction is essentially to be based on the Abstract observed score, plus the factor running through the nonverbal variables. The result is consistent with the view of Abstract Reasoning as a test requiring fluid adaptive ability. The small differences in weights for various tests do not deserve to be given a psychological interpretation until they are crossvalidated.

Table 10.5 displays the final weights for estimation equations for all

**TABLE 10.5.**  *Regression Weights for Predicting All Universe Scores, Obtained by Fitting All Predictors*

| Universe score to be estimated | Standardized regression weight[a] for | | | | | | | |
|---|---|---|---|---|---|---|---|---|
| | Ver | Num | Abs | Spa | Mch | Clr | Spl | Gra |
| Verbal | 0.6 | 0.1 | | | 0.1 | | 0.2 | 0.1 |
| Numerical | 0.1 | 0.7 | 0.1 | | | 0.1 | 0.1 | 0.1 |
| Abstract | 0.1 | 0.1 | 0.4 | 0.2 | 0.1 | 0.1 | −0.1 | 0.1 |
| Space | 0.1 | | | 0.8 | 0.1 | | | −0.1 |
| Mechanical | 0.2 | | 0.1 | 0.2 | 0.6 | | | −0.1 |
| Clerical | −0.2 | 0.2 | 0.3 | | −0.1 | 0.6 | 0.1 | |
| Spelling | 0.1 | 0.1 | −0.1 | | | | 0.8 | 0.1 |
| Grammar | 0.2 | 0.1 | | | | | 0.2 | 0.5 |

[a] Blanks represent weights in the range −0.05 to +0.05.

variables, rounded to emphasize major features of the equations. Few of the weights are difficult to rationalize; but substantive interpretations should be withheld until a relation is observed in further samples.

The reader has very likely noted that, in deriving equations from the correlation matrix, we have predicted each universe score on a scale where the standard deviation of the universe score is 1.00. If one wishes to express universe scores on a scale such that the universe-score standard deviation equals approximately $\sigma(_*\mu_p)$, he can simply multiply the scores estimated by our equation by the product $\hat{\rho}(_*X,_*\mu_p)\hat{\sigma}(_*X)$. The *estimated* universe scores will of course have a smaller standard deviation equal to $\hat{\rho}(_*\hat{\mu}_p,_*\mu_p)\hat{\sigma}(_*\mu_p)$. The whole issue of choice of scale for regression estimates is a vexing one that needs further thought. We remind the reader, finally, that for simplicity we have suppressed consideration of scale means; use of local means (along with variances and covariances for the local group) would of course change the equations.

### Reorganization of the profile

For a profile of $n_v$ scores there are $n_v$ universe scores, which can be thought of as defining a space of $n_v$ (or perhaps fewer) dimensions. The observed scores do not provide equally good information about all these dimensions, and, in fact, some of them may be very badly estimated. Where that is the case, it may be advisable to replace the original profile by a smaller set of variables, the universe score on each new variable being a composite of the original variables. (We shall use the word *composite* in a general sense; when all the $w_v$ except one are equal to zero, the composite coincides with one of the original variables.) The aim is to obtain a set of dimensions that are reliably measured and psychologically interpretable.

A suitable procedure for this follows the scheme described by Bock (1966). Using the mean squares and mean products as a starting point, he determines a series of composites. The first composite is that combination of the $_vX_{pi}$ for which the coefficient of generalizability is greatest. The second composite is orthogonal to the first, and has the largest coefficient of generalizability among the possible orthogonal composites. Each successive composite has a smaller coefficient than those previously extracted. The investigator can judge, on the basis of these coefficients, how many dimensions are reasonably well measured and should be retained.

The coefficients may indicate that the most prominent dimension in the universe-score space is measured with more precision than is necessary. If so, it should be possible to adjust test lengths so as to reduce the coefficient of generalizability for that dimension and to increase those for some dimension or dimensions that were inadequately observed with the original distribution of items.

Within the subspace defined by the selected composites, it is likely that correlated dimensions will be more interpretable than the orthogonal components determined mathematically. The test designer can rotate the orthogonal components into whatever set represents constructs he judges to be optimally interpretable. Universe scores on these variables can be estimated from the original observed scores.

Reducing and reorganizing the profile in this manner seems likely to retain all the information that can be reliably interpreted, and it reduces the interpreter's task by eliminating many differences within the profile that would not be confirmed if a second set of observations were made. It seems likely that for an instrument such as the Wechsler Scale, with its 11 subtests, as few as five dimensions may be sufficient to report a large proportion of the reliable information in the scores.

The Bock procedure is not the only method that can be used to examine the dimensions of the universe-score space and the accuracy with which each dimension is estimated. Nanda (1967) explored the application of Tucker's (1958) interbattery factor analysis. The vectors selected in this procedure are those whose projections into the universe-score space are largest, rather than those for which the coefficients of generalizability are greatest. Rao (1965) developed a technique of interbattery analysis, and in his early work on a "familial" correlation (1945, 1953) applied a procedure much like Bock's.

Abelson (1960) analyzed Semantic Differential data, where there is a score for each concept (e.g., mother) on each scale (e.g., strong-weak), given by each rater. Abelson employed discriminant analysis to form the composites of scales that are maximally generalizable over raters. His composite of scales is comparable to our composite of variables; in his analysis, concepts are the objects of inquiry where for us persons are ordinarily the objects of inquiry. Abelson's discussion is consistent with our separation of the facets over which one generalizes from the other bases for classifying data. Since Abelson's argument is developed in terms of the Cornfield–Tukey model, it can be directly extended into a multifacet approach. Abelson's interest in discriminant analysis reflects his awareness of a point made earlier by Tukey: that conventional factor analysis is distorted by linkage in the design. Factor analysis of covariance components for persons would apparently be consistent with Abelson's thinking.

The Tukey (1951) paper on "components of regression," even though two decades old, will necessarily be a point of departure for further development of the multivariate analysis discussed here. The paper is multifacet in principle. Among other points of interest, Tukey notes that the factors postulated in traditional factor analysis are composites of universe scores. However, if conditions for administering tests of a battery are linked (e.g., same occasion), nongeneralizable information will augment the test intercorrelations. In such an event, the factors accounting for intercorrelations will not be perfectly generalizable.

A major task of synthesis remains. The several formulations by Tukey, Tucker, Bock, Abelson, and others are essentially compatible, and compatible with our scheme. Yet the language of each argument is a bit different, and the point of view sometimes changes in subtle ways. Tucker's multimode factor analysis (see pp. 13) treats in symmetric fashion the several ways of classifying observations, whereas we (like Abelson) assign a distinctive place to persons and another distinctive place to variables. Still another variant appears in the "alpha factor analysis" of Kaiser and Caffrey (1965), which regards the $v$ as samples from a universe. One can obviously have further patterns that none of the present formulations is ready to deal with. For example, in the S–R Inventory of Anxiousness one can consider stimuli $i$

and modes-of-response $j$ as fixed, and replications or occasions as defining the universe of generalization; then one can embark on a multivariate examination of the $_{ij}X_{po}$.

Our formulation appears to generate no mathematical result that cannot be reached along the line suggested by Tukey and Abelson. Our way of stating the problem connects more explicitly the multi-way analysis with traditional problems in test theory, and so directs attention to particularly significant applications.

## B. *Relations among Universe Scores: the Multivariate "Attenuation" Problem*

The estimation of $\rho(_{v}\mu_{p}, _{v'}\mu_{p})$ was discussed in Chapter 9, where the classical correction for attenuation was modified to take multiple facets into account. We are now in a position to extend the correction in another way, estimating the multiple correlation $\rho(_{v}\mu_{p}, _{c}\mu_{p})$, $c$ being a "criterion" variable not included in **v**. We shall not go beyond the one-facet case.

$$(10.6) \qquad \mathbf{P} = \begin{bmatrix} \sum_{\mathbf{v}p} & \sigma(_{\mathbf{v}}p, _{c}p) \\ \sigma(_{\mathbf{v}}p, _{c}p)^{T} & \sigma^{2}(_{c}p) \end{bmatrix}$$

This is the variance–covariance matrix for the universe scores on the set **v** (regarded as "predictors") and the "criterion" $_{c}\mu_{p}$. $\sigma(_{\mathbf{v}}p, _{c}p)$ is a column vector of covariances of the form $\sigma(_{v}p, _{c}p)$. One can derive regression weights and the multiple correlation directly from **P**, but it may clarify matters to discuss the analysis in terms of correlations. For standardized variables, **P** becomes

$$(10.7) \qquad \mathbf{Q} = \begin{bmatrix} \mathbf{R} & \rho(_{\mathbf{v}}p, _{c}p) \\ \rho(_{\mathbf{v}}p, _{c}p)^{T} & 1 \end{bmatrix} = [\text{diag } \mathbf{P}]^{-1/2}\mathbf{P}[\text{diag } \mathbf{P}]^{-1/2}$$

This is a correlation matrix of order $n_{v} + 1$; diag $\mathbf{R} = \mathbf{I}$. Then

$$(10.8) \qquad _{c}\hat{\mu}_{p} = {_{\mathbf{v}}}\mu_{p} \cdot \boldsymbol{\beta}$$

where $_{\mathbf{v}}\mu_{p}$ is an $n_{p} \times n_{v}$ matrix of universe scores ánd $\boldsymbol{\beta}$ is a column vector of regression weights.

$$(10.9) \qquad \boldsymbol{\beta} = \mathbf{R}^{-1}\rho(_{\mathbf{v}}p, _{c}p)$$

Finally, the squared multiple correlation

$$(10.10) \qquad \rho^{2}(_{\mathbf{v}}\mu_{p}, _{c}\mu_{p}) = [\rho(_{\mathbf{v}}p, _{c}p)]^{T}\mathbf{R}^{-1}\rho(_{\mathbf{v}}p, _{c}p)$$

In practice, quantities in **P** are estimated rather than known. The off-diagonal entries may be filled with the covariances of the observed scores, or with estimates derived from a multivariate G study. Where there is linkage, it is necessary to employ $^{\circ}\hat{\sigma}(_vX,_{v'}X)$ and $^{\circ}\hat{\sigma}(_vX,_cX)$, not the linked covariances. The $\sigma^2(p)$ for the diagonal come from a G study. Unless all values entering **R** come from the same data, there is a risk that it will not be invertible. Reduced-rank methods may be employed in the analysis to reduce capitalization on chance.

The method may be extended to canonical correlations, simply by replacing the $c$ row and column of **P** or **Q** with matrices. Consider **c** to be not a single target variable but a set of variables (that ordinarily has no element in common with **v**). Then $\sigma(_vp,_cp)$ becomes $\sum_v p,_cp$ and $\sigma^2(_cp)$ becomes $\sum_cp$. The whole matrix is standardized and the usual canonical analysis is performed. Meredith (1964) showed that, in the absence of linkage, a squared canonical correlation for universe scores $\rho^2(_v\mu_p,_c\mu_p)$ is equal to

$$\rho^2(_vX,_cX)[\rho(_vX,_v\mu_p)\rho(_cX,_c\mu_p)]^{-1/2},$$

the squared correlation for the corresponding observed scores corrected by the Spearman attenuation formula.

Cochran (1970) pointed out the need for investigating the extent to which "errors of measurement weaken or vitiate the uses to which multiple correlation is put." (It is striking that the whole issue of correcting multiple correlations for attenuation has only so recently entered the literature, in view of the fact that the attenuation formula for simple correlation was introduced between 1904 and 1910, and that the multiple correlation was also coming into use at that time.) Cochran shows that the Spearman attenuation formula extends to the multiple correlation when predictors are uncorrelated; one need only substitute the weighted average of the predictor reliabilities for the usual predictor reliability $\rho(X,X')$. This average is the reliability of the best predictor of the criterion (or of its universe score). We can similarly substitute in the denominator of (9.31) a weighted average of the $\sigma^2(_vp)$. The weights are the squares of the standardized regression weights for combining the universe scores to predict the criterion (or its universe score). The numerator, of course, will be the covariance of the predictor (weighted composite) with the universe score of the target variable.

When predictors are correlated, no simple relation exists between the multiple correlation based on observed scores and that based on universe scores, according to Cochran. He was thus forced to confine attention to very limited cases, but our formulation is more powerful, and yields the desired multiple correlation in every case.

The essential difficulty Cochran encountered in seeking a straightforward conversion formula like Spearman's is this: the best combining weights for the fallible predictors assign quite different relative weights to the underlying universe scores than they have in predicting the same target directly from universe scores. That is to say, the corrected and uncorrected correlations report on the validity of distinctly different composites. A predictor that is inaccurately observed receives little or no weight in the observed-score composite even when the correlation of its universe score with the criterion is high.

Cochran emphasizes that conclusions about the relevance or irrelevance of variables will be incorrect if one attempts to interpret the regression weights in the observed-score equation substantively. Moreover, when the multiple correlation obtained from observed scores is low, the investigator may mistakenly conclude that further causal factors need to be unearthed, when inadequate observation of the *present* predictor variables is what holds down the correlation. The procedures described in this section enable one to learn about the contributions to the criterion of the predictor universe scores.

Ideally, an investigator who employs multiple-regression methods in an attempt to understand a system of variables will collect predictor and criterion data on the same large sample. He will replicate or subdivide the predictor observations in such a way that adequate estimates of the coefficients of generalizability and the intercorrelations can be made from the same sample. It is less critical to assess criterion generalizability, because the regression weights are not affected by it; but that coefficient is also useful.

## C. *Multivariate Estimation of Universe Scores for Composites*

The covariance of a weighted composite $_wX$ with any variable $_vX$ equals the weighted sum of the covariances of the elements of $_wX$ with $_vX$. This theorem enables us to derive covariances for $_w\mu_p$, and from the covariances one can develop a multiple-regression estimator of $_w\mu_p$.

If $n_v$ variables enter $_wX$, their linked covariances form an $n_v \times n_v$ matrix. One adds $n_v$ columns for the $^\circ\sigma(_vX,_{v'}X)$, which are interpreted as $\sigma(_vX,_{v'}\mu_p)$. Finally, one sums across the columns of this second matrix with weights $w_v$ to get a column of $\sigma(_vX,_w\mu_p)$. Suppose, for example, that $_wX$ is defined as $6\,_1X + _2X$. In the covariance matrix, the first two columns contain the variances and $^\bullet\sigma(_1X,_2X) = {}^\bullet\sigma(_2X,_1X)$; there are two columns for $_1\mu_p,\,_2\mu_p$, and a final column for $_w\mu_p$. The last column equals six times the $_1\mu_p$ column plus the $_2\mu_p$ column. Summing the elements of this column with weights $w_v$ gives $\sigma^2(_w\mu_p)$.

Using $_w\mu_p$ as the target variable and the $_vX$ as predictors in a regression analysis produces an estimating equation, plus a $\sigma^2(\varepsilon)$. It also generates a new coefficient of generalizability $\rho^2(_w\hat\mu_p,_w\mu_p)$, which is the multiple correlation of the $_vX_{pi}$ with $_w\mu_p$. It is greater than $\rho^2(_wX_{pi},_w\mu_p)$, though not necessarily by a worthwhile amount.

The covariances of $_w\mu_p$ with other variables, and the correlations and regression slopes for predicting other variables *from* $_w\mu_p$, are to be determined directly from the variance–covariance matrix, not from the estimated universe scores.

### Defining the universe score for a composite

Multiple-regression estimates of universe scores on composite variables depart substantially from those given by conventional test theory, even when classical assumptions hold. Take so simple a composite as the Full Scale (FS) score on the Wechsler test. This is the sum of Verbal and Performance scores. The reliability of the Full Scale score would customarily be determined either from two administrations of the test (with the same or different forms) or from an internal-consistency analysis (splitting within subtests). One could also carry out a multifacet study like those in Chapter 8, recognizing that subtests are nested within Ve and Pe scales.

Any such obtained coefficient estimates one kind of squared correlation of the observed Full Scale score with a universe score. An estimate of the person's "true" Full Scale score would be made by regressing the observed score toward the group mean. Such relatively conventional estimation procedures assign equal weight to the Verbal and Performance observed scores (means of scaled subtest scores). A multiple-regression procedure, however, will assign unequal weights. Which is to be preferred?

Perhaps the fundamental question is: if Wechsler had been fully conscious of the issues involved, how would he have defined the variable $_{FS}\mu_p$? (To sidestep questions of *universe* definition, we assume that generalization is over trials $t$, days $d$, items $i$, and subtests $j$. But for every reasonable alternative, the present argument would be much the same.) One imagines that Wechsler might have articulated either of two definitions for the Full Scale score (before conversion to IQ):

$$(i)\quad _{FS}X_{ptdIJ} = {_{Ve}}X_{ptdIJ} + {_{Pe}}X_{ptdIJ};$$

$$_{FS}\mu_p = \underset{p,t,d,I,J}{\mathscr{E}} {_{FS}}X_{ptdIJ}$$

$$= {_{Ve}}\mu_p + {_{Pe}}\mu_p$$

$$(ii)\quad _{FS}\mu_p = \frac{_{Ve}\mu_p}{\sigma(_{Ve}\mu_p)} + \frac{_{Pe}\mu_p}{\sigma(_{Pe}\mu_p)}$$

Definition (i) is perhaps the more obvious, because it embodies the classical concept of the true score as an expected value of observed score. But definition (ii) very likely comes closer to the construct Wechsler had in mind, because it assigns equal effective weights in the universe score to the verbal and performance aspects of intellectual performance. If Wechsler did not have such a balanced concept in mind, it is hard to understand why he assigned equal nominal weights in the Full Scale to the observed scores for Ve and Pe. (This he did by scaling and prorating subtests before calculating the Full Scale score.)

Clearly, an investigator must define carefully the universe score on a composite. This demand has consequences ranging far beyond the Wechsler scale. The true score of classical theory, which equals $\Sigma w_v \, _v\mu_p$, may not correspond to the construct an investigator has in mind. In the typical achievement test, an investigator seeks to achieve a certain balance among types of content, and so sets out a stratification plan that assigns a certain proportion of the test items to each stratum. What is meant by weighting of "influence" is obscure, because the number of possible score points gives one indicator of proportionate emphasis, a comparison of variances of stratum scores gives another, and a comparison of correlations of sections with the total provides a third. Here we raise comparable questions regarding the influence of the stratum *universe* scores in the universe score for the composite.

We need to reflect further on definition (ii). There is considerable appeal in the idea of combining standardized universe scores. Unfortunately, this will almost never be a viable definition. Standard deviations of universe scores such as $_{Ve}\mu_p$ vary from one subpopulation to another. To apply definition (ii) to the Wechsler, one would have to evaluate $\sigma(_{Ve}\mu_p)$ and $\sigma(_{Pe}\mu_p)$ within a specific population and define the person's universe score as a function of some population to which he belongs. Then two persons having the same universe scores $_{Ve}\mu_p$ and $_{Pe}\mu_p$ would be assigned different $_{FS}\mu_p$ if they belong, for example, to different age groups. (This is not a recurrence of the principle that the best *estimate* of the person's universe score differs from the estimate for another person with the same observed score, when the two persons belong to groups with different means. Here, we have been forced to *define* the universe score $_w\mu_p$ by two different functions of $_{Ve}\mu_p$ and $_{Pe}\mu_p$.)

While the combining weights for $_{Ve}\mu_p$ and $_{Pe}\mu_p$ for the 20–25 and 25–30 age groups, for example, will not differ much, this is merely the entering wedge of the problem. Once the proposal has been made to define universe score relatively, logic pushes toward a separate definition for men of age 20–25, or even for college graduates within that group. Where a construct is intended to describe performance of persons in different subpopulations, it is surely unwise to define it in terms of the parameters for each subpopulation.

The information of interest is contained in $_{Ve}\mu_p$ and $_{Pe}\mu_p$. It appears most important to estimate these separately and not to limit attention to the composite, however that is defined. But there will continue to be a need for composites. It appears that the composite should be defined so that weights of universe scores for the elements remain fixed, regardless of the subjects considered. This does not force Wechsler back upon definition (i). In place of (ii), he can employ a definition of the form:

$$(iii) \quad _{FS}\mu_p = w_{Ve} \, _{Ve}\mu_p + w_{Pe} \, _{Pe}\mu_p$$

It would be reasonable in deciding on the weights to take into account information on the standard deviations of $_{Ve}\mu_p$ and $_{Pe}\mu_p$ in one or another reference population. But there should be no pretense that any set of predetermined weights will cause component universe scores to have equal correlations with the Full Scale universe score in every population or subpopulation to which the test is applied. This is no more than a specialized version of the truism that the factorial makeup of any test is likely to change from one population to another. To the statistical considerations discussed in this section, considerations of construct validity must be added. These are discussed on p. 339.

### Difference scores

The multivariate regression approach in this chapter has previously been treated theoretically only in connection with measures of "change." Lord (1956) introduced the idea that when a certain variable is measured at time 1 and again at time 2, the best estimate of the "true" difference $_2\mu_p - _1\mu_p$ is a multiple-regression estimate.[1] The same logic applies to other kinds of difference scores, such as discrepancies between self-concept and self-defined ideal, or between Ve and Pe IQs.

Conventionally, the observed difference has been defined as $_2X_{pg} - _1X_{pi}$; we shall call this $_dX_{pi}$. The reliability of the difference score has been defined by a well-known and much-used formula. The rationale for the formula conceives of the D study as having the design $_1(i \times p)$, $_2(g \times p)$, $[i \circ g]$. Classical theory, assuming complete randomness of error, has not needed to distinguish between conditions or alternative designs. In our notation, the traditional formula is as follows:

$$(10.11) \quad {}^{\circ}\rho(_dX_{pi},_dX_{pi'}) = \rho^2(_dX_{pi},_d\mu_p)$$

$$= \frac{\sigma^2(_1p) + \sigma^2(_2p) - 2\,{}^{\circ}\sigma(_1X,_2X)}{\sigma^2(_1X) + \sigma^2(_2X) - 2\,{}^{\circ}\sigma(_1X,_2X)}$$

[1] No issue of standardization arises; one expects the variance of a posttest to differ from that of the pretest and wishes to retain that information.

As the formula is usually given, the expression $\rho(_1X_{pi},_1X_{pi'})\sigma^2(_1X_{pi})$ replaces the first term in the numerator, and a similar expression in $_2X_{pg}$ replaces the second term.

**Taking linkage into account.** A distinction between linked and independent covariances has to be made (Stanley, 1967). The classical model which leads to formula (10.11) requires that errors in measuring $v_1$ be uncorrelated with errors (departures from universe score) in measuring $v_2$. However, in short-term experiments appreciable correlation of error may arise when the same test form is used for pretest and posttest or if the same examiner makes both measurements. The pupil may recall certain answers, or the examiner may reinstate good or bad rapport such as was established during the initial measurement. The likelihood of "correlation of error" is even greater when one is examining the difference between two scores within a profile. The person's mental set in taking the Wechsler, his alertness, his reaction to the examiner, etc. will tend to produce a higher correlation between Ve and Pe scores on the same testing than between measures on different days with different examiners.

The following more general version of (10.11) should be employed wherever there is any possibility of linkage.

$$(10.12) \qquad \rho^2(_dX_{pi},_d\mu_p) = \frac{\sigma^2(_1p) + \sigma^2(_2p) - 2\,°\sigma(_1X,_2X)}{\sigma^2(_1X) + \sigma^2(_2X) - 2\,{}^\bullet\sigma(_1X,_2X)}$$

This formula is identical to (10.11) save that the denominator contains a linked covariance. Investigators in the past have at times made the mistake of employing a linked covariance in both the numerator and denominator of the classical formula. Assuming that $v_1$ and $v_2$ are positively correlated, this produces an underestimate of the generalizability of the difference score. Another implication of (10.12) is that an investigator ought, where possible, to use a linked design in estimating the difference score, because this raises the coefficient of generalizability provided that $v_1$ and $v_2$ have a positive relationship. Unwanted components such as $_1\mu_{pi}\sim$ tend to cancel out of the difference (see p. 94).

Estimates of the various terms of (10.12) can be obtained from a suitably linked G study; $°\sigma(_1X,_2X) = \sigma(_1p,_2p)$. While this equation holds for designs other than the crossed one, each design produces a different expected observed-score variance and, therefore, a different denominator. The situation is not appreciably modified by introducing additional facets; one needs only to make sure that the linkage in the G study permits one to estimate the co-variance that is defined by the linkage in the D study.

**Significance of an observed difference.** Classical theory concludes that the variance of the error of measurement for the difference score (in raw-score units) equals the sums of the variances of the errors for the two variables:

$$(10.13) \qquad \sigma^2({}_d\Delta_{pi}) = \sigma^2({}_1\Delta_{pi}) + \sigma^2({}_2\Delta_{pg})$$

[For many tests $\sigma^2({}_d\delta_{pi})$ will be much the same as this, but $\sigma^2({}_d\Delta)$ is the theoretically appropriate error measure for an absolute difference.) The commonly used overlap technique forms a confidence interval for ${}_1\mu_p$ symmetric about ${}_1X_{pi}$, its width being a multiple of $\sigma({}_1\Delta)$, and forms another such interval for ${}_2\mu_p$. If these two intervals, laid out on the same numerical scale, do not overlap, the decision is that ${}_d\mu_p$ differs from zero. The confidence interval ordinarily employed for each variable is $\pm 1\sigma({}_v\Delta)$. Such intervals will not overlap if $\left|{}_dX_{pi}\right| > \sigma({}_1\Delta) + \sigma({}_2\Delta)$. If the two error standard deviations are equal, and the errors are independent, this limit equals $1.41\,\sigma({}_d\Delta)$. The decision rule rejects the hypothesis that ${}_d\mu_p = 0$ with an $\alpha$ risk (two-tailed) of 0.16.

In reasoning such as this it is necessary to take linkage into account. It will be recalled that in a one-facet D study with $n_i' = 1$,

$$_1\Delta_{pi} = ({}_1\mu_i - {}_1\mu) + ({}_1X_{pi} - {}_1\mu_p - {}_1\mu_i + {}_1\mu)$$

and

$$_2\Delta_{pg} = ({}_2\mu_g - {}_2\mu) + ({}_2X_{pg} - {}_2\mu_p - {}_2\mu_g + {}_2\mu)$$

If $i \cdot g$, these components are not independent in the way the traditional argument requires. In calculating the variance within the person of the difference of these errors, i.e., of ${}_d\Delta_{pi}$, the crossproduct terms do not vanish, and in place of (10.13) we have

$$(10.14) \qquad {}^\bullet\sigma^2({}_d\Delta_{pi}) = \sigma^2({}_1i) + \sigma^2({}_2g) - 2\,{}^\bullet\sigma({}_1i,{}_2g) + \sigma^2({}_1pi,e)$$
$$+ \sigma^2({}_2pg,e) - 2\,{}^\bullet\sigma({}_1pi,e;{}_2pg,e)$$
$$= \sigma^2({}_1\Delta) + \sigma^2({}_2\Delta) - 2\,{}^\bullet\sigma({}_1\Delta,{}_2\Delta)$$

One may use ${}^\circ\sigma({}_d\Delta)$ or ${}^\bullet\sigma({}_d\Delta)$ directly to test the null hypothesis that ${}_d\mu_p$ equals some specified value: zero, or the mean ${}_d\mu$, or any other value $\kappa$. Whatever the value of $\kappa$, the procedure is to consider an interval ranging from $\kappa - a\hat\sigma({}_d\Delta)$ to $\kappa + a\hat\sigma({}_d\Delta)$. If ${}_dX_{p*}$ lies within the interval, the hypothesis ${}_d\mu_{p*} = \kappa$ cannot be rejected. When one rejects the hypothesis because the observed difference is outside the interval, the level of confidence depends on the value of $a$.

In Chapter 5 we questioned the formation of confidence intervals for the universe score symmetric about the observed score. The universe score is far more likely to lie within that interval than the interpreter believes, when the observed score is close to the mean, and far more likely to lie outside the interval when the observed score is far from the group mean in either direction.

Here we are dealing with a confidence interval symmetric about $\kappa$, and the objection considered previously does not apply. If in the universe the values of $_d\Delta$ for every subject are normally distributed with variance $\sigma^2(_d\Delta)$, the observed score for a person whose $_{d}\mu_p$ truly equals $\kappa$ will indeed fall outside the interval with probability equal to the risk the interpreter intends. Violation of the assumption of uniform $\sigma(\Delta)$ or of normality probably does not introduce great distortion.

In all confidence-limit procedures, one may misinterpret the finding that a statistic lies within the specified interval. The procedure is effective in making sure that one does not reject the null hypothesis too frequently, but it allows errors "of the second kind," of accepting the null hypothesis when other hypotheses are sound. Errors of the second kind are particularly numerous among persons having a universe score such as $\kappa + a\sigma(_d\Delta)$. As many as half of these persons will have observed scores falling within the interval, yet the investigator will interpret scores for these persons as if their $_{d}\mu_p = \kappa$. This is a consequence of the hypothesis-testing strategy, and not of special problems inherent in score interpretation. The investigator who wishes to avoid this kind of risk must turn to the point-estimation of $_{d}\mu_p$ or to some other statistical analysis that employs prior probabilities.

Even though the confidence interval symmetric around $\kappa$ serves its intended function, we have considerable reservations about the overlap technique. The overlap technique applies only where $\kappa$ is set equal to zero. It treats $_v\Delta$ and $_{v'}\Delta$ as if they were independent, where in most of the testing that yields plotted profiles, the two are linked. A further objection is that the plotting of bands suggests to the less sophisticated interpreter that one has formed a useful confidence interval for each $_v\mu_p$ considered by itself.

### Regression estimates of change scores and their use

Because of the widespread interest in the measurement of change or gain, we shall focus on that problem in this section, relying heavily on the earlier presentation of Cronbach and Furby (1970) and often quoting or paraphrasing it. The paper places particular emphasis on the estimation of $_{d}\mu_p$ by regression methods, but it also argues that it is usually less appropriate to fixate on the difference score in this manner than to emphasize the estimated universe score on the posttest. We shall return to that theme. It is also to be noted that every technical statement to be made regarding change scores applies to any difference between two variables or two distinguishable classes of variables. Many investigators have felt, for reasons good or bad, that their substantive questions required a measure of gain in ability or shift in attitude. "Raw change" or "raw gain" scores formed by subtracting pretest scores from posttest scores lead to fallacious conclusions, primarily because such scores are systematically related to any random error of measurement.

Although the unsuitability of such scores has long been discussed, they are still employed, even by some otherwise sophisticated investigators.

Just why gains or differences are thought to be worth estimating can perhaps be inferred from the studies where estimates of some sort have been made in the past. The following aims may be noted:

1. To provide an indicator of deviant development, as a basis for identifying individuals to be given special treatment or to be studied clinically.

2. To provide a measure of growth rate or learning rate that is to be predicted, as a way of answering the question, What kinds of persons grow (learn) fastest? Here, the change measure is a criterion variable in a correlational study.

3. To provide a dependent variable in an experiment on instruction, persuasion, therapy, or some other attempt to change behavior or beliefs.

4. To provide an indicator of a construct that is thought to have significance in a certain theoretical network. The indicator may be used as an independent variable, covariate, dependent variable, etc. An example is the interference score needed on the Stroop Color–Form Test.

*Selecting individuals on the basis of gain or difference scores.* Many who calculate difference scores are interested in making decisions about individuals—identifying underachievers for clinical attention or fast learners for special opportunities, for example. One can scarcely defend selecting such individuals on the basis of a raw-gain or raw-difference score of the type $_dX_{pi}$.

It is reasonable to estimate $_d\mu_p$ by regression methods. It was suggested by Lord, McNemar, and others that $_d\hat{\mu}_p$ be expressed as a function of the two observed scores $_1X$ and $_2X$. A multiple-regression estimator can be formed as was done earlier in this chapter for the Wechsler Performance score and for the DAT scores. We would take linkage into account; this possibility was not considered by earlier writers on this kind of estimation.[2] It may be noted that one can form the regression estimate in two ways. A direct estimate can be made of $_d\mu_p$, or one can form separate estimates of $_1\mu_p$ and $_2\mu_p$ and subtract the first from the second. The two procedures will give identical results.

Our broader multivariate conception leads us to bring predictors other than $_1X$ and $_2X$ into the picture. The statement of the problem as "the measurement of gain" or of "residual gain" implies a special affinity between

---

[2] Where subjects have experienced different treatments between time 1 and time 2, the equations ought to be determined from the correlations for each treatment group considered separately. There is no necessary reason for $\rho(_1X,_2X)$ or $\rho^2(_2X,_2\mu_p)$ to be the same under both treatments.

$_1X$ and $_2X$, the pretest and posttest, respectively. The two are seen as "the same variable" in some sense, but behavior is multivariate in nature and so is change.

Even when $_1X$ and $_2X$ are determined by the same operation, they often do not represent the same psychological processes. At different stages of practice or development, different processes contribute to performance of a task. Nor is this merely a matter of increased complexity; some processes drop out, some remain but contribute nothing to individual differences within an age group, some are replaced by qualitatively different processes. Purely empirical studies of changes in the operationally defined variable may be justified; to assess such changes, even when one cannot describe them qualitatively, may be practically important. But one must not fall into the trap of assuming that the changes are in a particular psychological attribute.

Instead of confining attention to two specific variables, we consider two classes of variables. Assume that some point of time or some time interval is selected as defining the onset of treatment and the termination of treatment. All measures the investigator collects regarding the person's status before that time are pretest measures, considered collectively as a vector of observed scores $_IX$. All measures following treatment are considered together as a set of $_{II}X$. The measures may be test scores, but they may be school marks, demographic indicators, or other information (e.g., enrolled in honors section of algebra, or went to college). Division of the time continuum to form two classes of variables does not directly allow the study of change as a process extended over time, but successive cuts can be used to extend the model.

We find a conditional notation helpful. The symbol $_2\hat{\mu}_p \mid _2X$, for example, refers to a regression estimate of the universe score on $v_2$ from the observed score $_2X$; the following refer to particular multiple-regression estimates: $_2\hat{\mu}_p \mid _1X, _2X$ or $_2\hat{\mu}_p \mid _{II}X$. The logic developed earlier in the chapter leads us to recommend, then, that a person who wishes to estimate $_d\mu_p$ proceed to set up a variance–covariance matrix for all the observed scores and the universe scores $_1\mu_p$ and $_2\mu_p$. The computational procedures previously demonstrated may then be used to form $_1\hat{\mu}_p \mid _IX, _{II}X$ and $_2\hat{\mu}_p \mid _IX, _{II}X$. The difference between these is the desired estimate of "true change." It would be slightly more direct to form $_d\hat{\mu}_p \mid _IX, _{II}X$, but the two separate scores are likely to be worth examining. In any procedure of this sort, one must recognize the role sampling errors play in the determination of regression weights, and probably should employ a reduced-rank procedure to minimize capitalization on chance.

A simple example will perhaps clarify why such a procedure has advantages. Suppose one wishes to teach recognition of the letters of the alphabet to kindergartners. A direct pretest on the alphabet might serve as $v_1$, and a posttest as $v_2$. Consider as an auxiliary pretest a good individual test of general ability; call this $v_3$. Now the set $_1X$ consists of $_1X$ and $_3X$. The quantity $v_1$ may be a useful test for identifying those few children who have indeed mastered most of the alphabet, but for children who have only hazy knowledge of a few letters, a simple test with one trial per letter will be quite unreliable, especially when attention and work habits are as poorly developed as they are at the start of kindergarten. The correlation of $v_3$ with $_1\mu_p$ seems likely to be appreciable. We suggest that the best estimate of true knowledge of the alphabet at the beginning of the training is the regression equation that combines $v_1$, $v_2$, and $v_3$ with whatever weights the data justify. It seems unlikely that there will be a large enough correlation between $v_2$ and $_1\mu_p$ to produce an appreciable weight for $v_2$ in the regression equation; but calculating the weight allows for the possibility that $v_2$ adds useful information. We also propose that $_2\mu_p$ be estimated from $v_1$, $v_2$, and $v_3$, where these are the only facts at hand; again, the data will tell whether $v_1$ and $v_3$ add appreciably to the accuracy of the estimate. To extend the example just a bit further: we would look with favor on employing a report on parents' education as a further predictor variable $v_4$. Among children with the same observed score on the unreliable alphabet test, it seems quite likely that those whose parents are more educated have given them more experience with the alphabet and that if tested more thoroughly they would earn better scores than those with less-educated parents. However, this is only a working hypothesis; if parents' education does not improve the estimate, it will be assigned no weight in the end.

Having described a logical approach to the estimate of the universe score $_d\mu_p$, we must still question whether there is any purpose in such a measure. We see no practical reason for selecting high gainers and low gainers for clinical study or for using the gain score as a basis for decisions about the individuals. For most decisions, $_2\mu_p$ is more pertinent than $_d\mu_p$. A case can be made for estimating some type of "residual gain" measure. Persons who have gained less than others who started at the same level often should be made the subject of clinical study or singled out for some special treatment. The residual gain, however, is indistinguishable from the posttest score with pretest information partialled out. Thus $_dX \cdot _1X \equiv _2X \cdot _1X$, and $_d\mu_p \cdot _1\mu_p \equiv _2\mu_p \cdot _1\mu_p$. So we shall speak of an adjusted posttest score, not of residual gain.

The observed adjusted posttest score is $_2X \cdot _1X$ or $_2X_{pi} - \sigma(_1X,_2X)_1X_{pi}/\sigma^2(_1X)$. But this score is fallible, and one might suspect it of being biased in the same sense as the observed difference score is. Actually, as is demonstrated by Cronbach and Furby, when one estimates the universe score $_2\mu_p \cdot _1\mu_p$ from $_1X$ and $_2X$ jointly, the estimate is proportional to the observed partial variate. Consequently, if one relies entirely on the $v_1$ and $v_2$ data, he identifies the same persons as having unexpectedly high or low universe scores on the posttest when he interprets the observed residual gain as when he forms a regression estimate of the residual gain in universe score. If, however, auxiliary predictors can be used in estimating the universe score on the partial variate, the regression estimate will select different persons as having exceptional posttest universe scores.

The focus on $_2\mu_p \cdot _1\mu_p$ is open to question. One really wishes to know which persons profited less from the treatment than others having similar initial status. Pretest status is multivariate, and it appears that one ought to identify persons having exceptional performance on the basis of $_2\mu_p \cdot _1\mu_p$. This takes into account all relevant aptitudes rather than just the pretest that is the operational counterpart of the posttest $_2X$.

To estimate $_2\mu_p \cdot _1\mu_p$ requires a complicated algorithm whose stages follow:

1. Estimate all variances and covariances among the $_1\mu_p$ and $_2\mu_p$.
2. Identify by regression analysis the weights $w_v$ such that $_w\mu_p = \sum_{v \in I} w_v \ _v\mu_p$ best predicts $_2\mu_p$. The $w_v$ are unstandardized regression weights.
3. Calculate $\sigma(_w\mu_p, _2\mu_p)$ and $\sigma^2(_w\mu_p)$. By definition, $_2\mu_p \cdot _1\mu_p = _2\mu_p \cdot _w\mu_p = _2\mu_p - \sigma(_w\mu_p, _2\mu_p)_w\mu_p / \sigma^2(_w\mu_p)$.
4. Calculate all variances and covariances among and between the $_IX$ and the $_{II}X$, and estimate their covariances with $_2\mu_p \cdot _w\mu_p$.
5. From these, calculate a regression equation for estimating $_2\mu_p \cdot _w\mu_p$ from the $_IX$ and $_{II}X$ jointly.

The reader setting up a computational procedure should note these points:

$$\sigma(_v\mu_p, _{v'}\mu_p) = {}^\circ\sigma(_vX, _{v'}X)$$

$$\sigma(_*X, _2\mu_p \cdot _w\mu_p) = {}^\circ\sigma(_*X, _2X) - \frac{\sigma(_w\mu_p, _2\mu_p)}{\sigma^2(_w\mu_p)} \sum_{v \in I} w_v {}^\circ\sigma(_*X, _vX)$$

Linked covariances of observed scores are required at stage 4. Reduced-rank methods should usually be employed in the regression analyses.

Posttest status as well as pretest status is multivariate. One could estimate an individual's adjusted posttest universe score on every variable in **II**, just as we did for $v_2$. This vector of scores, however, would not provide an obvious basis for selecting cases for special treatment. If one wishes to select cases on the basis of unexpectedly high or low posttest status, taking all posttest variables into account, it appears that he has to define a particularly significant composite $_{w'}\mu_p$ of the variables in **II**. The calculations suggested above with respect to $_2\mu_p$ can be carried out with respect to $_{w'}\mu_p$.

*Criteria in correlational studies.* Correlational studies are often intended to investigate a question such as this: What pretest attributes distinguish persons who profit most from the treatment? This asks about the correlation of each variable in **I** with the posttest or, perhaps, with a gain score. For inquiries such as this, one should not calculate gain scores, nor should one form a regression estimate of such a score. One cannot obtain the correct correlation of any universe score with another variable by correlating the *estimated* universe score with that other variable.

The most expeditious way to investigate correlational questions is to work from a variance–covariance matrix. One determines the variances of $_1\mu_p$ and $_2\mu_p$ and their covariances with each other and with each observed variable in **I**. The covariance of any variable with $_d\mu_p$ (if that is the "criterion" of interest) equals its covariance with $_2\mu_p$ minus its covariance with $_1\mu_p$. (Note that an estimate of variance for a universe score is not the variance of estimated universe scores; similarly for a covariance.) Dividing covariances by the appropriate standard deviations gives a complete set of predictor intercorrelations, and a set of correlations of the predictors with $_v\mu_p$. If a question is asked about correlations of the adjusted posttest (i.e., the residual gain), one forms its variance and covariances and proceeds similarly.

*Gains as a consequence of treatments.* Investigators often think of an experiment as testing whether a treatment produces gains, but a proper analysis of the separate pretest and posttest scores themselves serves at least as well as any treatment of gain scores (Engelhart, 1967).

Where an experiment has been performed on cases allocated randomly into two or more treatment groups, there is no need to use a measure of change as a dependent variable and no virtue in using it.[3] In testing the null hypothesis that two treatments have the same effect, the essential question is whether posttest scores $_2X$ vary from group to group. Consequently, $_2X$ is an entirely suitable dependent variable.

Analysis of covariance takes variation in $_1X$ into account. [If $\rho(_1X,_2X) <$ 0.4, blocking on $_1X$ is probably to be preferred, according to Elashoff, 1969.] The adjustment estimates the scores $_2X$ expected under the null hypothesis and then expresses each observed $_2X$ as a deviation from the estimate. It is desirable to base the adjustment, not on $_1X$, but on whatever linear function of $_1X$ best predicts $_2X$ within groups.

Where within-treatment regressions differ in slope, the effect of the treatment depends on the level of pretest variables. The most meaningful scientific report consists of regression functions describing the relation of $_2\mu_p$ to the $_1\mu_p$. These can be computed for each treatment group with the aid of the within-group covariance matrix for universe scores.

---

[3] In the one-group experiment, if only $v_1$ and $v_2$ are measured, gains scores give the same result as analysis of the paired pretest and posttest scores.

Lord (1960) proposed that universe scores on the pretest be estimated and that these be used as a covariate instead of the observed scores. The Lord procedure would not alter one's conclusions when the groups have been formed at random and when all information from which the universe scores might be estimated is taken into the set of covariates. In a quasi-experiment, however, where treatment groups are not formed at random, no analysis is satisfactory. To be sure, Lord's technique can be applied, and augmented by using data other than $_1X$ to estimate $_1\mu_p$. Since publication of the Cronbach–Furby paper, however, new issues have arisen. The covariance adjustment, we find, may overestimate or underestimate the difference in population means on posttest, depending on what variable is selected as covariate. Bias is likely regardless of what one does to improve the estimate of scores on the covariate. We cannot digress to examine these unsettled problems. Perhaps the last word will be Lord's statement made in 1967: "There simply is no logical or statistical procedure that can be counted on to make proper allowances for uncontrolled preexisting differences between groups."

The findings of such a study can be summarized by calculating within-group regression functions relating $_2\mu_p$ to the $_1\mu_p$, using the covariance matrix for universe scores. What cannot be done is to interpret the difference in means, adjusted or unadjusted, as a treatment effect.

***Differences and gain scores as constructs.*** One of the most common uses of difference scores is to operationalize a concept: For example, self-satisfaction is sometimes defined as the difference between the rating of self and ideal-self on an esteem scale. One might likewise think of a gain score as reflecting "learning ability" on a certain task. Operational definitions will often take the form of linear combinations of operations.

But there is little *a priori* basis for pinning one's faith on $_2\mu_p - _1\mu_p$ as distinct from the more general $_2\mu_p - a_1\mu_p$. Just what weight to assign the "correcting" variable is an empirical and theoretical question. To arbitrarily confine interest to $_d\mu_p$ (which means that $a$ is fixed at 1.00) is to rule out possible discoveries. This argues, then, for discovering what function of $_2\mu_p$ and $_1\mu_p$ has the strongest relationships with variables that theory suggests should connect with the construct. The statement applies equally to the defining of additive composites such as the Wechsler FS score.

The claim that an index has validity as a measure of some construct carries a considerable burden of proof. There is little reason to believe and much empirical reason to disbelieve the contention that some arbitrarily weighted function of two variables will properly define a construct. More often, the profitable strategy is to use the two variables separately in the analysis to allow for complex relationships.

## EXERCISES

**E.1.**  Given: $\rho(_1X,_2X) = 0.50$; $\rho^2(_2X,_2\mu_p) = 0.70$.

a. If $_2X$ were taken as the sole predictor of $_2\mu_p$, what would be the value of $\rho^2(_2\hat{\mu}_p,_2\mu_p)$?

b. What would be the value of $\rho^2(_2\hat{\mu}_p,_2\mu_p)$ if both $v_1$ and $v_2$ were used as predictors? (Use Figure 10.1.)

c. What would be the value if $\rho^2(_2X,_2\mu_p)$ were 0.90?

**E.2.**  The following matrix (Bouchard, 1968) gives correlations between scales of two instruments, an adjective checklist and a self-rating schedule for three attributes: Dominance, Endurance, and Order. The within-scale correlations are indicated as linked. Test–retest coefficients from a second sample are given in the diagonal. Using Figure 10.1, judge as well as you can whether any combination of two scores is likely to estimate one of the six universe scores well enough to make $\rho^2(_*\hat{\mu}_p,_*\mu_p)$ exceed $\rho^2(_*X,_*\mu_p)$ by 0.05 or more?

|  | Adjective checklist | | | Self-rating schedule | | |
|---|---|---|---|---|---|---|
|  | Do | End | Ord | Do | End | Ord |
| **Adjective checklist** | | | | | | |
| Do | 0.76 | •0.48 | •0.26 | 0.51 | 0.31 | 0.02 |
| End |  | 0.74 | •0.88 | 0.17 | 0.46 | 0.42 |
| Ord |  |  | 0.63 | 0.09 | 0.44 | 0.50 |
| **Self-rating schedule** | | | | | | |
| Do |  |  |  | 0.76 | •0.18 | •0.04 |
| End |  |  |  |  | 0.67 | •0.54 |
| Ord |  |  |  |  |  | 0.81 |

**E.3.**  The following data come from a study by Costin (1968). Scores were obtained before and after a psychology course. Treat data on the assumption of independent sampling of conditions. Entries in the diagonal are internal-consistency coefficients.

| Test | Standard deviation | Correlation | | | |
|---|---|---|---|---|---|
|  |  | 1 | 2 | 3 | 4 |
| $v_1$: Principles test (pretest) | 4.2 | 0.52 | 0.25 | 0.28 | 0.18 |
| $v_2$: Principles test (posttest) | 6.6 | 0.25 | 0.78 | 0.36 | 0.51 |
| $v_3$: Misconceptions test (pretest) | 5.0 | 0.28 | 0.36 | 0.57 | 0.42 |
| $v_4$: Misconceptions test (posttest) | 4.4 | 0.18 | 0.51 | 0.42 | 0.65 |
| SCAT (with arbitrary s.d.) | 5.0 | 0.42 | 0.38 | 0.27 | 0.34 |

a. Calculate the coefficient of generalizability for the difference between $v_1$ and $v_2$ observed scores. What is the universe of generalization? (Hint: In problems such as this series, first form the variance–covariance matrix.)

b. What is the correlation of the raw-gain score $_2X - _1X$ with $_2\mu_p$? with $_1\mu_p$?

c. What combination of these five (or fewer) variables gives the best available estimate of $_1\mu_p$? (If a computer is not available, make a reasonable inference as to which variables would enter a regression equation; do not calculate weights.)

d. What is the correlation of each variable with $_d\mu_p(= _2\mu_p - _1\mu_p)$?

e. What combination of $_1X$ and $_2X$ gives the best estimate of $_d\mu_p$?

f. What combination of five (or fewer) variables gives the best available estimate of $_d\mu_p$?

**E.4.** The data given in Table 10.E.1 report correlations for subtests within the same administration of the test. For the sake of this example, assume that the covariance arises entirely from the person component (i.e., linkage is negligible).

*TABLE 10.E.1. Intercorrelations within Form A, and Estimated Alternate-Form Retest Correlations for Subtests of the Metropolitan Readiness Test* (*Test Manual, 1966, pp. 12, 14*)

| Subtest | Mean | Standard deviation | Correlation with |  |  |  |  |  |
|---------|------|--------------------|------|------|------|------|------|------|
|  |  |  | 1 | 2 | 3 | 4 | 5 | 6 |
| 1. Word Meaning | 8.67 | 3.1 | 0.70 | 0.49 | 0.43 | 0.46 | 0.55 | 0.39 |
| 2. Listening | 8.89 | 2.8 |  | 0.50 | 0.42 | 0.40 | 0.50 | 0.36 |
| 3. Matching | 7.49 | 4.0 |  |  | 0.79 | 0.53 | 0.60 | 0.49 |
| 4. Alphabet | 9.39 | 4.7 |  |  |  | 0.85 | 0.64 | 0.45 |
| 5. Numbers | 12.02 | 4.7 |  |  |  |  | 0.81 | 0.53 |
| 6. Copying | 6.82 | 3.9 |  |  |  |  |  | 0.82 |

For intercorrelations, means, and standard deviations, $N = 12{,}225$. Diagonal entries are correlations with Form B given a few days later ($N = 278$); these have been adjusted to allow for the fact that the standard deviation in the retest sample was smaller for most subtests than the standard deviation for the sample giving intercorrelations.

a. What is the coefficient of generalizability (over items and occasions within a week) of the total score (formed by adding the subtest raw scores)? Hint: First form the variance–covariance matrix.

b. Which subtest contributes most to the universe-score variance for the total score?

c. When regression equations were formed for estimating the universe scores for the first three subtests, the standardized regression weights were as follows:

| Universe score predicted | \multicolumn{6}{c}{Subtest to which weight[4] is assigned} | $\rho^2(_*\hat{\mu}_p,_*\mu_p)$ | $\rho(_*X,_*\mu_p)$ |
|---|---|---|---|---|---|---|---|---|
|  | 1 | 2 | 3 | 4 | 5 | 6 | | |
| Word meaning | 0.6 | 0.1 | — | — | 0.2 | — | 0.78 | 0.71 |
| Listening | 0.3 | 0.4 | 0.1 | — | 0.2 | — | 0.75 | 0.50 |
| Matching | — | — | 0.7 | — | 0.1 | — | 0.83 | 0.81 |

Does this indicate that regression weighting would be useful where the subtest scores are to be used to diagnose the child's specific weaknesses?

d. The raw-score regression equations for Word Meaning and Listening take this form (when very small terms are ignored):

$$_1\hat{\mu}_p = 0.52(_1X) + 0.14(_2X) + 0.09(_5X) + 1.85$$

$$_2\hat{\mu}_p = 0.19(_1X) + 0.25(_2X) + 0.06(_3X) + 0.10(_5X) + 3.36$$

Consider the following three persons, all of whom are well below the mean and therefore would be candidates for diagnostic interpretation. Is the impression given in each case by the unregressed scores for variables 1 and 2 similar to the impression given by the regression estimates? (The means for the two variables are near 9, and the raw score standard deviations near 3; see Table 10.E.1.)

| Person | \multicolumn{4}{c}{Raw score on subtest} |
|---|---|---|---|---|
|  | 1 | 2 | 3 | 5 |
| 1 | 6 | 3 | 4.5 | 4.5 |
| 2 | 4.5 | 4.5 | 4.5 | 4.5 |
| 3 | 3 | 6 | 4.5 | 4.5 |

**E.5.** Table 10.E.2 gives data for subtests of the Analysis of Learning Potential (two forms, given a short time apart; these data supplied by Harcourt, Brace, and Jovanovich, Inc.). Examine the following (haphazardly selected) pairs of subtests to answer the question: Is there appreciable linkage between subtests administered at the same time?

a. 2, 9 Figure Series, with 3, 10 Number Fluency.
b. 2, 9 Figure Series, with 5, 12 Number Series.
c. 4, 11 General Information, with 7, 14 Story Sequence.

**E.6.** A "reading composite" is made up of subtests 4, 6, and 7, which presumably are most related to aptitude for reading instruction. The observed score is the sum of standard scores on the three subtests.

---

[4] Weights less than 0.10 are not given, although these terms were allowed to affect the multiple correlation.

TABLE 10.E.2. Correlations for Subtests of Analysis of Learning Potential in Grade 3 (N = 551)

| Subtest within form[a] | 1 A1 | 2 A2 | 3 A3 | 4 A4 | 5 A5 | 6 A6 | 7 A7 | 8 B1 | 9 B2 | 10 B3 | 11 B4 | 12 B5 | 13 B6 | 14 B7 | Standard deviation |
|---|---|---|---|---|---|---|---|---|---|---|---|---|---|---|---|
| 1 A1 | | 0.28 | 0.30 | 0.47 | 0.31 | 0.61 | 0.25 | 0.70 | 0.18 | 0.31 | 0.41 | 0.24 | 0.54 | 0.20 | 5.11 |
| 2 A2 | | | 0.51 | 0.46 | 0.68 | 0.37 | 0.54 | 0.33 | 0.67 | 0.43 | 0.37 | 0.57 | 0.32 | 0.45 | 3.67 |
| 3 A3 | | | | 0.42 | 0.57 | 0.42 | 0.41 | 0.28 | 0.36 | 0.76 | 0.32 | 0.48 | 0.39 | 0.36 | 5.07 |
| 4 A4 | | | | | 0.50 | 0.56 | 0.46 | 0.51 | 0.32 | 0.41 | 0.63 | 0.43 | 0.50 | 0.40 | 3.30 |
| 5 A5 | | | | | | 0.46 | 0.52 | 0.32 | 0.55 | 0.51 | 0.39 | 0.78 | 0.44 | 0.45 | 4.78 |
| 6 A6 | | | | | | | 0.38 | 0.56 | 0.28 | 0.39 | 0.44 | 0.36 | 0.75 | 0.31 | 4.46 |
| 7 A7 | | | | | | | | 0.33 | 0.43 | 0.36 | 0.34 | 0.48 | 0.35 | 0.66 | 6.03 |
| 8 B1 | | | | | | | | | 0.26 | 0.34 | 0.47 | 0.32 | 0.57 | 0.30 | 4.60 |
| 9 B2 | | | | | | | | | | 0.42 | 0.35 | 0.68 | 0.32 | 0.52 | 3.49 |
| 10 B3 | | | | | | | | | | | 0.38 | 0.53 | 0.40 | 0.38 | 5.06 |
| 11 B4 | | | | | | | | | | | | 0.41 | 0.42 | 0.38 | 2.97 |
| 12 B5 | | | | | | | | | | | | | 0.44 | 0.50 | 4.88 |
| 13 B6 | | | | | | | | | | | | | | 0.31 | 4.53 |
| 14 B7 | | | | | | | | | | | | | | | 5.54 |

[a] 1 = Word Picture Association, 2 = Figure Series, 3 = Number Fluency, 4 = General Information, 5 = Number Series, 6 = Word Meaning, 7 = Story Sequence.

a. Calculate a coefficient of generalizability to the universe score defined by various forms of the test (all with the same subtests) and by various occasions within a week or so.

b. The authors apply Kuder–Richardson formula 20 (internal-consistency of items) to each subtest. They use a traditional formula for the reliability of an equally-weighted composite score, to obtain a coefficient for the reading test of 0.91. The formula is

$$(\Sigma r_{vv} + \Sigma r_{vv'})/(n_v + \Sigma r_{vv'})$$

where $n_v$ is the number of subtests, $r_{vv}$ is a subtest reliability coefficient, and the $r_{vv'}$ are intercorrelations of subtests within the battery. Why is the coefficient obtained in *a* above more valuable to the test user?

c. One might generalize over the universe of subtests of which 4, 6, and 7 may be regarded as a sample. What would be the coefficient of generalizability, taking sets of random subtests, and occasions, as randomly sampled?

E.7. What is the regression equation for estimating the universe score on the reading composite from the standardized observed scores on subtests 4, 6, and 7 together? Assume subtests fixed, but let forms of each subtest, and occasions, vary in the universe. How much is gained from using multiple predictors in place of the observed score on the reading composite?

There are several ways to carry out the required calculations. It is suggested that one set up a symmetric matrix of within-battery correlations for the three subtests. For example, fill cell 4, 6 with the average of the correlations of 4A with 6A and 4B with 6B. Set up also a symmetric between-battery matrix, by averaging, for example, the correlation of 4A and 6B with that of 4B and 6A. The entries in this matrix are also covariances of the subtest universe scores, and covariances of the subtest observed scores with the subtest universe scores. The sum of this matrix is the variance of the subtest universe score. Summing the three entries in a row gives the covariance of the subtest observed score with the reading-composite universe score.

### Answers

**A.1.**   a. 0.70
b. About 0.74.
c. Negligibly different from 0.90.

**A.2.**   No other score correlates highly enough to produce such an improvement for adjective checklist (ACL) Dominance (where an intercorrelation of about 0.65 is required, according to the figure). Similarly, no combination improves the estimate for self-rating schedule (SRS) Dominance by the amount desired.

If the data are taken at face value, it would appear that combining adjective checklist End and Ord would make $\rho^2$ for estimating $_{End}\mu_p$ close to 1.00. One is suspicious of the data, however, because the retest correlation comes from a different sample than the within-scale correlations. Moreover, the presence of linkage means that a conclusion would have to be based on (10.5) rather than (10.4);

Figure 10.1 is based on the latter. More complete analysis comparing correlations with occasion common to correlations with occasion different would provide the basis for constructing and evaluating a two-variable estimator of $_{End}\mu_p$.

Similar arguments apply to the adjective checklist Ord scale (estimated from ACL Ord and SRS Ord, and also from ACL End if linkage is not too great) and the Self-rating Schedule End scale (estimated from SRS End plus ACL End, plus perhaps SRS Ord).

**A.3.** Variance–covariance matrix, extended

|       | $_1X$ | $_2X$ | $_3X$ | $_4X$ | SCAT | $_1\mu$ | $_2\mu$ | $_d\mu$ |
|-------|-------|-------|-------|-------|------|------|------|-------|
| $_1X$   | 17.64 | 6.93  | 5.88  | 3.33  | 8.82  | 9.16 | 6.93  | −2.23 |
| $_2X$   | 6.93  | 43.56 | 7.56  | 9.42  | 12.54 | 6.93 | 33.98 | 27.05 |
| $_3X$   | 5.88  | 7.56  | 25.00 | 9.24  | 6.75  | 5.88 | 7.56  | 1.68  |
| $_4X$   | 3.33  | 9.42  | 9.24  | 17.64 | 7.48  | 3.33 | 9.42  | 6.09  |
| SCAT  | 8.82  | 12.54 | 6.75  | 7.48  | 25.00 | 8.82 | 12.54 | 3.72  |

a. $\widehat{\sigma^2}(_1p) = 9.16$, $\widehat{\sigma^2}(_2p) = 33.98$, $\hat{\sigma}(_1p,_2p) = 6.93$
Using (10.11),

$$\widehat{\rho^2}(_dX,_d\mu_p) = \frac{9.16 + 33.98 - 2(6.93)}{17.64 + 43.56 - 2(6.93)} = \frac{29.28}{47.34} = 0.62$$

b. $\hat{\sigma}(_2X - _1X,_2\mu_p) = \hat{\sigma}(_2X,_2\mu_p) - \hat{\sigma}(_1X,_2\mu_p) = 33.98 - 6.93 = 27.05$
$\hat{\sigma}(_2X - _1X,_1\mu_p) = -2.23$

The raw gain score then depends very heavily on the posttest but has almost no relation to the universe score on the pretest. The reason for this is the comparatively large universe-score variance for $v_2$.
c. The three best predictors give this equation:

$$_1\hat{\mu}_p = 0.39 \,_1X + 0.09 \,_2X + 0.18 \text{ SCAT}. \qquad R = 0.796.$$

$\widehat{\rho^2}$ goes from 0.52 to 0.63. (The standardized regression weights are 0.55, 0.15, and 0.31.) Very small weights could be added for the two other variables, but these add negligibly to the prediction.
d. $\sigma^2(_d\mu_p) = 29.28$ (from a). $\sigma(_d\mu_p) = 5.41$.
Correlations are: for $_1X$, $= -0.10$; $_2X$, 0.76; $_3X$, 0.06; $_4X$, 0.26; SCAT, 0.14.
e. $_d\hat{\mu}_p = 0.69(_2X) - 0.40(_1X)$, $R = 0.816$; $\rho^2$ goes from 0.58 to 0.67. (The standardized weights are 0.84 and −0.31.)
f. A five-variable equation is

$$_d\hat{\mu}_p = 0.77(_2X) - 0.34(_1X) - 0.17(_3X) - 0.12(_4X) - 0.04(\text{SCAT}).$$

$R = 0.839.$

A three-variable equation gives $R = 0.834$. ($\rho^2$ rises from 0.58 to 0.70). Perhaps the most interesting aspect of that equation is that the standardized regression weights are nearly equal for $_1X$ and $_3X$. I.e., the Misconceptions pretest estimates gain in Principles as well as does the Principles pretest. The equation in terms of standard scores is

$$_d\hat\mu_p = 0.89(_2Z) - 0.27(_1Z) - 0.19(_3Z).$$

**A.4.  a.**

|  | Universe score variance | Observed score variance |
|---|---|---|
| Sum off-diagonal entries in Table 10.E.3 | 220.88 | 220.88 |
| Sum diagonal entries in parentheses | 72.43 | |
| Sum diagonal entries not in parentheses | | 92.84 |
| Totals | 293.31 | 313.72 |

Coefficient of generalizability $= 293.3/313.7 = 0.935$

b. Subtest 4 (whose entries into the numerator have the largest total).

c. The increase in the accuracy of generalization is tiny for the Matching universe score. The increase for Word Meaning cuts the error variance by one-fourth, and might be sufficient to repay the effort of weighting. The weighting cuts in half the error variance in the estimate of the Listening score, and is clearly profitable.

The Listening equation places considerable weight on variables other than Listening ($v_2$) because $\rho^2(_2X,_2\mu_p)$ is low.

**TABLE 10.E.3.** *Variance–Covariance Matrix$^a$ (Based on Table 10.E.1)*

|  | 1 | 2 | 3 | 4 | 5 | 6 |
|---|---|---|---|---|---|---|
| 1 | 9.61 (6.73) | 4.25 | 5.33 | 6.70 | 8.01 | 4.71 |
| 2 | 4.25 | 7.84 (3.92) | 4.70 | 5.26 | 6.58 | 3.93 |
| 3 | 5.33 | 4.70 | 16.00 (12.64) | 9.96 | 11.28 | 7.64 |
| 4 | 6.70 | 5.26 | 9.96 | 22.09 (18.78) | 14.13 | 8.25 |
| 5 | 8.01 | 6.58 | 11.28 | 14.13 | 22.09 (17.89) | 9.71 |
| 6 | 4.71 | 3.93 | 7.64 | 8.25 | 9.71 | 15.21 (12.47) |

$^a$ Parenthetical entries are estimated universe-score variances.

d. We arrive at these estimates:

| Person | Observed score $v_1$ | $v_2$ | Regression estimate $v_1$ | $v_2$ |
|---|---|---|---|---|
| 1 | 6 | 3 | 5.7 | 6.0 |
| 2 | 4.5 | 4.5 | 5.2 | 6.1 |
| 3 | 3 | 6 | 3.6 | 6.2 |

There is, then, a considerable change in the profile shape for Person 1. The very low observed score on $v_2$ is regressed toward the mean and the false impression that $_2\mu_p$ is less than $_1\mu_p$ for Person 1 is erased. Likewise, the profile shape for Person 2 changes, though not greatly.

**A.5.** a. Within-battery (linked) correlations are 0.51 and 0.42. Independently administered, the subtests correlate 0.43 and 0.36. Linkage raises correlations within the battery by about 0.07.
b. Linked: 0.68, 0.68. Independent: 0.57, 0.55. About 0.12 of the linked correlation is attributable to person-occasion linkage.
c. Linked: 0.46, 0.38. Independent: 0.40, 0.34. Linkage for this pair is weaker (0.05).

**A.6.** a. For each form–occasion combination there is a within-form matrix of correlations of subtests 4, 6, and 7, with 1.00 in the diagonal. The sum of the matrix is the observed-score variance for the reading composite; the average for the two forms, 5.51, is the estimate of expected observed-score variance.

There is also a between-battery matrix of subtest correlations for 4 with 4, 4 with 6, etc. These nine correlations add to 4.38, which estimates the universe-score variance.

The ratio 4.38/5.51 or 0.795 is the coefficient of generalizability.
b. The analysis by the Kuder–Richardson method allows person-occasion and person–subtest–occasion variance to enter the estimate of universe-score variance. But an interpreter of an aptitude test of this sort is almost invariably interested in generalizing over occasions of testing, at least within a limited time period. Consequently, our analysis considers a more pertinent universe; because this universe is broader, the coefficient drops.

Another answer, which amounts to the same thing, is that we use independent, rather than linked, correlations in the numerator of our formula.

As a matter of fact, the test manual follows its report on the Kuder–Richardson coefficient with a directly calculated correlation between reading composites on two forms on two occasions, speaking of this as "the most rigorous measure of test precision." The value of 0.80 agrees with our coefficient of 0.795. There is probably no legitimate argument for calculating and reporting the Kuder–Richardson reliability.
c. The estimated observed-score variance is the same as in $a$: 5.51. The universe-score variance now has to be estimated as nine times the average of the $^\circ\rho(_vX,_{v'}X)$ between unlike subtests. This is 3.81 and the coefficient is 0.691.

A.7. A larger correlation matrix is presented here than is necessary for the question posed.

| | | Subtest observed scores (linked) | | | | | Reading-composite universe scores | |
|---|---|---|---|---|---|---|---|---|
| | 1 | 2 | 3 | 4 | 5 | 6 | 7 | Subtests fixed | Subtests random |
| 1 | 1.00 | 0.27 | 0.32 | 0.47 | 0.32 | 0.59 | 0.28 | 0.61 | 0.66 |
| 2 | | 1.00 | 0.47 | 0.40 | 0.68 | 0.34 | 0.53 | 0.54 | 0.58 |
| 3 | | | 1.00 | 0.40 | 0.55 | 0.41 | 0.40 | 0.53 | 0.57 |
| 4 | | | | 1.00 | 0.46 | 0.49 | 0.42 | 0.70 | 0.65 |
| 5 | | | | | 1.00 | 0.45 | 0.51 | 0.61 | 0.66 |
| 6 | | | | | | 1.00 | 0.34 | 0.74 | 0.62 |
| 7 | | | | | | | 1.00 | 0.65 | 0.54 |

Standardized regression weights[5] were calculated with three predictors and seven predictors:

| Predictor | 4 | 6 | 7 | 1 | 2 | 3 | 5 | $\rho^3(\hat{\mu}_{p},_{w}\mu_{p})$ |
|---|---|---|---|---|---|---|---|---|
| Fixed subtests in the universe | 0.33 | 0.46 | 0.35 | | | | | 0.80 |
| | 0.27 | 0.35 | 0.29 | 0.14 | 0.08 | 0.04 | 0.06 | 0.82 |
| Random subtests in the universe | 0.38 | 0.36 | 0.28 | | | | | 0.64 |
| | 0.21 | 0.10 | 0.11 | 0.35 | 0.21 | 0.12 | 0.14 | 0.81 |

The answer to the question posed is that regression weighting of the three subtests, where the universe is defined by the three fixed subtests, has very small effect, raising $\rho^2(_{w}\hat{\mu}_{p},_{w}\mu_{p})$ from only 0.795 to 0.798. Allowing the four other subtests to take on weights does not make an appreciable difference.

The universe score defined by an indefinitely large number of tasks like General Information, 4, Word Meaning, 6, and Story Sequence, 7 is much less well predicted by the three-variable weighted composite ($\rho^2 = 0.64$; compared to 0.62 for prediction from the unweighted observed score). Not only do the additional predictors raise the coefficient markedly with the broader universe (to 0.81), but two of the predictors with highest loadings are outside the original composite! To be sure, these precise loadings would not be replicated in another sample.

This is an unusually striking example of the possibility of using auxiliary predictors for a universe score. However it suggests that the distinction between tests in the universe for the reading composite, and other tests, was ill-conceived, and that 4, 6, and 7 do not represent a particularly distinctive variable. In fact, the reading composite score originated out of the test developers' hope that they could form diagnostically useful "reading" and "arithmetic" composites. Selected subtests

[5] These weights do not take into account $\sigma(_{w}\mu_{p})$, which is arbitrarily taken as 1.00.

were combined on the basis of correlations with subsequent achievement tests, and it was concluded that 4, 6, and 7 were most relevant to reading. Further work showed that the two diagnostic composites do not differentiate well enough to be profitably used side by side. The reading composite is recommended for use when prediction of only reading performance is desired, in which case 4, 6, and 7 are to be administered by themselves, as a short form.

A question of considerable interest, which cannot be answered from the data treated here, is whether the universe score defined by fixed subtests, or the universe score defined by random subtests, is more informative about subsequent achievement in reading. This amounts to asking whether the specific factor in such a subtest as Story Sequence makes a valid contribution to prediction or is a source of error.

While the broader universe score (where subtests are regarded as random) is considerably harder to predict than the narrow one when the observed score or the three-variable weighted composite are used as predictors, when seven variables are used it is as easy to predict as the narrower universe score with subtests fixed.

# CHAPTER 11
# Contributions and Controversies— a Summing Up

A statistician complains, in a recent book review, that "students so frequently come away from the literature with the feeling that no unresolved problems exist. Philosophy, controversy, and the true complexity of issues tend to be left out. Intelligent discussion can reveal the exciting problems, and uncertainties, and the wide range of opinions and approaches in a field." (Elashoff, 1970, p. 104). This chapter is dedicated to making sure that such a complaint is not made about the present book.

For the reader who has lost his bearings amid the twists and branches of our argument, and for the reader whose habit is to turn first to a technical book's concluding chapter, we start with a summary of the highlights of the system and of the more striking implications of the argument. The second section of the chapter treats the technical limitations of the system and points toward work remaining to be done. We then move to a more fundamental level of criticism and interpretation, discussing pertinent challenges that have been offered and the connection between generalizability theory and the theory of test validity. Finally, we evaluate the approach.

## A. The Model and Its Implications

In principle, an investigator could make numerous observations of a person's performance with respect to any variable. These observations are equivalent in that the investigator would take each of them as a sample of the same kind of information—they fit the same operational definition. This does not imply that the observations are statistically equivalent. Our theory is closely

350

related to what Lord and Novick report as theory for "imperfectly parallel measurements."

The investigator would like to know the person's universe score, the mean over the whole set of admissible observations. Practically, he is limited to making an inference from a sample of observations. Very likely he should make a regression estimate of the universe score, instead of taking the observed score as the estimate (p. 102 ff.). While this proposal is to be found in virtually all major works on test theory, it has been given little theoretical attention and has rarely been carried over into practical test interpretation.

Multiple-regression methods are applicable; mysteriously, treatises on classical test theory have failed to develop this possibility. Instead of inferring the universe score on a variable from observations on that variable alone, information on additional observed variables can and probably should be used (p. 312 ff.). The universe score on the Wechsler Performance Scale, for example, is better estimated from a weighted combination of the observed Verbal and Performance scores than from the Performance score alone. A still better estimate can be made by weighting subtests separately. A few persons working along classical lines have previously suggested this kind of multiple-regression estimate, but only for the limited purpose of measuring "change."

This work also has touched on the measurement of change. We conclude that data collected prior to and following a treatment can be profitably analyzed in quite untraditional ways. To estimate a "change score," however, is rarely appropriate (p. 334 ff.).

Any universe score may be estimated from a multiple-regression equation when relevant data supplementary to the observed score are available. The procedure is a simple application of the algorithm for predicting a criterion from a best linear combination of predictors. Because the criterion here is a universe score, its correlations with the observed scores that serve as predictors have to be estimated rather than directly calculated from data. Under strictly classical assumptions, estimation is quite easy, but we find it necessary to add a distinction between linked and independent observations. The strict independence assumed in classical theory rarely holds when several scores are collected in the same setting (p. 268 ff.).

The multivariate estimation of a single universe score obviously extends to the estimation of a whole profile. A profile of estimated universe scores is likely to differ in shape from that for the observed scores, and gives a different impression of the person. In general, such estimated profiles will be flatter as differences arising from errors of observation tend to be suppressed. One can go further, reorganizing and simplifying profiles so that only the more generalizable information is reported (p. 323 ff.).

The "true score" concept has served to conceal ambiguities. With regard specifically to a composite score, for example, the true or universe score can be defined in conflicting ways. Perhaps the investigator is really interested in the expected value of the Wechsler Full Scale score, over all measurements of the person that might be made. This expected value is interchangeable with the sum of the Verbal and Performance universe scores. The variance of Verbal universe scores is greater than that of Performance universe scores, however, and consequently the sum of the two gives heavier weight to individual differences in Verbal abilities. A better score might be a composite that gives about equal weight to Verbal and Performance components. How to define a universe score to represent a construct adequately raises questions for which measurement theory has no answers. This is just one of the many hidden cracks beneath the much travelled surface of established test theory.

We classify conditions of observation with respect to facets; for example, test forms, observers, and occasions. This does much to sharpen the definition of the universe of generalization and brings to attention the importance of the universe definition. Investigators often choose procedures for evaluating reliability that implicitly define a universe narrower than their substantive theory calls for. When they do so, they underestimate the "error" of measurement, that is, the error of generalization.

A given observation belongs to a number of universes of generalization. The investigator has to make conscious decisions about his choice of universe; about which facets shall be considered sources of "wanted" information and which contribute to "error." The observed score generalizes well to some universe scores and poorly to others. Classical theory is unable to deal formally with the fact that a given score can be generalized in various ways (over forms, over days, over scorers, etc). Our model requires the investigator to make an explicit choice of universe. The system then generates a plan for data collection and analysis that is properly matched to the universe chosen.

The facet model lends itself naturally to analysis of variance and to estimation of the magnitude of score components. While one-facet analysis of variance was brought into test theory before 1940, multifacet techniques have been largely neglected. In a complex study, components of variance and covariance are easier to work with than correlations, and make fuller use of the data. They readily separate one kind of variation from another, and this separation casts considerable light on the nature of effects that are lumped together in typical reliability coefficients.

Correlations between universe scores are undoubtedly significant for substantive theory and for practical applications of tests. The multifacet model brings to the surface ambiguities in the common correction for attenuation. Because any score belongs to several universes, "the universe-score correlation" between it and another variable can be defined in many ways. Each of the "corrected" correlations has its own substantive meaning

(p. 290 ff.) Persons employing the traditional correction formula have not infrequently interpreted the result incorrectly. Indeed, at times they have made wholly unsound applications, in effect using conflicting definitions of "error" in evaluating the numerator and the denominator of the formula. Our argument from components of covariance is more straightforward, less subject to mishandling, and more likely to be correctly interpreted.

The statistical concepts of traditional reliability theory reappear in our system, but with altered interpretations and emphases. Least altered, perhaps, is the concept of true-score variance, which is seen here as a universe-score variance. This, we argue, is a property not of the measure and the population alone, but of the chosen universe of generalization as well. For each universe within which the procedure fits, there is a different universe-score variance. Such distinctions as that between coefficients of scorer agreement and parallel-form coefficients (which were a verbal supplement to classical reliability theory rather than part of the model) are directly imbedded in the model for generalizability theory.

Observed-score variance is a property of the measuring procedure and the population (as classical theory recognizes), and also of the experimental design by which the procedure is applied. This variance will be larger, for example, when a different judge rates each subject than when the judge is constant over subjects. This increase occurs whenever each observer has his own constant error; classical theory assumes equivalence of observer means, and cannot acknowledge such sources of variance. Abandoning equivalence assumptions makes it necessary for the generalizability study to estimate an *expected* observed variance. This is the average to be expected when a design is applied many times to samples of persons from the same population. Not knowing what test forms, occasions, or judges will be selected when the procedure is applied, the person analyzing the generalizability data can do no better than report an average value for the procedures encompassed by the stated design (p. 90 ff.)

The coefficient of generalizability is the ratio of universe-score variance to expected observed-score variance. This is, approximately, the average of the values that would be obtained if the ratio of universe-score variance to actual observed-score variance were known for each application of the design. The ratio, which we denote by $\mathscr{E}\rho^2$, is an intraclass correlation (pp. 75, 80, and 97 ff.). Such correlations have appeared in various guises in earlier psychometric writings: Rulon's split-half formula, Horst's formula for reliability with multiple observations, Kuder–Richardson formulas 20 and 21, and the Hoyt–Cronbach alpha coefficient. All are intraclass correlations for one or another design. Each of these is now identified as a specific case within a general structure. We have shown how to obtain a far wider variety of coefficients, with different numerical values and distinctive meanings.

Traditional studies of test reliability have had limited aims. The generalizability study has more flexible purposes. Traditional studies assume that the design of a measuring procedure has been fixed, with the possible exception of the number of items, and that one wants a numerical index of the precision of the device. The report gives a reliability coefficient and a standard error of measurement. Beyond this, one may, if he chooses, form a regression equation for estimating true score from observed score. The generalizability study is best carried out in the preliminary phase of instrument development. The developer will thereby gain a general understanding of the causes likely to create discrepancies among observations. Such disturbing influences can be brought under control either by modifying the operations performed (e.g., preparing clearer rules for scoring), or by drawing a larger sample of whatever conditions are particularly serious sources of discrepancy.

Deciding upon test length with the aid of the Spearman–Brown formula is an illustration of the kind of pilot work we recommend. Our methods go far beyond that technique, extending to all kinds of observations that yield numerical scores, and to all the variable aspects of the procedure. Altering the number of observations to be made is only one of the options open to the investigator. Where the universe is multifaceted, many different designs can be applied to select the conditions for the observations from which future decisions will be made.

The G (generalizability) study is distinguished from the D (decision) study. One chooses the design for the D study in the light of its purposes, the costs of collecting information, and the information the G study provides regarding generalizability from any of the possible designs. That is, a G study carried out according to one design—perhaps a rather elaborate one— indicates the effectiveness of a whole panorama of alternative designs among which the investigator can choose. What design produces the best results will vary with the decision intended. Where the observations are to produce a group mean, for instance, an item-sampling design may be best. Where individuals are to be compared, crossing certain facets with persons will usually be advisable. Where the study is to yield an absolute score that permits a decision about the person without reference to norms or to the performance of competitors, very "weak" designs with joint sampling of conditions from various facets are advantageous. Also one will design a testing procedure differently, depending on whether he wishes to generalize over subtest tasks, over days, or both.

When the oversimplifications of classical theory are abandoned, the concepts of error variance and standard error are seen to be ambiguous. There is an error $\delta$ that has the property that its variance [the "group specific error variance" of Lord and Novick (1968, p. 178)] is the difference between observed-score variance and universe-score variance. The "generic" error of

Lord and Novick, our $\Delta$, equals the difference between universe score and observed score. In classical theory, these errors are identical, but with weaker assumptions $\sigma^2(\Delta) \geqslant \sigma^2(\delta)$. Once the components of variance have been determined, either of these variances can be estimated for various designs that may be employed in the future (p. 84 ff.).

The standard error $\sigma(\delta)$ can properly be applied to determine whether one of two persons tested under the same conditions has a significantly higher universe score than the other (p. 93 ff.). That is, it is relevant where a decision depends on a ranking (and where there is no basis for making the decision save the observed score). We conclude that the time-honored standard error of measurement is of rather limited importance.

The index $\sigma(\Delta)$ indicates how far measures are likely to depart from their "true" values; i.e., from the person's universe score. It gives an indication of the precision of the measuring procedure (p. 84 ff.). When we take multiple facets into account instead of regarding error as undifferentiated, we frequently find that errors are larger than conventional analyses suggest. For example, when taped interviews of patients are rated on the Spitzer–Burdock Mental Status Schedule, the standard error is 6.5, if one considers only the person–rater interaction as a source of error. When the component for rater main effects is considered in the error also, the standard error rises to 8.3. In general, multifacet G studies encourage increased caution in generalizing and encourage more elaborate designs for collecting data, so as to reduce error of generalization.

Closely related to $\sigma(\Delta)$ is the standard error for a group mean, useful in evaluating the generalizability of an experimental result. Unlike the usual standard error of a mean, our index takes sampling of conditions as well as sampling of persons into account (p. 96). The index $\sigma(\Delta)$ also plays a part in sequential testing, where one seeks to determine which persons have universe scores above (or below) a certain standard, and is willing to test borderline cases further until a confident decision can be reached about each one. Still another use of $\sigma(\Delta)$ is in testing the significance of differences within a profile, but the rationale and computing procedure are more complicated than older theory has recognized (p. 332 f.).

One important use of the standard error of measurement has been to "establish a confidence interval" for the person's "true" score. The confidence-interval procedure for locating a population mean is a sound one, and the interval reached via generalizability theory, which recognizes sampling of conditions, is a distinct improvement on the conventional procedure that recognizes only the sampling of persons. Applied to scores of individuals, the interval-estimation procedure is almost inevitably misleading. We concur in the recent warnings of Lord and Novick, and point out additional reasons for distrusting the procedure. The intervals stated for the universe score are frequently incorrect among persons whose observed scores are far from the group mean.

When one wants to move from general statements about the overall adequacy of a design for collecting data, toward statements about each individual's universe score, the regression technique for estimating the universe score is probably advisable. But the nonequivalence of conditions entertained in generalizability theory makes it impossible to estimate parameters for the genuine regression equation. When conditions are not equivalent, and conditions are crossed with persons in the D study, there is (in theory) a regression equation for the particular conditions used. The parameters of the equation will vary with the set of conditions. Our methods yield an estimation equation that is an approximation to the regression equation for typical conditions. The investigator runs some risk in applying that equation when conditions are not equivalent, especially when the number of conditions in the D study is small. Those conditions may be atypical, so that the genuine regression equation is not close to the estimation equation.

Associated with the genuine regression estimate is an error $\varepsilon$. This error will, on the average, be less than the error $\delta$. The value of $\sigma(\varepsilon)$ is a rough indication of the precision with which our estimation procedure estimates universe scores, but various logical and statistical problems prevent exact interpretation of it.

The foregoing is only a superficial summary, but perhaps enough has been said to make clear why we regard generalizability theory as powerful. The work is far from finished—if finishing so open-ended an exploration is indeed possible. Theoretical problems remain. Illustrative studies in great profusion are needed, because every new application clarifies our thinking and may bring new puzzles to attention. There is need for study of the sampling errors of the statistics obtained from G studies. Above all, there is need for open discussion of the system. Apart from whatever technical criticisms may be warranted, our proposals for change in the design, analysis, and interpretation of tests have to be debated in the light of professional values and purposes. Debate will clarify whether testers do indeed wish, for example, to generalize over verbal tasks when they interpret a Verbal IQ, and whether they do wish to reduce batteries to differential information that can be generalized. Our presentation has brought issues such as these into the open. The community of test users now has to decide what detailed version of the model best fits scientific and practical requirements.

### B. *Technical Limitations of Generalizability Theory*

Any mathematical model is vulnerable to criticism. Being an idealization, it leaves out of account some aspects of the real world. Occasionally, the model may even be flatly contradictory to plain observation. To appreciate a model for what it can do, one must be thoroughly sensitive to its limitations. Criticism directed to this end supports the development and proper use of the model.

### *"Weakness" in the model*

We have attempted to work with a model far less restrictive than the classical theory. Classical theory embodies what we would call a universe of observations on a subject, but regards those observations as indistinguishable. If scores are obtained by observing each person in a population, every such collection will, according to classical theory, have the same mean and standard deviation; pairing of two such collections will always yield the same correlation. Writers have weakened this model by one or another isolated amendment, but nothing like an alternative system has hitherto been put forth.

We started boldly, placing no restriction upon the universe. The assumption that conditions of observation are identifiable and classifiable is a nominal one; the argument applies when the identification is arbitrary.

The one central assumption made is that conditions are randomly sampled when data are collected. This assumption is shared with Lord's work on randomly parallel, stratified-parallel, and item-parallel tests. The merits of the sampling assumption will be argued below.

From this beginning it follows that a collection of scores will have different statistical properties depending on the experimental design proposed. A design in which sampling of conditions is carried out separately for each person will yield score collections that fully conform to the classical requirements even when the universe itself is substantively and statistically heterogeneous.[1] With crossed designs, however, the model allows for varying means, varying standard deviations for observed score, varying correlations among observations, and varying correlations of observations with the universe score.

As is usual with a highly generalized weak model, it can be strengthened by adding assumptions and so brought into line with many simpler models. If one retains every aspect of our formulation, but reduces the number of facets to one (or assumes that all facets are confounded in the design), the model becomes identical to Lord's for "random parallel" tests (1955; see Lord & Novick, 1968, p. 234 ff.). If one imposes the further assumption that conditions are single-factored [i.e., that $\sigma^2(v_{pi})$ defined in (5.9) is zero] one has

---

[1] This statement must be qualified in one respect. Even if conditions of facet $i$ are fully equivalent, and conditions of facet $j$ are fully equivalent, when a design such as $(i \times j):p$ is used two collections of scores obtained with the same $i$ within the person (and different $j$) will correlate to a greater degree than scores with different $i$ and $j$ within the person.

"congeneric tests" (their p. 217). One can retain the multifacet feature and impose strict assumptions of single-factoredness or equivalence of conditions on each facet and arrive at a valuable intermediate model that copes with distinctions among kinds of measurement error. Finally, one can impose the full classical assumptions; classical true-score theory is the extreme simple case of generalizability theory.

It may be worthwhile to comment on the relation between this general model and the distinction between systematic and random error that physical scientists and engineers typically make. Systematic errors arise from particular conditions of observation (e.g., personal constants of observers, idiosyncrasies of individual machines, departures of temperature from a standard value). These can be assessed and brought under control by mechanical adjustment or calibration. What is left over is considered to be random fluctuation. We deal in one way with systematic error when we treat a condition as fixed. If the WISC scale as it now exists is to be used in all observations of IQ, any idiosyncrasy in its content or standardization can be disregarded so long as one limits his generalizations to WISC IQs. The standardization procedure is intended to remove such systematic "error" as arises from the selection of harder or easier items; it is a calibration process. Many score components that can be regarded as systematic error if we know what specific conditions will be employed in the D study become random sources of variation when we anticipate D studies for which the conditions have not yet been specified. The constant error of a rater is, from a philosophical point of view, systematic; from a practical point of view, if a procedure will be carried out by an observer of a certain kind who has not yet been selected, his constant error must be regarded as a random error. Because it enters into the observations of many persons, however, we do not assume that the errors of observing those persons are independent.

*Indefiniteness of results.*    It was possible to derive a great many conclusions within our weak model. The conclusions all have a certain indefiniteness, in that they refer to expected values and never to particular values. The classical theory can say, after a suitable generalizability study, "If you apply any one of these parallel tests to a large sample from this same population, 96% [for example] of observed-score variance will arise from the universe score." Ordinarily the conclusion from generalizability theory must be: "Over all investigations with tests drawn from this universe and applied to samples from this population, the variance arising from the universe score will be 96% of the *average* observed-score variance." Similarly, our estimated value of $\sigma(\Delta)$ applies over all observations that might be formed from the universe, but does not ordinarily apply to the set for any particular person or the set for any particular study.

The weakness of these conclusions means that any reasoning from a generalizability study must be most tentative. Having a correlation between Form A of Test 1 and Form Q of Test 2, we sometimes want to estimate the correlation of universe scores $_1\mu_p$ and $_2\mu_p$. For this purpose we need

$$\rho(_1X_{pA}, _1\mu_p)$$

but the only available estimate is an average correlation of various forms of Test 1 with $_1\mu_p$; similarly for Test 2. Unless we introduce the strong assumption that coefficients of generalizability are equal for all forms of Test 1, and equal for all forms of Test 2, the estimate is a very loose one. Classical theory makes that strong assumption and arrives at what appears to be a much tighter conclusion: that the correlation of the two universe scores *is* a certain number. This number may be the same number that we report from the same data—though this depends on how the generalizability study and the correlational study were carried out. The difference between the two conclusions is less in the way the information is handled than in the statements we allow ourselves to make at the end. Classical theory, by denying the existence of certain kinds of variation, brings hypothetical "information" into the solution of the problem. If generalizability theory were to accept the same hypothetical values [e.g., $^{\bullet}\sigma^2(_1pi, _2pg) = \sigma^2(_1pi) = \sigma^2(_2pg) = 0$] it could arrive at the same strong statements. The fact that we have not specified several zero values *a priori* leads us to look for empirical information on them, and this information often leads us to doubt one or all of the suggested zero values. The assertion we make about our calculated value is not so strong as the one classical theory makes about its zero values. The values are subject to considerable uncertainty, and we fall back on an average value or an indefinite statement.

The issue was discussed helpfully in the famous "bridge" metaphor of Cornfield and Tukey (1956, pp. 912–913):

> In almost any practical situation where analytical statistics is applied, the inference from the observations to the real conclusion has two parts, only the first of which is statistical. A genetic experiment on *Drosophila* will usually involve flies of a certain race of a certain species. The statistically based conclusions cannot extend beyond this race, yet the geneticist will usually, and often wisely, extend the conclusion to (a) the whole species, (b) all *Drosophila*, or (c) a larger group of insects. This wider extension may be implicit or explicit, but it is almost always present. If we take the simile of the bridge crossing a river by way of an island, there is a statistical span from the near bank to the island, and a subject-matter span from the island to the far bank. Both are important.
>
> By modifying the observation program and the corresponding analysis of the data, the island may be moved nearer to or farther from the distant

bank, and the statistical span may be made stronger or weaker. In doing this it is easy to forget the second span, which usually can only be strengthened by improving the science or art on which it depends. Yet a balanced understanding of, and choice among, the statistical possibilities requires constant attention to the second span. It may often be worth while to move the island nearer to the distant bank, at the cost of weakening the statistical span—particularly when the subject-matter span is weak.

In an experiment where a population of $C$ columns was specified, and a sample of $c$ columns was randomly selected, it is clearly possible to make analyses where

(1) the $c$ columns are regarded as a sample of $c$ out of $C$, or
(2) the $c$ columns are regarded as fixed.

The question about these analyses is not their validity but their wisdom. Both analyses will have the same mean, and will estimate the effects of rows identically. Both analyses will have the same mean squares, but will estimate the accuracy of their estimated effects differently. The analyses will differ in the length of their inferences; both will be equally strong statistically. Usually it will be best to make analysis (1) where the inference is more general. Only if this analysis is entirely unrevealing on one or more points of interest are we likely to be wise in making analysis (2), whose limited inferences may be somewhat revealing.

But what if it is unreasonable to regard $c$ columns as any sort of a fair sample from a population of $C$ columns with $C > c$. We can (at least formally and numerically) carry out an analysis with, say, $C = \infty$. What is the logical position of such an analysis? It would seem to be much as follows: We cannot point to a specific population from which the $c$ columns were a random sample, yet the final conclusion is certainly not to just these $c$ columns. We are likely to be better off to move the island to the far side by introducing an unspecified population of columns "like those observed" and making the inference to the mean of this population. This will lengthen the statistical span at the price of leaving the location of the far end vague. Unless there is a known, fixed number of reasonably possible columns, this lengthening and blurring is likely to be worth while.

In an extensive discussion of samples, universes, and hypothesis formulation in psychology, de Groot (1969, pp. 182–197) develops much this same position.

One would like to estimate, for example, the specific coefficient for an identifiable condition. Thus, for Test A of p. 359 one would like to estimate $\rho^2(_1X_{pA},_1\mu_p)$ more specifically than we do when we employ $\widehat{\mathscr{E}\rho^2}(_1X_{pi},_1\mu_p)$. If we could do this, we would have retained the weaker, more defensible model and yet would have a very firm final statement to make. Indeed, where we have developed our theory in terms of the expected value of $\rho^2(X_{pi}, \mu_p)$ Lord and Novick are inclined to favor estimation of the coefficient for a particular condition $i$ and they present a formula for this (their p. 210). As we have virtually ignored such possibilities in this report, a word of explanation is in order.

If a G study employs condition $i^*$ along with at least two other conditions randomly selected from the universe, it is possible in theory to estimate the coefficient of generalizability for the specific condition $i^*$ (p. 101). Such estimates were indeed made by Burt in a study of graders of essays as early as 1936. Our attempts to apply this kind of formula have been discouraging. When there is enough variation among the coefficients for different conditions to make the specific estimate important, any one coefficient can be estimated accurately only by employing a large number of conditions in the G study. This effort can be justified when one or more of those particular conditions will again be used in D studies. This rarely seems to be the case, except for well-standardized tests, and these are so nearly equivalent that it is unnecessary to estimate specific coefficients. We have mentioned (p. 102) the proposal to apply item-sampling designs to estimate the coefficient. Those methods may prove to be practically useful.

The "weak" $\widehat{\mathscr{E}\rho^2}$ and $\widehat{\sigma^2}(\Delta)$ appear likely to serve well enough for gross evaluation of a technique and for planning the design of a D study. When one tries to reach a conclusion about a particular person's universe score, this weakness becomes distressing. Where a regression estimate of $\mu_p$ is to be made, as we have recommended, one would very much like to have a specific regression formula for the conditions that in the D study are crossed with persons. It is not too difficult to correct for variations in the mean observed score, but there is no way, at present, to allow for the variation in slope, from condition to condition, that our model anticipates.

Because of the difficulties of estimation, our discussion has abandoned the attempt to retain the full flexibility of the model in developing linear estimators. We have written as if the estimation equation based on $\widehat{\mathscr{E}\rho^2}$ or $\widehat{\mathscr{E}\sigma^2}(X)$ were the genuine regression equation for whatever condition is used. To have pursued the implications of the weaker model by qualifying all our statements about $\hat{\mu}_p$ and the associated error would have made an already complex and novel argument much too confusing.

The multivariate argument has been simplified even more drastically, by assuming full equivalence and eventually restricting attention to one-facet universes and crossed designs. This extreme simplification seemed to be necessary for an initial presentation of the model. Most of the argument would hold if the argument were developed from the more elaborate weak model. In multivariate regression estimates, the equivalence assumption becomes crucial.

We have some confidence that the estimation equation based on an average coefficient of generalizability will give better information about the universe score than the alternatives so far known. Nevertheless, it should be remembered that this method of generalization brings in hazards of unknown magnitude.

***Need for extensive data.***    In principle, a weak model can reach conclusions just as strong and just as precise as those from a strong model. The stronger model assumes the values of certain parameters that have to be estimated in the weak model. The weak model leaves more questions to resolve, and the investigator has to collect more data to reach equally firm conclusions. Often, the weak model demands far more data than it is practicable to obtain.

Components of variance and covariance estimated from limited data are subject to considerable sampling error. This is not to argue that one should fall back upon a stronger model. The estimates reached from the stronger model have a smaller error *only so long as the strong assumptions are satisfied*. Staying within the stronger model, the investigator can make no test of its added assumptions.

Investigation of sampling errors in generalizability studies is much needed. The literature on sampling errors to date has limited itself severely, usually to one-facet studies and quite often to homogeneous universes. In this report we have barely touched on sampling errors. References to pertinent statistical sampling theory are given in Chapter 2. The literature is quite limited in its applicability but it is growing fast.

We reported one empirical study previously (p. 180), a reorganization of some Endler–Hunt data from successive samples. In that study there was reasonable consistency among estimates of components from various samples. Two other studies of the same general character were reported by Burdock, *et al.* (1963). Components of variance for ward observers applying a behavior rating scale to patients, and components for residual, were estimated in six institutions. There was remarkable agreement of results from four institutions even though data were obtained from only two to four raters. Conversely, the components dropped markedly (the residual component being cut about in half) at the two remaining institutions. Burdock attributes this at least in part to superior training of observers in those settings. A study with an instrument employed in rating interviews gave rather similar findings, with samples of three to five raters per study. The estimated inter-action component was quite consistent from one study to another. In both studies, the component for raters (which is sampled $n_j$ times rather than $n_p n_j$ times) was much more vulnerable to sampling errors.

Much can be done with Monte Carlo methods, in which the parameters for a universe and population are specified and the computer is asked to generate sample data. For a certain G-study design, one sample after another is formed and analyzed; this indicates how widely estimates of (for example) $\mathscr{E}\rho^2$ vary, and how they relate to the value that ideally should have been obtained. One can ultimately draw conclusions about the effect of changing $n_p$, $n_i$, $n_j$, etc. on the goodness of estimates with various kinds of design. In an earlier phase of our work, we carried out a number of one-facet studies of this type (Cronbach & Azuma, 1962; Cronbach, Ikeda, & Avner, 1964; Cronbach, Schönemann, & McKie, 1965). We have not recapitulated them, because they do not contribute much information on multifacet designs. They do, however, illustrate a powerful and flexible technique. Research is required to learn what number of conditions need to be drawn to make G studies of various types dependable. No simple and general answer is to be expected, because the desirable $n_i$, for example, will depend on the magnitude of $\sigma^2(pi)$ and other parameters.

We would not be surprised to learn that G studies of customary size—for instance, 40–100 subjects, 2 forms, 2 occasions—are insufficiently accurate. If so, this is not a limitation of generalizability theory. Rather, it would demonstrate the power of generalizability theory to bring to light the inadequacies of studies carried out under traditional models that deliberately oversimplify.

### The steady-state requirement

The model assumes that the person's universe score, the interactions $\mu_{pi}\sim$, and other components "exist." Presumably, $p$ stands ready to give certain performances when the measuring procedure is applied to him. Because we do not control all conditions, and we do not control the fluctuations of the person's mood and physiological state, a residual $e_{pij}$ comes into the observation also. While it is not assumed that $p$ is completely stable during the period to which the universe definition applies, it is taken for granted that $p$'s characteristics fluctuate around a typical value. That is, $p$ is regarded as being in a steady state such that the $e_{pij}$ fluctuate irregularly around zero.

Sometimes the universe is defined to cover a short time span, the admissible observations being thought of as nearly simultaneous. The model, in effect, assumes that the expected score under any condition is the same, no matter where in the series of observations within the time span that condition is placed.

Something similar is implied when the time span is longer and time intervals are considered as samples. It is mathematically sound to define the universe score as the average over the time span, and to use randomly selected moments for observation. The model is satisfactory as long as there is no regular trend in performance. If there is a trend, however, systematic sampling of occasions spaced regularly over the time span will not fit our random-model arguments.

The whole concept of a universe score is of dubious value if the universe stretches over a period when the person's status is changing regularly and appreciably. An average height in inches, over the period when a boy goes from age 12 to age 15, can be determined—but that value would not be of much interest. An average height within the month after his twelfth birthday is a useful construct (for example, as a dependent variable in a study of nutrition, where it is impractical to test every boy on his exact birthday). Even though the growth trend continues during the month, the change is not enough to be disturbing.

Because our model treats conditions within a facet as unordered, it will not deal adequately with the stability of scores that are subject to trends, or to order effects arising from the measurement process. This is a limitation common to all reliability theory. A large contribution will be made by the development of a model for treating ordered facets. For many variables, adjacent scores agree more closely than scores that are further separated. Some kind of growth or decay function will be needed, in place of the present universe score, to describe a process that is strongly time-dependent. A development along these lines will amplify, not replace, our model.

### Other topics for further work

At a number of places we have restricted our development. The matters neglected will no doubt be important to some users. There should be systematic attention to certain of the neglected problems; it will be more profitable to work out ad hoc solutions for others to fit a particular application than to seek a broad theory. A summary of neglected areas follows:

1. Our formal developments have been restricted to universes in which facets are crossed with each other and with persons. We have, however, encountered cases where one facet is nested within another in the universe. While we were able to cope with these, no general formulation covering this extension was offered.

Moreover, we have given no consideration to universes where all the admissible conditions are nested within the person. That is, the universe of generalization may be unique to the individual, covering his behavior settings or his friends, etc. Personality theory now has a strong concern for the person's interaction with and interpretation of *his* environment; this is notable in Mischel's work (1968), derived in part from G. A. Kelly. To pursue measurement in this vein appears to require different questions for (or about) each individual, and a personalized universe of generalization. Generalizability theory should, nevertheless, be readily adaptable to such measuring procedures.

2. The random-sampling model derived from Cornfield and Tukey yields much information not available by other procedures. Maxwell (1968) is critical of the application of estimation of variance components in the usual $i \times p$ internal-consistency study because he anticipates that residual components will tend to be correlated. He uses a simplified model that omits the "cell effect" $\mu_{pi}\sim$, and defines an $e_{pi}$ equal to $X_{pi} - \mu_p\sim - \mu_i\sim + \mu$. Then, he points out correctly,

$$\mathscr{E}_p \, \sigma(e_{pi}, e_{pi'}) = \mathscr{E}_p \, \sigma(X_{pi} - \mu_p - \mu_i + \mu, X_{pi'} - \mu_p - \mu_{i'} + \mu) \neq 0.$$

We acknowledge the existence of such a covariance, positive or negative, but place weight on the fact that its expected value over pairs of conditions is zero. Maxwell argues that the presence of the covariance in the sample data tends to give too great a $\widehat{\sigma^2}(p)$, but he appears to assume that the covariance will be positive. Further study of the criticism is needed.

3. We have given little attention to the sort of facet that includes only a limited number of conditions. The methods for dealing with fixed facets, or facets with $N$ finite and greater than $n$, follow along lines mentioned briefly in our earlier chapters.

4. Attention has been restricted to designs where the number of conditions $n_i$ (or $n_i n_j$, or $n_i'$, etc.) is the same for all persons, but there will be studies where one has more data for some persons than others. There are also designs where the number of $j$ nested within $i$ varies from $i$ to $i'$, etc. For the sake of simplicity, in this book we have "squared off" all designs, discarding observations if necessary. However, in principle one can estimate $\mathscr{E}\sigma^2(X)$, etc. for irregular designs.

Earlier papers (Cronbach, *et al.*, 1963; Rajaratnam, *et al.*, 1965) have examined certain one-facet designs, discussing stratified tests where strata are assigned different numbers of items. Such a stratified test can now be handled as a special type of composite to which the theory of Chapter 9 applies. This theory can cope with multifacet studies of stratified tests where the number of items varies from stratum to stratum.

Another type of design that we have virtually neglected is the blocked design where, for example, randomly assembled items are organized into two or more forms (see pp. 39 and 217). No doubt there are circumstances where this design will be a superior design for a G study.

5. The possibility of carrying out economical G studies by various forms of item sampling needs to be explored. Knapp (1968) has shown that with $n_p = $ ca. 400 one can obtain good estimates of $\mathscr{E}\rho^2$ and $\mathscr{E}\sigma^2(X)$ for an $i \times p$ design using a balanced incomplete-block design in which each person takes only seven items. Multifacet G studies pose more stringent requirements, but it is often easier to replicate a study with additional persons than to

:rease the number of observations $n_i n_j$. Lord and Novick indicate that em-sampling designs are capable of evaluating the specific parameters $\mu_{i*} - \mu$, $\rho^2(X_{pi*}, \mu_p)$, etc. for a condition $i^*$ that will be used in the D study. (See p. 10 ff. Condition $i^*$ must be included in the G study.) This proposal urgently requires development; if successful, it would in many studies provide additional data and yield stronger conclusions.

6. It has been recommended that observed scores be regressed toward subpopulation means. While the logical justification for this is clear, the worth of such estimates depends on both the size of the differences among subpopulation means and the size of the subsamples. This is a variant of the shrinkage problem in multiple regression. Studies are needed to make clear how small a subgroup it is wise to treat separately.

7. Where the D-study sample has an observed-score variance different from that in the G-study sample, we face a "restriction of range" problem more complex than the classical one. We have adopted the classical assumption that only the variance component for persons has changed. Alternative assumptions could perhaps be defended; the topic merits further attention.

8. Bayesian statistical theory needs to be exploited systematically. It appears likely that developments now available in the statistical literature could, in some problems, profitably replace the methods of estimating variance components that Chapter 2 relies on. Also, whereas we obtain all estimates from the G study, one could, by Bayesian methods, take into account the additional information offered by the D study to reach final conclusions about the generalizability of the D data.

Traditional analytic procedures are most dependable when conditions are close to equivalence, or the number of conditions in the G study is large. But generalizability theory has been especially concerned with nonequivalent conditions, and to sample several facets extensively is rarely practical in a G study. Bayesian statistics offer some hope of dealing with the cases that present the greatest difficulty for traditional methods (Novick, 1971).

### C. *Defining Universes*

The central concept of generalizability theory is that the observation is a sample from a universe, or, more formally, that any condition is sampled from a universe of conditions. This in itself is probably not likely to be disputed, but arguments do arise regarding the way universes are defined and about the appropriateness of the formal sampling model. Two preliminary comments may help to avoid minor objections.

First, while the mathematics assumes strict random sampling, this does not necessarily mean blind and planless sampling. The model explicitly takes into account the cross-classification of observations in the universe with respect to forms, testers, and other facets. A model referring to random sampling of conditions is considerably more sophisticated than a model that considers a true score plus an error drawn from an undifferentiated distribution of random errors. Second, our model can deal with systematic samples. Therefore, one can classify test items with respect to content (or any other aspect), and can call for random sampling of items within these strata. This plan can be accommodated entirely within the mathematics of random sampling.

### Operational definition as a requirement

The theory we have presented employs the mathematical model of probability theory. There is an aggregation from which samples are drawn. From the sample, inferences are made about the aggregation. Critics have often pointed out the hazards of inferences of this sort even in traditional applications of statistical method. A common practice is to observe a group of persons who are conveniently available to the investigator and then to generalize to a population of persons "like these." Such an expression is indeed loose; moreover, it is used to mask the fact that the investigator intends to generalize over all persons in a certain age range, but realizes that such an unqualified generalization would be criticized. Scientists have found it better to apply statistical inference to samples obtained haphazardly than to refuse to use information from those samples or to take the sample data as purely descriptive and relevant only to the sample in hand (or at most to other samples assembled in the same locality by the same haphazard process). The justification for applying sampling theory to studies where the sampling does not conform to the model is essentially that quoted above from Cornfield and Tukey.

No doubt a reference to "the population of persons like these" is based on much clearer thinking than a reference "to the universe of tasks like these." (See critical remarks, p. 376 ff.). We have suggested, for example, that the interpreter of a Wechsler Verbal IQ probably generalizes, at least implicitly, to a domain of verbal tasks of which Wechsler's six subtests are representative. While this does indeed seem to be the way test interpreters think, the "domain of verbal tasks" is an unexamined, crude notion. Loose inference will undoubtedly continue to be made, with or without help from generalizability theory. It appears less profitable to discuss the iniquities of loose inference than to ask what rigorous inference regarding universes of conditions would be like. As the possibility of rigor becomes clearer, it will be increasingly possible to formulate statements and investigations that permit better inference.

The probability model requires that the elements in the population be discrete, as are balls in an urn. In behavioral observation there are exceptional situations where the conditions of a facet are distinct from each other in the same way that each person in a population is separate from the next. There is, for example, the universe of addition problems. The problem $15 + 6$ is clearly a different problem from $24 + 7$. Small quibbles arise, e.g., about whether to distinguish $15 + 6$ from $6 + 15$ and from $\begin{smallmatrix} 15 \\ \underline{\phantom{1}} 6 \end{smallmatrix}$, but defining rules to cover these cases are not hard to prepare. Where raters constitute one facet of an observation, there is no more difficulty in distinguishing one rater from another than there is in any application of inference to samples of persons.

Where conditions blend into one another, elements are difficult to define. One may speak of sampling items on, for instance, the physical geography of North America, from a universe of such items. But what separates one item from another is far from obvious. "What river runs from Cairo to Memphis?" Is this item different from "What river runs from Cairo to New Orleans?" As the content of instruction becomes more theoretical (e.g., "the conservation of matter"), it is even less obvious what constitutes an element. Often the elements to be observed are more-or-less arbitrary segments cut out of a continuous stream. This is true, for example, when we try to sample behavior so as to learn whether a person is "typically" outgoing. This does not prove to be a very troublesome problem. It is necessary to impose some arbitrary divisions, but this is a familiar practice in "area" sampling for opinion polling, and in time sampling of continuous processes. The fact that different items may measure "the same" content is not ordinarily troublesome. However, if the overlapping versions of the same content amount to a sizeable fraction of the universe, simple random sampling without replacement may not be an adequate analogue to the test-construction process.

The essential requirement for reasonable application of sampling theory is that the universe be defined clearly enough that one can recognize elements belonging to it. This is a requirement of operational definition (de Groot, 1969, pp. 170, 240). A semantically and logically adequate definition will enable the consumer of research to consider possible candidates for membership in the universe, to judge whether the candidate is indeed included by the definition, and to reach the same conclusion as other interpreters of the definition.

Some writers on scientific method seem to regard an operational definition as somehow describing one unique and unvarying operation, but it is better to regard it as defining a class of equipment and rules for action. Often the members of the class are very nearly identical and will give measures that are indiscriminable from one another. Even so, to say that measures are indiscriminably different is not to say that they, or the pieces of equipment themselves, are identical. In the early stages of development of a measuring technique there may be considerable variation between successive applications of the same procedure, either because the pieces of equipment differ in unidentified ways, or because the investigator has not yet learned what ambient variables need to be controlled. To define operationally is only to state clearly the identifying characteristics of the measurement procedure; it is not necessary that the procedure be accurate. Kaplan (1964, pp. 40–41) criticizes the strict operationism that does not allow the definition to be a *class* description.

What justifies the assumption that the operation I perform is the same as the one you carry out? The operationist principle is that different operations define different concepts. Without the assumption, therefore, no two scientists could ever understand any scientific idea in the same way, and mutual criticism or corroboration would become impossible. The difficulty arises even for a single scientist: each performance of the operations is different in some respects from any other. Unless these differences are dismissed as irrelevant, it is impossible to replicate even one's own experiments. As Gustav Bergmann has pointed out an extreme operationist would presumably refuse 'to "generalize" from one instance of an experiment to the next if the apparatus had in the meantime been moved to another corner of the room.'

Nash (1957, p. 242) makes it clear that what is to be controlled (i.e., explicitly represented in the definition) is a matter of judgment and an expression of the current state of substantive knowledge.

[The investigator] "will expend most of his time and effort in attempts to achieve the effective duplication of just those conditions that, in the light of the conceptual scheme or working hypothesis that has suggested the experiment to him, appear to be capable of significantly affecting the results. He will be able to spare little or no effort to secure the duplication of those factors that appear to be irrelevant to the outcome of the experiment. Thus it is seen that the whole design will inevitably depend on the conceptual outlook of the experimenter. Long delays may ensue whenever this outlook encourages the investigator to regard as 'trivial,' and to leave uncontrolled, some factor that may actually be capable of contributing to the production of anomalous and misleading experimental results."

To specify a universe of conditions will be easy in some studies and difficult in others. It should not be difficult, for example, to specify the universe of examiners over which one proposes to generalize, by indicating the amount of training expected and, if relevant, their age, sex, color, or other demographic identification. Apart from a few cases with marginal training, different referees should agree very well in deciding which examiners are and are not within the universe. It will be noted that this is not a universe all of whose members can be listed in a roster; it is most unlikely that anyone knows who all the qualified persons are. Hence, while the universe of generalization is well defined, a genuine random sample cannot be drawn.

How would one investigate whether a universe is "clearly characterized?" Essentially, by presenting various judges with the universe definition and a broad aggregation of conditions from a facet. If the facet of examiners is under discussion, one would present, say, 200-word descriptions of each of 50 persons who might be asked to give the test, and require the judge to indicate which ones fit the universe definition. The examiners who are described should include many who were not employed in the G study, and a reasonable number who fall outside the universe. The better the definition, the more nearly the judges will select the same examiners as admissible. If the universe is defined only as a universe "of examiners like these" (i.e., like the two or three examiners used in a certain G study) we expect it to fail the clarity-of-description test.

### Homogeneity, heterogeneity, and stratification

One has a choice between making the universe definition very restrictive, so that relatively few conditions of a given kind qualify, or of leaving the bounds open. There is nothing in generalizability theory itself, for example, that restricts the universe of admissible Wechsler testers to persons trained in the traditional manner, who have administered a certain number of tests under supervision. If one wants to define the universe of generalization to include any person who has read the WAIS manual, a study can easily be set up to determine how well results obtained by one such person are likely to agree with the mean result over all such persons. If one wants to restrict the universe to persons who hold the Diploma in School Psychology, that too is an adequately explicit definition. (To investigate how well the two universe scores agree is a useful investigation of a different kind.)

This point requires considerable emphasis because it is so easily misconceived. Generalizability theory is an abstract model that generates procedures for testing a working hypothesis or claim. The investigator sets forth a universe definition. He carries out a study with two or more exemplars of the definition. He learns how well he can generalize, on the average, from the observed score given by any one exemplar to the universe score. The machinery will work whether the elements of the universe are identical in all discernible respects or are manifestly diverse. The one requirement is that the universe be clearly characterized. If that is not the case, the statement about generalizability is essentially meaningless. For a test it is necessary to specify the kinds of items to be admitted to the universe; the universe might, for example, be limited to completion (constructed-response) items. There may be a variety of editorial rules. For a science test one might direct that questions be made as easy to read as possible, by using simple vocabulary and short sentences. Per contra, one might direct that the technical terminology and conventions adopted in the Chemical Bond Approach to highschool chemistry be used.

The definition should encompass the instructions to the subject and other administrative procedures, and the rules to be followed by observers or scorers. These aspects of the description refer to the testing operations; most of them have the effect of fixing some aspect of the procedure. It remains possible to generalize over collections of items fitting the definition, and over such aspects of the testing procedure as are not pinned down by the operational definition. That is to say, a class of tests is defined, over which one will ordinarily generalize. Class membership is identified not just by item content and form but by the procedural rules as well.

Essentially the same is to be said about other types of instruments. Consider a rating scale to be used to evaluate drug effects in psychiatric patients. It is necessary to define a concept such as state of excitement, very likely by providing descriptions of behavior at different points on the scale. These "anchors" tell the observer what to look for, and guide him in recording what he observes. The experimental plan has to tell what period of time and what range of behavior settings the universe encompasses. The observer needs to know whether to attend to the typical intensity of the symptom or the level of its most extreme manifestation during the period. The class of admissible observations is thus defined by the scales printed on the blank and by the whole set of instructions that guide the observing operation (Gleser, 1968).

Whether it is wise or unwise to restrict a universe to uniform elements is a question to which we shall return. It may be beneficial, however, to indicate what might be taken into account by a person who wants to achieve a high degree of equivalence of observations. The question is most often raised in discussing the construction of tests made up of items. The test constructor will attain a higher degree of equivalence among tests if he considers the universe of admissible items to be subdivided into strata, each of which is itself a universe. The stratum may be defined in terms of the topical coverage of the items, or it may be further restricted by imposing a constraint on item difficulty.

The procedures by which the test is assembled from the universe must be specified as part of the universe definition. If there are strata, there will be distribution rules to indicate (at least approximately) how many items will be taken from each stratum. For further discussion of specifications for test construction, see Cronbach, 1971.

When a universe is loosely specified, the elements admitted are likely to be diverse in character. Even a rubric that is rather narrowly defined may prove to be heterogeneous. Consider use of *-ir* verb forms in French as an example; narrow though this task is, recognition and production of the correct graphic forms perhaps develops faster or slower than proficiency with spoken forms.

The more heterogeneous the universe, the larger is the sample of observations required to estimate the universe score with a desired degree of accuracy. For example, we may want to determine what impression a prospective teacher makes on pupils. It is reasonable for the universe to include pupils with various backgrounds, abilities, and attitudes. The fact that the teacher will make a favorable impression on some and an unfavorable one on others is a fact of life. For an accurate estimate of the universe score, the investigator will have to obtain a suitably large and representative sample of pupils. Narrowly specified universes can be examined with smaller samples.

Specificity reduces the scientific and practical significance of the universe. Practical decisions typically call for a score that refers to a broad range of content or situations. This would be true in a professional qualifying examination, in an examination used for gross evaluation of the effectiveness of instruction in French, or in a measure used as a criterion for a selection decision.

The scientist may use constructs defined in terms of narrow operations, or constructs defined more broadly. If homogeneity were the only criterion, one would reduce the broad construct "self-esteem" to constructs as specific as "favorable self-report on questions having to do with proficiency in physical games, worded so that a YES response expresses a favorable view of the self." There is a stage in research where it is necessary to measure so specifically, to learn how large the interaction components involving activity and form of question may be. For research on the antecedents and consequents of self-esteem, however, one would certainly move to a larger universe of generalization. Activity and form of question could remain as stratifying variables within the universe.

To sample within a narrow universe just because one can then generalize accurately from a small sample is often to ask the wrong question. In the context of drug evaluation, for example, patients are interviewed and their adjustment judged (Gleser, 1968). The investigator has enough resources to assign two raters to each patient in his D study. A traditionally trained investigator wanting to get "reliable" scores might conduct two reliability studies. In one study two interviewers would separately examine the same patient. If components for persons, person × occasion, and person × rater were 2.60, 0.80, and 0.60, for example, the correlation of scores for these interviews would be 0.65, and the "reliability" for the two scores together would be 0.79. In the second study, a single interview would be rated by two persons, one making his judgments from an observation room. In this study the person–occasion interaction would not contribute to error variance though it would enter the observed-score variance. The "reliability" of a single score would be 0.85, and that for two raters would appear to be 0.92. This might encourage the investigator to use the available rater hours to obtain two scores for a single interview on the patient, because such data "are reliable." But the coefficient of 0.92 indicates degree of generalizability to a universe with the interview fixed. To generalize over raters *and* interviews, rater time is best used to cover twice as many interviews, with one rating of each. Narrowing the universe often increases the coefficient of generalizability at the expense of validity. Lehmann's words (1960) are pertinent:

> The reliability of a method is not of necessity positively related to its validity. In fact, in the behavioral sciences we sometimes find a negative correlation between validity and reliability, as the validity of certain results often decreases when we try to control all experimental and environmental factors to such an extent that the test–retest reliability is raised to a maximum.

As any field comes to be better understood the definition of what is to be measured evolves (Kuhn, 1961). Certain facets of the measuring procedure are found to have an influence on scores and other facets not. Then the definition can be liberalized by removing the constraint imposed on conditions of the noninfluential facet. (Jokes aside, who puts the thermometer in the patient's mouth does not affect the reading, and the "examiner" facet can in this instance be left unrestricted.) Among the influential facets, some are considered to be nuisance variables and they are brought under control, perhaps by fixing one condition of such a facet. Fixing conditions is the usual way of handling test instructions, time limits, and many other procedural details. The condition fixed upon is, of course, the one regarded as likely to maximize validity. (Temperature will not be taken just after the patient drinks hot coffee.) With respect to other influential facets, one will find it necessary to form subclassifications or strata, and to substitute a stratified sampling plan for the original random sampling of conditions.

### Universes in instrument design: concrete examples

The notion that test constructors lay out a series of test specifications and write items to conform to them has been criticized as hypothetical and unrealistic, but a number of investigators have actually constructed tests from explicit universe specifications.

**Measures of interpersonal perception.** Brunswik's (1947) emphasis on systematic design, coupled with his interest in perception of persons, led to research on person perception that has used sampling concepts in various ways. Palmer (1960a, b) quite explicitly laid out a universe for test construction. He wished to assess the subject's knowledge about other persons with whom he had the opportunity to interact. He set forth the following preliminary specification:

> The domain of interpersonal knowledge may be thought of as being defined by all possible questions which might be asked about another person. The following incomplete list of general categories will indicate the kinds of personal data which may provide suitable questions: personal habits and personal qualities; food, drink, and smoking; travel; hobbies, recreations, and leisure; organizations, clubs, and teams; education; skills and knowledge; medical, dental, optometric, and related information; personal property; financial affairs; job and career; family, marriage, and courtship. Many questions can be developed within each category. Conservatively estimated, 5,000 to 10,000 questions can be developed without repetition of the same content.

Palmer's omission of more subjective and affective items is deliberate. He discusses the weighting among categories. There is a fairly elaborate list of specifications for drafting questions within each of the rubrics. (Information on the outcome of Palmer's effort is not available.)

Content somewhat similar to that catalogued by Palmer was defined in a more empirical manner by Belson (1956). He was planning a questionnaire on interests of the adult population, as affected by television viewing. As a base for selecting questionnaire items he carried out some 3000 interviews of persons in various kinds of neighborhoods to learn the full range of interests that might be relevant. In Belson's view, the final list of topics from which questionnaire items were selected was far more representative of "interests in general" than a list that might have been set down *a priori* by the research worker. An *a priori* definition is required, of course, to fix the question posed to the interviewee; but this question can be less restrictive than a formal definition of the universe of topics would otherwise be.

The "David" test of Soskin (1954) was in a sense a face-valid test of interpersonal understanding. It was to be used to test the clinician's ability to draw correct inferences from tests, observation of filmed expressive behavior, or other information on a subject. Soskin studied David's life history thoroughly. Incidents from it were converted into test items, by describing a conflict situation and listing plausible courses of action. The question put to the clinician (who had been given a file of information on David) was, "Which course of action do you think David chose?" The correspondence between the judgment and the historical record of what David actually had done thus became a measure of the clinician's skill in interpreting whatever sort of data he had been handed. All the episodes in David's life history were potential test items. It is possible that Soskin chose episodes of a relatively dramatic character (and conceivably atypical ones); but strict representativeness could have been assured. A psychologist using the test might generalize from the score over items about David to the universe of items about David. Or he might generalize over a universe of target persons on whom a test could be prepared. One might also think of the information presented to the clinician as a sample from a universe of information. Studies could be designed to appraise the amount of variance arising from each such facet: target person, information, test items, etc. The series of studies would go far to make clear what the instrument measures and what influences performance.

*Achievement tests.* An example of a quite different sort is provided by Hively, Patterson, and Page (1968), who faced the task of measuring achievement in "basic calculation" among Job Corps trainees. It was important to have many tests. These men often claim to have mastered a unit of instruction prematurely, and have to be tested more than once. To design a family of tests on subtraction, the domain was subdivided into such categories as "basic subtraction fact, minuend $> 10$," "borrow across zero," and "equation with missing subtrahend." An example for the last category is $42 - \underline{\quad} = 25$, or, in general, $A - \underline{\quad} = B$. A set of formal rules for generating items was prepared. For this third category, it was specified that $A$ be a two-digit number, that $B$ be a one- or two-digit number, and that $B < A$. Something like 5000 distinct exercises can be formed by randomly sampling digits for the first and second positions of $A$ and $B$.

One randomly generated item of each type went into a test. The 27-item subtraction tests so constructed have a coefficient of generalizability of 0.880 (generalizing over forms and occasions, and assuming that test forms and occasions differ from person to person). Even though difficulty of items was not directly equated, the tests came close to equivalence.

This kind of test construction has been discussed by Hively in other unpublished reports, and also by Osburn (1968) and Bormuth (1970). Osburn thinks of outlining topics and subtopics in relatively complex subject matter. At the third or fourth level in this hierarchy, each entry refers to a kind of problem. A list of many concrete instances of the problem is prepared. In an example Osburn gives, the problem requires calculation of probabilities. Osburn draws one of several statements of problem-in-context (for example, a sentence on market research) and draws numbers from a list to particularize the data to be interpreted; an item has been constructed. (Bormuth's approach is similar but makes use of linguistic transformations to generate items.) Osburn apparently visualizes outlining a chapter or more of text, drawing categories, subcategories, and item types within subcategories by a sampling plan, and then generating automatically the particular problem to be solved. Each kind of problem could be represented by a predetermined number of items, or one could sample from the list of problems. Formally, this is a choice between regarding strata as fixed or as random.

One may not be able to speak of Osburn's tests as "representing" the universe of content, because the number of subcategories and concrete kinds of items is a somewhat arbitrary function of the material and the tester's ingenuity. But, having once mapped the universe of content, one can assign whatever weights seem reasonable at the category level or the subcategory level. This defines a universe of admissible tests that can be constructed as required. Osburn suggests that such testing is feasible in mathematics, statistics, and, possibly, engineering and physical science.

### D. *Some Questions Raised by Critics*

Papers by Tryon and Lord introduced the theme of constructing tests by sampling into mathematical test theory in the 1950's, though Tryon's mathematical argument had a concealed assumption of strict equivalence. We began to report on our emerging views in 1960. Criticisms of these previous papers merit attention here, though we like to think that the theory as it now stands overcomes the objections expressed.

R. L. Thorndike found the notion of a universe "puzzling and somewhat confusing."

> As soon as we try to conceptualize a test score as a sample from some universe, we are brought face to face with the very knotty problem of defining the universe from which we are sampling. But I suppose this very difficulty may be in one sense a blessing. The experimental data-gathering phase of estimating reliability has always implied a universe to which these data corresponded . . . . Perhaps one of the advantages of the sampling formulation is that it makes us more explicitly aware of the need to define the universe . . . .
>
> When we are dealing with the typical aptitude or achievement test, . . . the conception of the universe from which we have drawn a sample becomes . . . fuzzy . . . . Some of the recent discussions seem to imply a random sampling of tests from some rather loosely and broadly defined domain—the domain of scholastic achievement tests, or the domain of reading-comprehension tests, or the domain of personal adjustment inventories. Clearly, these are very vague and ill-defined domains . . . .
>
> . . . Cronbach [and his coworkers offer] . . . the single term "generalizability" to cover the whole gamut of relationships from those within the

most restricted universe of near-exact replications to those extending over the most general and broadly defined domain, and develops a common statistical framework which he applies to the whole gamut. Recognition that the same pattern of statistical analysis can be used whether one is dealing with the little central core, or with all the layers of the whole onion, may be useful. On the other hand, we may perhaps question whether this approach helps to clarify our meaning of "reliability" as a distinctive concept.

A third context in which the random sampling notion has been applied to the conceptualization of reliability has been the context of the single test item . . . .

. . . But here, again, we encounter certain difficulties. These center on the one hand upon the definition of the universe and on the other upon the notion of randomness in sampling. In the first place, there are very definite constraints upon the items which make up our operational, as opposed to a purely hypothetical, universe. If we take the domain of vocabulary items as our example, . . . . First, there is typically a constraint upon the format of the item—most often to a five-choice multiple-choice form. Second, there are constraints imposed by editorial policy—exemplified by the decision to exclude proper names or specialized technical terms, or by a requirement that the options call for gross rather than fine discriminations of shade of meaning. Third, there are the constraints that arise out of the particular idiosyncrasies of the item writers: their tendency to favor particular types of words, or particular tricks of distracter construction. Finally, there are the constraints imposed by the item selection procedures . . . . Thus, the universe is considerably restricted, is hard to define, and the sampling from it is hardly to be considered random.

Presumably we could elaborate and delimit more fully the definition of the universe of items. Certainly, we could replace the concept of random sampling with one of stratified sampling, and indeed Cronbach has proposed that the sampling concept be extended to one of stratified sampling. But we may find that a really adequate definition of the universe from which we have sampled will become so involved as to be meaningless. We will almost certainly find that in proportion as we provide detailed specifications for stratification of our universe of items, and carry out our sampling within such strata, we are once again getting very close to a bill of particulars for equivalent tests . . . . Frequently a test is fairly sharply stratified—by difficulty level, by area of content, by intellectual process. When this is true, correlation estimates based on random sampling concepts may seriously underestimate those that would be obtained between two parallel forms of the test, and consequently the precision with which a given test represents the stratified universe. [Thorndike, 1967, pp. 288–289].

Rozeboom (1966) is an even more sympathetic critic; indeed, his discussion of reliability might well be ours, restated in a more rigorous style.

> ... the definition of a particular testing procedure $X$ may be considered to determine a specific probability distribution over potential scores on $X$ for each individual $p$ to whom a test is applicable, while the $X$-score, if any, actually received by $p$ is a sampling of this distribution.
>
> ...
>
> Now, to define a testing procedure is *partially* to identify the causally relevant circumstances which are operative ... , and hence to ask what score *would* a particular individual $p$ receive on test $X$ if $p$ *were* to be tested by $X$ is to speculate about what could be deduced from the laws governing such situations given exhaustive information about what $p$ is like, but only so much information about test circumstances as is included in the definition of the test. [Rozeboom, 1966, p. 383; notation altered.]

As Rozeboom develops the basic notions of reliability theory, he apologizes for the "mysterious" nature of the hypothetical true score. It enters "only because we have attempted openly to confront—not to solve, but at least to make our discomfiture explicit—certain fundamental conceptual problems of test theory which lie unrecognizably smothered in the more ornate flourishes of most traditional accounts of reliability." (p. 391). A brief characterization of classical theory is followed by a characterization of the random-sampling model.

Rozeboom sees it as "rather powerful mathematically ...." "Mathematical potency is something which can be achieved only at the expense of postulational risk, however, and how legitimate the sampling assumptions are in any given application is always questionable. Further limitations of this approach are its failure to provide a satisfactory definition for the reliability of noncomposite tests or single items, and the difficulties which arise in attempting to specify the domain." (p. 393).

The critical argument emerges in full force under the heading of "content validity," with which Rozeboom properly identifies much of our model. Space does not permit full representation of Rozeboom's logic or his eloquence, but his spirit of discontent can be communicated. He thinks of the universe score as a composite criterion variable:

> ... a fundamental but outstandingly neglected prerequisite for judging a test's content validity is specification of just what the composite criterion *is*. The composition of the test itself does little to clarify this, for there are an unlimited number of potential composite criteria whose components are simultaneously sampled by the test .... [An illustration follows.] Moreover, it takes much more than a few commonplace phrases ... to

specify the domain with any useful precision, and definitions of composites which determine . . . exactly what variables do or do not fall within their scope virtually never occur in practice." (pp. 198–199).

Rozeboom goes on to further examples and into a technical argument whose burden is essentially that a universe score cannot be estimated unless the elements in the universe are positively, if not uniformly, intercorrelated. His critique—which we have been able to represent only by much-too-short excerpts—ends on the note that an attempt to appraise performance on a universe from a sample "is in actuality being surreptitiously construed to measure a *theoretical* variable hypothesized to unify the composite's domain. Hence consideration of content validity ineluctably feeds into the problems of *construct validity*." (p. 205). As Rozeboom goes on, he clarifies that he thinks well of construct interpretations, and of construct validation as a scientific rationale.

We turn next to Loevinger, who is also noted for her advocacy of construct validation of tests. Her Presidential address to psychologists concerned with measurement (Loevinger, 1965) objects to applying the row-and-column symmetry of analysis of variance to persons-and-tests.

A person retains recognizable identity through slight superficial changes, while a test may not. We are always clear whether we are confronted by one person or two people. People do not, in front of one, shade off into one another imperceptibly. Tests, on the other hand, may differ in minor ways whose significance is doubtful. [p. 145.]

My objection to all psychometric developments that assume random sampling of items or tests is in the first instance that they grossly misrepresent the actual case, which is almost invariably expert selection rather than random sampling. But there is also implied in my argument a subtler and deeper point. The term population implies that in principle it is possible to catalogue or display or index all possible members . . . . Tests and items are not that sort of thing. There is no meaning to talking about populations of tests or items. No system is conceivable by which an index of all possible tests could be drawn up; there is no generating principle. [p. 147.]

To be sure, science has often advanced by breaking out of old meanings and opening new possibilities . . . . Obviously, population has also been progressively redefined . . . . Perhaps what it now means is a class of objects or events, usually hypothetical, that can be randomly sampled . . . . Thus the burden of definition is shifted to the term random . . . . Crucial points for me are the difference between random and expert selection and that sampled objects should maintain a continuing identity. [pp. 153–154.]

We concur with the critics that if one claims that his observations represent a universe he ought to define the universe clearly. His readers should agree that the conditions appearing in his G study fit within the universe and that they are reasonably distributed over its whole range, not confined to some narrow subuniverse. The critics see possible conflict between generalizability and construct validity, yet we agree with them that it is well to measure any one concept by operations that are phenomenally diverse (Campbell & Fiske, 1959; Cronbach & Meehl, 1955). We cannot satisfy those critics who insist that statistics derived from the random-sampling model may be applied only to data collected by random sampling. On this point, we have already aligned ourselves with Cornfield and Tukey.

Much of the doubt expressed regarding the reference to universes apparently relates to the idea of random sampling of items from an aggregation of rather diverse content. While this theme was dominant in many early papers, the concept of stratification of item content was present even in the papers of the 1950's and is now far more prominent. Stratification does not loom large in the *mathematics* of generalizability theory, because random sampling within a stratum is still random sampling from a universe of items. The investigator proposes to assemble data from fixed strata into a composite observation, using predetermined weights. Accordingly, the stratified test can be examined with the aid of "univariate" generalizability theory, treating items as nested within fixed strata in the universe, or one can apply multivariate generalizability theory, considering each stratum as defining a new variable. The objections critics have made regarding random sampling seem not to apply, or to apply with very little force, to sampling from the universe of stratified tests.

A proposal to sample items from a broad domain at random is generally but not always a sign that one's understanding is crude. That is, one is employing a crude construct, and will willingly move on to purer variables when possible. But even when theory is highly refined one will use relatively global constructs where they are adequate for a proposed inquiry. In many studies of nutrition, for example, it may be sufficient to speak of intake "of fats," rather than to record separately the "saturated fats" and "unsaturated fats" or to move to several still finer constructs such as "olein," "lanolin," etc. Purified constructs, even when available, are not used in every scientific conclusion where they might apply. After one learns what is aggregated in his global construct, he still retains it—especially in applied measurement and research.

That generalizability theory blurs the distinction between reliability and validity distresses Thorndike and Rozeboom. We, on the other hand, feel that generalizability theory—even though it is a development of the concept of reliability—clarifies both content and construct validation. It is important to pursue this issue. A paper by Campbell and Tyler (1957) seems pertinent. At that date they were writing without reference to our theory, but they were concerned with reliability over observers and occasions, not merely over test forms.

A given scientific construct has multiple operational specifications. If, as sampled, these operational specifications concur, the construct and the sampled measurement techniques have validity. Constructs for which diverse operational specifications persistently fail to agree are in the long run modified or abandoned. In the physical sciences, . . . methodologically independent operational specifications of the same construct may agree on the order of .99 . . . over a population of instances . . . .

Construct validity thus becomes the correlation among two or more independent measures as conceptually identical in their referents as possible. But the distinction between reliability and validity is still a very important one to retain. Insofar as the measures share the same apparatus or the same approach, they tend to share correlated error variance, or common variance due to features of the apparatus which are irrelevant to the construct in question . . . . *Reliability* is epitomized by the correlation of two specifications of a construct through maximally *similar* approaches. *Construct validity* is epitomized by the correlation between two or more specifications of a construct maximally *different* in apparatus or method. [p. 91.]

Our position differs little from the position all the critics mentioned accept. The operational definition of an indicator or a body of content to be sampled defines a universe. Generalizability analysis indicates how closely different realizations of the definition correspond to each other and to the universe score. The more definitively the developer of a measuring procedure describes the universe he has in mind, the greater the degree of generalizability he can hope to attain with a sample of practicable size. The question remains whether he has defined a universe particularly worthwhile investigating. To answer this, one asks for convergence of *diverse* indicators. One does not move far toward construct validation until he has established convergence of indicators of different kinds, each representing its own universe of admissible operations (Cronbach, 1971).

A universe can encompass diverse situations. Recall our suggestion that the subtests of the Wechsler Verbal Scale represent a larger domain of verbal tasks. One who wishes to restrict his interpretation to the fixed subtest tasks, singly or together, may do so. It is arguable whether it is more parsimonious to attribute properties of the data to the features of the several subtests or to assume that their unique features can be neglected and that "verbal ability" is a suitable interpretative construct. A critic may object to random sampling, and propose to define verbal ability in terms of subcategories from which tasks could be systematically sampled. This would seem to be a forward step, as generally occurs in the evolution of constructs. Similarly, a critic might propose to replace the universe score (which is something like a centroid factor of the subtests), with a factor defined by some other dimension in the same space. This too is acceptable within generalizability theory.

Because of the latitude allowed for universe definition, Ross and Lumsden (1968) consider generalizability theory to have "abandoned" the problem of reliability. They want to consider tests that "measure an attribute," in that the observed score is a monotone function of a latent unidimensional score plus a random error. As they point out, our readiness to investigate *any* operationally defined universe means that we deal with many a universe that does not, for them, appear to fit an acceptable theoretical construct. Ross and Lumsden directly pursue the problem of inferring values of the latent variable as a nonlinear function of the observed variable. They see this as a more interesting problem than the one we have addressed: "The outcome of the attempt to set up true score as an ideal is, in point of fact, the position reached by Cronbach, *et al.*, which attributes to the test as many true scores as universes which may be found for it. In brief, then, while true score may suffice as a conceptual tool with which to examine the exterior properties of a test (its relations to other tests and criteria) it will not suffice for examining the intrinsic and singular interior properties of the test." Such faint damns appear to us as praise, and we cheerfully assent to the characterization.

We do see generalizability analysis as having some value for theory. Generalizability analysis indicates which facets contribute strong main effects or strong interactions with persons. Suppose, for example, it is found that in peer ratings there is a substantial subject–rater interaction component; then one may generalize to a universe of undifferentiated raters, but must employ many raters to obtain adequate generalizability. However, the finding should impel the investigator to ask what rater characteristics contribute to the interaction. This would lead him to subuniverses representing types of raters. Whether he reassembles the types by some stratified sampling plan is less important than the fact that the study has increased his understanding of social perception.

We see research as gradually improving the specification and mapping of the universe of interest. As a construct becomes better understood, it becomes possible to define the admissible observations so that they will be highly comparable. Stratified sampling plans and various types of calibration or correction serve just this purpose. When the investigator's generalizability study shows that he is successfully controlling a large number of unwanted influences, that in itself is partial evidence of validity. A score can, alternatively, have a high degree of generalizability without having a high degree of validity, either because the universe has been narrowed without scientific or practical justification, or because a very large sample of conditions has been drawn.

## E. *Evaluation of Generalizability Theory*

### *Applied testing and social policy*

A mathematical model is socially neutral, even though the flight of cannon-balls was Galileo's starting point. Any discussion of social values in a technical monograph would have been considered out of place two decades ago, but times have changed. Today it is obvious that even to publish a report dealing with behavioral measurement is to express a conviction that measurement can be socially beneficial. For writers to pretend that they have no views about the application of the devices they describe is disingenuous and patronizing of the reader. Worse, it is self-destructive, because the writer who conceals his social views leaves himself open to the critic who may impute a malign policy to him. Discussion of policy is imperative in the present monograph since the technical devices we suggest may have re-gressive social consequences if applied thoughtlessly.

Testing of all kinds has been under heavy attack in recent years. A good part of the criticism has been well motivated and well informed. Since the critics have made many points testers must take seriously, it is foolish to envision a return to the status quo ante. New values are in the ascendancy, and political, educational, and business institutions are being reshaped. The victories that progressive social critics win over the old system will have to be celebrated with some sacrificial victims, and among the casualties will be some uses of tests. Because of minority-group objections some American school boards have greatly curtailed the administration of general mental tests. In a more violent confrontation abroad, the Japanese New Left threw over an innovative and liberalizing program of college aptitude tests, but one that symbolized the Establishment to them.

Much of the attack of tests represents an egalitarian doctrine. Equality of opportunity, dignity, and justice is to be promoted by every possible means. But to insist that equality requires identity and that there is in fact no differ-ence among persons is to deny reality. One can make tests taboo, if they embarrass one's doctrine by continually making unwelcome differences evident; but to do so is only to force decisions of employers, teachers, physicians, and policy makers back onto biased, inaccurate impressions.

Social planning, evaluation of social programs, and planning treatments for individuals require information. Some of the information is of the generalized sort that comes from scientific research, and some is local and personalized. The better the information of either type, the sounder the planning and the greater the social benefit. The social order that is evolving will have as much need for social and behavioral measurement as did that of the 1940's, though it probably will ask different questions. Already, the educational romantics are beginning to ask how they can collect evidence to evaluate their neo-Summerhills. Computer-aided instruction in reading seems to be particularly beneficial to children coming from poor, non-reading families—and it makes intensive use of week-by-week measurement. If the much-talked-of tuition vouchers come into being, so that each family can pick and choose among schools for its children, the very diversity of opportunity will make the parent an avid consumer of facts about the results schools achieve. Such diversification of education must also renew interest in differential aptitudes.

Application of generalizability theory should operate ultimately to increase the accuracy of test interpretations. It will make interpretation more cautious as the inadequate generalizability of a procedure becomes recognized, and it will encourage the development of procedures more suitable for generalized interpretation. Item-sampling designs in evaluation, for example, make it far more practical to collect evidence on diverse outcomes of a course. With such data one can recognize when an innovative program is truly producing effects qualitatively different from those of a traditional course. Equally, such data enable one to detect undesirable side effects. The reorganization of aptitude and interest profiles in terms of estimated universe scores will suppress some pseudoinformation, and so make guidance more intelligent. Moreover, it will lead to a redistribution of effort, because some instruments now invest much effort in measuring a verbal–educational core of ability that is already reasonably well estimated by ordinary school records. Once multivariate estimation of universe scores makes this redundancy clear, effort will be redirected to the study of other educationally significant outcomes.

Almost all the uses of tests that are now attracting increasing interest demand absolute, content-referenced, or criterion-referenced measurement. It is the level of the graduates' performance, and not their standings relative to each other, that one requires to evaluate early childhood education. It is absolute measurement that one uses to prescribe next week's instructional activities in computer-based instruction, and in testing to see if a standard is met. Any effort to facilitate the learner's self-evaluation should emphasize an absolute multidimensional report on his performance. In another area, there is a role for instruments that report interest profiles in absolute terms (Cronbach, 1970, pp. 486–488). The emphasis in generalizability theory is on the universe score itself and not on the comparisons of individuals stressed by classical test theory.

Point estimates of universe scores are pertinent to either absolute or comparative decisions. Such estimates are higher than observed scores for some persons and lower for others. If subgroups are used, the fact that the person moves toward the mean of his own demographic group can have dramatic consequences. Persons with the same observed score will obtain different universe-score estimates, the higher estimates going to persons whose group has the higher mean.

Scores of blacks and whites are likely to be differentially affected. Similarly, children whose parents have little education are likely to do worse than children of educated parents. Regressing scores will lower the apparent standing of any child in the former group whose observed score is unusually high for that group. To be sure, regression will also tend to bring down the very high scores among the children of educated parents, and to bring up scores in the lower end of the distribution for children of undereducated parents. Accordingly, on the average, regressing scores does not alter the group difference. But it does alter the apparent standings of children from different backgrounds in that region of the scale where their scores are most likely to overlap, and where they are most likely to be compared with each other.

Although regressing scores would have unfavorable consequences for children or adults who are above-average members of disadvantaged groups, it is not a distortion of the available evidence. The statistical logic is clear: if exhaustive measurement of each individual were carried out, there would be the same kind of rearrangement of ranks, and the same kind of reduction in the overlap of the groups. If thorough measurement is to be regarded as biased because it reports a difference that exists, then the only procedure that could be called unbiased is a completely inaccurate measurement; scores determined wholly by chance exhibit no group difference. We cannot defend a measurement procedure that knowingly conceals a difference. The regression procedure gives as accurate a report as we can give.

There is something repugnant about a correction procedure that seems to penalize a person by virtue of his color or his parents' education. The feeling is a consequence of the view that high scores are beneficial to the individual. Scores should not be used in such a way that attaining a high score *in itself* confers benefits. If we were to look on the score not as a measure of merit but as a measure of need, then the more accurate (but lower) regressed score would be conferring a benefit on the individual from the disadvantaged group. If the person with the greater need is to have the greater educational resources expended on him, full disclosure of his present educational deficit would be to his advantage.

In personnel decisions, it is not of central importance to estimate the universe score on a predictor variable. The real question is what outcome is to be expected, if the person is accepted and assigned to a job, or accepted for a particular instructional program. The decision rule ought to be established on the basis of the joint distribution of test score and criterion score within each group, the cutting score being located so that the persons at the cutting score have an acceptable probability of reaching satisfactory criterion performance. The cutting score is likely to be different in the various groups. The rule can be expressed in terms of the observed score or in terms of the estimated universe score, but if properly determined, the two rules will select the same members of the group, and hence it is pointless to regress the observed score toward the group mean.

The justifiability of the regression procedure will vary from context to context. Because a serious attempt to apply this technique has never been made, despite its long years of good standing in test theory, regressed scores are likely to be misunderstood even by professional test users. A period of exploratory use of both regressed and unregressed scores side by side may be the best way to arrive at a sensible policy.

The task of measurement theory is to improve the quality of information. Score information alone does not necessarily dictate what action should be taken. Even if, after all pertinent statistical corrections are made, the information about two persons is the same save for their group membership, it may be appropriate to treat them differently. Such a strategy may involve "bias," but we see no objection to bias so long as it is open and based on thoughtful policy. Preferential hiring, quotas, and other devices intended to redress social balance ought to be open and explicit, where they can be recognized for what they are. To use poor estimates of scores in place of better estimates as a covert way of accomplishing social balance seems unlikely to prove benign in the long run.

### Scientific implications

Generalizability theory is both familiar and exotic. Its basic ideas have been well known to behavioral scientists at least since Fisherian analysis of variance came into prominence late in the 1940's. Even though the applications of these ideas to measurement theory have with rare exceptions been limited to simplified cases, many of those applications are well known and embody most of the concepts with which this monograph deals. However, in the attempt to build a relatively comprehensive system, with assumptions somewhat more realistic than those of previous developments, we have constructed a maze of argument in which one can easily lose himself, and which is very difficult to see as a whole. Complicated as our argument is, it by no means goes into all the matters that will ultimately have a place in the theory.

A large part of our development is a restatement of classical theory in more general terms, displaying how the argument would look if derived from weaker assumptions. In the end, we are often forced back upon some of the classical simplifications in order to reach one or another conclusion. Fortunately, the reader who wishes to use traditional formulas and traditional language may do so because classical theory remains, as a special case, entirely consistent with our model. We do not expect the words "reliability," "true score," and "standard error of measurement" to disappear from the tester's vocabulary. But the reader who has grasped the perspective of generalizability theory will thereafter see those concepts in a different way. He will recognize that error and true score can be defined in almost innumerable ways. He will recognize that any sentence that refers to "the" reliability coefficient is simplistic, and almost certainly wrong. Even if he carries out only single-facet, single-variable studies of instruments, applying the most conventional of formulas, generalizability theory should induce him to formulate his question more thoughtfully before proceeding to collect data.

In the course of what once set out to be a simple free-style retelling of a familiar plot, we found the familiar topics taking on a most unfamiliar aspect. The "true IQ" is a hero-figure as well known to us as the Lone Ranger; we try to tell about it and suddenly realize that not even Wechsler himself knows what the "true Full Scale IQ" might mean. The correction for attenuation, we find, takes on as many identities as the Old Man of the Sea. The regression estimate of the universe score has always been in the cast of the psychometrician's ritual, but has never been given lines to speak; in the present theory it finds itself thrust to the center of the stage, as fraught as Hamlet with grand messages and grander uncertainties. The legend of the confidence interval, which once flowed so smoothly from the tongue, has to be cut from the plot as implausible. Surely the story of reliability theory will never again have its old meaning for our readers.

Generalizability theory offers endless possibilities for elaborate studies to be carried out in the course of instrument development. To exploit the full possibilities of these investigations, the investigator will have to develop his own skill through experience. Even though we have tried to present methods in cookbook form, virtually never does one of our illustrative studies prove to be a routine application. Every measuring device has its idiosyncrasies that must be considered in designing the study and in reflecting on the data. Analysis of variance components is much like factor analysis in this respect; how one designs his study, what decisions he makes at various points in the analysis, and what psychological background he brings to the interpretation do much to determine what he will get out of the inquiry. Cookbook procedures can be offered, but sophisticated inquirers will depart from them in almost every application.

Generalizability information offers welcome possibilities for improving instruments, possibilities that could not be attained through a simpler formulation. First, there is the possibility of identifying the conditions largely responsible for inconsistency from one observation to another. This both explains the character of what is measured and allows one to reduce the inconsistency by taking a better sample. Second, there is the power of multifacet design to adapt one's procedure for data collection to suit his purpose. Instead of attempting to create an instrument that will "have high reliability" forever after, he adjusts the crossing and nesting of facets, the extent of sampling, etc., to fit the decisions for which he will use the scores. Item sampling is just one of the variants on this theme. Third, the use of approximate regression estimates seems to promise more accurate conclusions than would be drawn by alternative methods. This is *a fortiori* the case when a profile or a composite score is to be interpreted.

Our theoretical ideas have changed during the long period in which we have been trying to put them into a coherent form. The statements offered here differ, in emphasis and sometimes in content, from our formulation of three or more years ago. Often it has seemed as if the theory were imposing itself upon us. Certainly we never "set out to solve" many of the problems this monograph deals with; rather, the solution and the problem tumbled simultaneously out of the machinery. We have found some of our reasoning incorrect, usually because we were unwittingly employing old habits of thought to interpret elements of the new model. It is possible that inconsistencies remain in the present argument. Even if everything we say now stands up under criticism, generalizability theory will change as measurement specialists begin to apply it and sense possibilities that have not called themselves to our attention.

Today's reader, coming to a fully elaborated generalizability theory for the first time, no doubt finds it forbidding. As measurement specialists become accustomed to its language and its ways of treating data, this strangeness will pass. As the theory is put in different words by successive writers, it will be rounded into smoother form. As it becomes better integrated with other recent developments in error theory, and with the validation theory of which it is a part, it will become inseparable from the measurement theory of the next generation.

## EXERCISES

**E.1.** Tuddenham obtained ratings of adolescents, given by two raters each of whom studied a different file of background material on the subjects. In a later study Kagan and Moss also obtained ratings of adolescents; their various raters all worked from the same files of material. Kagan and Moss regarded their technique as

superior because they obtained higher reliability coefficients (rater agreement) than Tuddenham.
a. Use the language of generalizability to discuss the choice between techniques.
b. Comment on the remark of Honzik (1965) that the Tuddenham technique is "more akin to a validity coefficient."

**E.2.** Illustrate, with respect to a simple test of typing speed to be used in an employment office, the following:
a. A vague universe definition.
b. A clear universe definition where the universe is broadly inclusive.
c. A clear definition of a narrow universe.

**E.3.** Show that rifle marksmanship can be defined in terms of broad or narrow universes. For what purposes is it useful to make the universe extremely narrow?

**E.4.** A reading test, used to advise students whether or not they should consider taking a remedial reading course, presents selections of about 200 words from college texts and asks four questions on each selection to test comprehension.
a. Should the universe of generalization be defined as: i. selections from the whole range of college textbooks and questions on these selections, or ii. as the universe of questions that might be asked about the selections in the test?
b. Assuming that an internal-consistency G study is conducted, how does the answer to *a* affect the procedure or analysis?
c. Assuming that a parallel-form G study is conducted, how does the answer to *a* affect the procedure or analysis?
d. What arguments can be offered for and against compiling separate tests for commerce majors, science majors, and humanities majors? (Confine your attention to considerations of generalizability and validity.)
e. Suppose that there is a single test for all fields. It is found that means of students in different majors vary considerably. Is it fair to use the point-estimation technique, regressing to the mean of students entering that field and reporting estimated universe scores rather than observed scores?

**E.5.** The following statement is made by R. B. Cattell (Cattell, Eber, & Tatsuoka, 1970, p. 32), who measures such personality traits as humility or conservatism by assembling a number of items that represent rather different aspects of the trait into a score.

> If one wishes to create high homogeneities (and call them reliabilities!) as some test handbooks do, it is easily possible to do so by multiplying the writing of very similar items. But any broad and important personality trait has to be assessed across a wide variety of areas and forms of expression. Furthermore, even from a purely psychometric point of view, the highest multiple-*R* validity is obtained by finding items that correlate consistently with the factor, but trivially with one another.

Comment on the passage from the standpoint of generalizability theory. Can Cattell's ideas be expressed in terms of this theory, or is there a contradiction?

**E.6.**  Where there is an excellent indicator of a construct, one often carries out a concurrent validation study to evaluate a substitute (usually a cheaper) indicator. Accordingly a household thermometer may be checked against a well-calibrated thermocouple. In such a study, the latter is referred to as the criterion. Defend or criticize the statement: "A universe score may be regarded as a criterion score, as that term is used in concurrent validation."

**E.7.**  Consider the following passage. Is it true that to maximize validity one may be required to reduce the generalizability from observed score to the universe score of interest?

> [With regard to reliability and validity:] in order to maximize one, you may be required to reduce the other . . . . When one is developing a test it is natural to retain those items which correlate highly . . . with the total score. This has the effect of making the test more homogeneous and more reliable.
>
> However, it is almost always the case that the criterion behavior, which is to be correlated with the test scores, is not a homogeneous, single-factor behavior . . . .. To measure these skills, one needs to retain such diverse items in the predictor test. By retaining such items, the reliability will undoubtedly be reduced." [Dick & Haggerty, 1971, p. 137.]

**E.8.**  "For the test interpreter, the universe of generalization is a construct." Which of the following comments best amplifies and defends this statement?
a.  The universe consists of observations that have not been made.
b.  The universe is hypothetical; its elements could not all be listed, even by exhaustive effort.
c.  The universe definition reflects someone's working hypothesis that sentences embodying that concept will provide a useful description of reality.
d.  The universe definition implies a theoretical assertion that all observations in the universe measure the same thing; that is, all are influenced by the same actions or characteristics of the subject.

**E.9.**  Nunnally (1971, Chapters 5 & 6) identifies generalizability with both reliability and validity. Any construct, he says, has to be identified by the theorist with a specified "domain of observables." One would verify the adequacy of the concept by demonstrating that the several observables, each of which is determined by a distinct kind of measurement, perform similarly. For the construct of "fear," one would check whether, under given conditions, persons' scores on the several measures correlate. Also, one would determine if the mean scores on the measures show similar trends when the level of threat is manipulated experimentally.

Is Nunnally's view (much abbreviated here) consistent with our concept of generalizability?

### Answers

**A.1.**  a. Tuddenham used a G-study design $(r,f) \times p$ in which raters $r$ and files $f$ were confounded. His coefficient treats both sources of error together, and investigates the adequacy of generalization over the universe of raters and that of files of material. Kagan and Moss held files constant. That is, their design could be described as $r \times p$ with files as a hidden facet; or as $r \times f \times p$ with $n_f = 1$. Their analysis gives information only on the $r$ and $pr$ components of variance within the file, and, in effect, investigates how well one can generalize over raters when the files are of the type used in their study. The more limited universe of generalization is sufficient to account for the greater size of their coefficient. A more complex design, with at least two raters for each file, and with more than one file per subject, would be needed to arrive at any comparison of the two rating methods.

b. It makes sense to use the term validity rather than generalizability for the Tuddenham study if the two files of material on the subject are of different kinds; for example, one representing psychological tests and one representing notes by an observer who visited the school. If the two files are regarded as reasonably similar in character (e.g., each set includes a sample of test data and a sample of reports from acquaintances), the term generalizability seems more natural. Perhaps the easiest way to make this discrimination is to ask: If there were three sets of files, would it make sense to consider the three intercorrelations separately because the sets are so distinctive in character, or would a single intraclass correlation for the three files give the pertinent information? It is in the latter case that generalization over the universe of further files is a plausible idea.

**A.2.**  a. Samples of English text to be typed.

b. Measures of speed in typing material of the kind assigned to typists in office typing pools, using one of the common office machines, with performances that involve more than two errors per hundred words not being counted. (This of course could be specified still more precisely.)

c. Measures of speed in typing outgoing letters in the wholesale grocery business, using an IBM Model Q typewriter. (Error control to be added as in *b*. One might think of stratifying to produce sets reflecting different kinds of correspondence.)

**A.3.**  The basic purpose of training may be to prepare the rifleman to hit the sorts of targets he will encounter in the field, where lighting conditions, terrain, and target motion vary. This is a broad universe. A narrow definition would specify one kind of target, one distance, a modest range of lighting conditions, and one position of fire. It probably would specify the model of rifle to be used. The narrow definition would generally be preferred in any comparative assessment, for example, an experiment comparing two training techniques, comparing two models of rifle sight or testing the effect of lack of sleep on steadiness. Also, for a competition in marksmanship (though a stratified universe including more than one narrowly specified condition would usually be preferred).

For these purposes there is likely to be some merit in selecting whatever particular conditions are most favorable to markmanship, and fixing them. But the broader universe would be preferred when training is being evaluated, because one wants to forecast the overall operational efficiency of graduates in the field, where conditions require adaptability. In general, the greater the stress on realism in the decisions to be made, the less satisfactory a narrow universe will be

**A.4.**    a. It is hard to see any basis for preferring ii as the universe.

b. Selections should be taken as the unit of analysis, or the analysis should treat the data as having a (questions:selections) × persons design. One could also split-score the test by dividing selections into alternate halves. Person–selection interaction must be treated as one of the sources of error.

c. The two forms must use different selections, not different questions on the same set of selections.

d. If one assumes that students in these fields read such different books that comprehension will vary considerably depending on the kind of book used in the test, separate forms are advisable. The universe seemingly ought to be the books students are likely to have to read, and students in different majors overlap considerably in the courses they are expected to take. This argues against separate tests. Even if there were little overlap, it might reasonably be argued that comprehension of texts is a general skill that generalizes over categories of subject matter.

e. By the rationale of this chapter, regressing toward group means is fair. The mean square error of generalization is reduced, and the student who by luck earns a score well above the mean for his group is not given false encouragement. It would be well, of course, to interpret the score with relation to expectancy tables determined for the separate groups; if that is done and all persons take a single test, regression is unnecessary.

**A.5.**    Generalizability theory recognizes a coefficient obtained from an $i \times p$ design as information about generalizability over randomly sampled items. If an investigator uses a different procedure for assembling items into a test, it is possible to apply that procedure to two or more items and to form separate tests; that investigator clearly wants to generalize over tests formed in that manner, not over randomly generated tests. Where Cattell is using something akin to a multiple-regression procedure to decide what items to retain, the two forms for the G study would have to be formed by applying the method to two separate but randomly parallel item pools. Generalizability theory is neutral with respect to the question as to whether an instrument should measure a broadly defined trait; the machinery can be adapted to Cattell's purposes and also to the purposes of a person theorizing about a very narrow trait that coincides with a homogeneous universe of items.

**A.6.**    The statement is acceptable only if the universe represents what the investigator truly desires to measure. If he is interpreting his result as a measure of "temperature," the universe score for thermocouples is a good approximate criterion. To be sure, it is an unobserved and unobservable criterion, and it is not a perfect criterion, because there are still better (and more expensive) laboratory devices to measure temperature. The universe score for the rather questionable household thermometers, which may be biased in various ways, is definitely not regarded as a criterion. In general, it appears that to call the universe score a criterion is inadvisable.

**A.7.**    This passage is sound, within the limits of traditional ways of talking about internal-consistency coefficients as representing "the reliability." This passage is in accord with Cattell's view (**E.5**), but more traditionally worded. The coefficient of generalizability for a test that is deliberately constructed to cover a broad universe (whether the construction plan is stratified or not), ought to refer to that universe, not to the universe of tests from which low-correlating items have been culled. To be sure, a heterogeneous test of length $n_i$ will ordinarily correlate less with the universe score for tests randomly parallel to it than a homogeneous test of length $n_i$ with the universe score for a set of such homogeneous tests. But these two numbers answer different questions, and no meaningful comparison can be made. (The comparison is only a shade less absurd than asking "Did it rain more than it was windy, last night?") This is precisely the issue raised in **E.1** above.

**A.8.** A construct is a term that (together with other constructs) describes how a person "construes" a phenomenon. It is a part of his theory (which may be elaborate or vague). He uses it because he finds that he can make statements that summarize his experience and help in analyzing new events. Accordingly, **E.8.c** is closest to the heart of the meaning, and it is true that universes are chosen for this reason.

Choices **E.8.a** and **E.8.b** are irrelevant. Under certain circumstances, a construct is a grouping of events or phenomena that have already been observed. ("The French Revolution" is a construct, though not a scientific one. So is "the gross national product in 1932.")

Comment **E.8.d** may or may not be largely true. If the construct is something like "Jones' reputation among the American public," one does not expect homogeneity of cause or description, yet this may be a most useful construct for the behavioral scientist. Even with a strictly scientific construct, it is understood that we are dealing with an abstraction, not a "thing." Consider "blood sugar level" as measured by any standard procedure. Clearly, despite the purity of the concept, the phenomena it reflects in different persons and at different times are quite diverse.

**A.9.** We tend to restrict the term generalizability to the case where the operations are phenomenally similar. With regard to "fear" we would not speak of generalizing over a universe that embraces measures of skin conductivity, self-report, heart rate, and observers' ratings of disorganization. But we would willingly consider a universe of measures of skin conductivity taken simultaneously at different places on the body—though, in line with Nunnally's logic, we would have to revise that working construct if we found that different parts of the body respond differently to experimental stress. We do not use the term generalizability as broadly as Nunnally, but the disagreement is merely terminological.

# Appendix

**Proof regarding covariance components**

Each expression in (8.1) and (8.2) is a special case of what might be written as:

$$\mathscr{E}\sigma(X_{pij}, X_{pi^+j^+}) = \sigma^2(\mu_p\!\sim) + \sigma(\mu_{pi}\!\sim, \mu_{pi^+}\!\sim) + \sigma(\mu_{pj}\!\sim, \mu_{pj^+}\!\sim)$$
$$+ \sigma(\mu_{pij}\!\sim, \mu_{pi^+j^+}\!\sim) + \sigma(e_{pij}, e_{pi^+j^+})$$

where $i^+$ may be interpreted as either $i$ or $i'$, and $j^+$ as either $j$ or $j'$. Each of the four possible interpretations gives rise to either (8.1) or (8.2), because components of covariance for $i$ with $i'$ or $j$ with $j'$ vanish in the expectation, and a covariance of like components is a variance. The following is a proof for the case where $i^+$ is interpreted as $i'$ and $j^+$ is interpreted as $j$.

(1) $$\mathscr{E}_{\substack{i \neq i' \\ j}} \mathscr{E}\sigma(X_{pij}, X_{pi'j}) = \mathscr{E}_{\substack{i \neq i' \\ j}} \mathscr{E}\mathscr{E}_p (X_{pij} - \mu_{ij})(X_{pi'j} - \mu_{i'j})$$

(2) $$X_{pij} = \mu + (\mu_p\!\sim) + (\mu_i\!\sim) + (\mu_j\!\sim)$$
$$+ (\mu_{pi}\!\sim) + (\mu_{pj}\!\sim) + (\mu_{ij}\!\sim) + (\mu_{pij}\!\sim, e)$$

(3) $$\mu_{ij} = \mu + (\mu_i\!\sim) + (\mu_j\!\sim) + (\mu_{ij}\!\sim)$$

(4) $$X_{pij} - \mu_{ij} = (\mu_p\!\sim) + (\mu_{pi}\!\sim) + (\mu_{pj}\!\sim) + (\mu_{pij}\!\sim, e)$$

Similarly for $X_{pi'j} - \mu_{i'j}$.

(5)    $(X_{pij} - \mu_{ij})(X_{pi'j} - \mu_{i'j})$

$$= (\mu_p\sim)[(\mu_p\sim) + (\mu_{pi'}\sim) + (\mu_{pj}\sim) + (\mu_{pi'j}\sim,e)]$$
$$+ (\mu_{pi}\sim)[(\mu_p\sim) + (\mu_{pi'}\sim) + (\mu_{pj}\sim) + (\mu_{pi'j}\sim,e)]$$
$$+ (\mu_{pj}\sim)[(\mu_p\sim) + (\mu_{pi'}\sim) + (\mu_{pj}\sim) + (\mu_{pi'j}\sim,e)]$$
$$+ (\mu_{pij}\sim,e)[(\mu_p\sim) + (\mu_{pi'}\sim) + (\mu_{pj}\sim) + (\mu_{pi'j}\sim,e)]$$

(6)    $\underset{p}{\mathscr{E}}(\mu_p\sim)(\mu_p\sim) = \sigma^2(p);$     $\underset{p,j}{\mathscr{E}}\,(\mu_{pj}\sim)(\mu_{pj}\sim) = \underset{p,j}{\mathscr{E}}\,(\mu_{pj}\sim)^2 = \sigma^2(pj)$

All remaining expected products in (5) reduce to zero.

(7)            $\underset{i\neq i'}{\mathscr{E}}\,\sigma(X_{pij}, X_{pi'j}) = \sigma^2(p) + \sigma^2(pj)$

Two examples of the argument regarding reduction of products to zero should suffice:

(8)      $\underset{p,i'}{\mathscr{E}}\,(\mu_p\sim)(\mu_{pi'}\sim) = \underset{p}{\mathscr{E}}\,[(\mu_p\sim) \cdot \underset{i'}{\mathscr{E}}(\mu_{pi'}\sim)] = \underset{p}{\mathscr{E}}\,[\mu_p\sim \cdot 0] = 0$

(9)      $\underset{p}{\mathscr{E}}\,\underset{i,i'}{\mathscr{E}}\,(\mu_{pi}\sim)(\mu_{pi'}\sim) = \underset{p}{\mathscr{E}}\left[\underset{i}{\mathscr{E}}(\mu_{pi}\sim) \cdot \underset{i'}{\mathscr{E}}(\mu_{pi'}\sim)\right] = 0$

# Bibliography

Abelson, R. P., "Scales Derived by Consideration of Variance Components in Multi-Way Tables," in *Psychological Scaling: Theory and Applications*, pp. 169–181. Edited by H. Gulliksen and S. J. Messick. John Wiley & Sons, Inc. New York, 1960.

Belgard, M., Rosenshine, B., and Gage, N. L., "The Teacher's Effectiveness in Explaining: Evidence on Its Generality and Correlation with Pupils' Ratings and Attention Scores," in *Research into Classroom Processes*, pp. 182–209. Edited by I. Westbury and A. A. Bellack. Teachers College Press, New York, 1971.

Belson, W. A., "The Effects of Television upon the Interests and the Initiative of Adult Viewers." Paper presented to the British Association for the Advancement of Science, 1956.

Blalock, H. M., Jr., "Estimating Measurement Error Using Multiple Indicators and Several Points in Time," *Am. Sociol. Rev.* Vol. 35, pp. 101–111, 1970.

Blalock, H. M., Jr., and Blalock, A. B. (Eds.), *Methodology in Social Research*. McGraw-Hill Book Co., Inc., New York, 1968.

Block, J., "The Equivalence of Measures and the Correction for Attenuation," *Psych. Bull.*, Vol. 60, pp. 152–156, 1963.

Bock, R. D., "Multivariate Analysis of Variance of Repeated Measurements," in *Problems in Measuring Change*, pp. 85–103. Edited by C. W. Harris. University of Wisconsin Press, Madison, Wis., 1963.

Bock, R. D., "Contributions of Multivariate Experimental Designs to Educational Research," in *Handbook of Multivariate Experimental Psychology*, pp. 820–840. Edited by R. B. Cattell. Rand McNally, Chicago, Ill., 1966.

Bock, R. D. and Wiley, D. E., "Quasi-Experimentation in Educational Settings: Comment," *School Rev.*, Vol. 75, pp. 353–366, 1967.

Bormuth, J. R., *On the Theory of Achievement Test Items*. University of Chicago Press, Chicago, Ill., 1970.

Boruch, R. F., and Wolins, L., "A Procedure for Estimation of Trait, Method, and Error Variance Attributable to a Measure," *Educ. Psych. Meas.*, Vol. 30, pp. 547–574, 1970.

Bouchard, T. J., Jr., "Convergent and Discriminant Validity of the Adjective Check List and the Edwards Personal Preference Schedule," *Educ. Psych. Meas.*, Vol. 28, pp. 1165–1171, 1968.

Box, G. E. P., and Tiao, G. C., "Bayesian Estimation of Means for the Random Effects Model," *J. Am. Stat. Assoc.*, Vol. 63, pp. 174–181, 1968.

Brunswik, E. *Systematic and Representative Design of Psychological Experiments*. University of California Press, Berkeley, 1947.

Bulmer, M. G., "Approximate Confidence Limits for Components of Variance," *Biometrika*, Vol. 44, pp. 159–167, 1957.

Burdock, E. I., Fleiss, J. L., and Hardesty, A. S. "A New View of Inter-Observer Agreement," *Personnel Psych.*, Vol. 16, pp. 373–384, 1963.

Buros, O. K. *Schematization of Old and New Concepts of Test Reliability Based Upon Parametric Models*. Gryphon Press, New Brunswick, N.J., 1963 (dittoed).

Burt, C., "The Analysis of Examination Marks," in P. Hartog and E. C. Rhodes, *The Marks of Examiners*, pp. 245–314. The Macmillan Company, London, 1936.

Burt, C., "Test Reliability Estimated by Analysis of Variance," *Brit. J. Stat. Psych.*, Vol. 8, pp. 103–118, 1955.

Cahen, L. S., Romberg, T. A., and Zwirner, W., "The Estimation of Mean Achievement Scores for Schools by the Item Sampling Technique," *Educ. Psych. Meas.*, Vol. 30, pp. 41–60, 1970.

Campbell, D. T. and Fiske, D. W., "Convergent and Discriminant Validation by the Multitrait-Multimethod Matrix," *Psych. Bull.*, Vol. 56, pp. 81–105, 1959.

Campbell, D. T., and Tyler, B. B., "The Construct Validity of Work-Group Morale Measures," *J. Appl. Psych.* Vol. 41, pp. 91–92, 1957.

Cattell, R. B., Eber, H. W., and Tatsuoka, M. M., *Handbook for the 16 Personality Factor Questionnaire*. Institute for Personality and Ability Testing, Champaign, Ill., 1970.

Cattell, R. B., and Warburton, F. W., *Objective Personality and Motivation Tests*. University of Illinois Press, Urbana, Ill., 1967.

Cleary, T. A., and Linn, R. L., "Error of Measurement and the Power of a Statistical Test," *Brit. J. Math. Soc. Psych.*, Vol. 22, pp. 49–55, 1969.

Cochran, W. G., "Some Effects of Errors of Measurement on Multiple Correlation," *J. Am. Stat. Assoc.*, Vol. 65, pp. 22–34, 1970.

Coffman, W. E., and Kurfman, D. G., "A Comparison of Two Methods of Reading Essay Examinations," *Am. Educ. Res. J.*, Vol. 5, pp. 99–107, 1968.

Collins, J. R., *Jackknifing Generalizability*, Doctoral dissertation. University of Colorado, Boulder, Col., 1970.

Cornfield, J. and Tukey, J. W., "Average Values of Mean Squares in Factorials," *Ann. Math. Stat.*, Vol. 27, pp. 907–949, 1956.

Costin, F., "Dogmatism and the Retention of Psychological Misconceptions," *Educ. Psych. Meas.*, Vol. 28, pp. 529–534, 1968.

Cronbach, L. J., "Test 'Reliability': Its Meaning and Determination," *Psychometrika*, Vol. 12, pp. 1–16, 1947.

Cronbach, L. J., *Essentials of Psychological Testing*. Third Edition. Harper and Row, New York, 1970.

Cronbach, L. J., "Test Validation," in *Educational Measurement*, pp. 443–507. Edited by R. L. Thorndike. American Council on Education, Washington, 1971.

Cronbach, L. J., and Azuma, H. "Internal-Consistency Reliability Formulas Applied to Randomly Sampled Single-Factor Tests: an Empirical Comparison," *Educ. Psych. Meas.*, Vol. 22, pp. 645–665, 1962.

Cronbach, L. J., and Furby, L., "How We Should Measure Change—or Should We?" *Psych. Bull.*, Vol. 74, pp. 68–80, 1970. See also: Errata, *ibid.*, Vol. 74, p. 218, 1970.

Cronbach, L. J., and Gleser, G. C., "Assessing Similarity Between Profiles," *Psych. Bull.*, Vol. 50, pp. 456–473, 1953.

Cronbach, L. J., and Gleser, G. C., *Psychological Tests and Personnel Decisions*, Second Edition. University of Illinois Press, Urbana, Ill., 1965.

Cronbach, L. J., Ikeda, H., and Avner, R. A., "Intraclass Correlation as an Approximation to the Coefficient of Generalizability," *Psych. Reports*, Vol. 15, pp. 727–736, 1964.

Cronbach, L. J., and Meehl, P. E., "Construct Validity in Psychological Tests," *Psych. Bull.*, Vol. 52, pp. 281–302, 1955.

Cronbach, L. J., Rajaratnam, N., and Gleser, G. C., "Theory of Generalizability: a Liberalization of Reliability Theory," *Brit. J. Stat. Psych.*, Vol. 16, pp. 137–163, 1963.

Cronbach, L. J., Schönemann, P., and McKie, T. D.,"Alpha Coefficients for Stratified-Parallel Tests," *Educ. Psych. Meas.*, Vol. 25, pp. 129–312, 1965.

Das, Rhea S., "Some Models for Assessment of Intelligence and Scholastic Attainment," *J. Psych. Res.*, Vol. 11, pp. 77–90, 1967.

Dayhoff, E., "Generalized Polykays, an Extension of Simple Polykays and Bipolykays," *Ann. Math. Stat.*, Vol. 37, pp. 226–241, 1966.

de Groot, A. D., *Methodology*, Mouton, The Hague, 1969.

Dick, W., and Hagerty, N., *Topics in Measurement: Reliability and Validity*. McGraw-Hill, New York, 1971.

Ebel, R. L. "Estimation of Reliability of Ratings," *Psychometrika*, Vol. 16, pp. 407–424, 1951.

Eber, H. W., Cattell, R. B., and Delhees, K. H., "Improvement of Factor Scale Validities by 'Computer Synthesis' Progams, Using 'Allocation of Variance,' Illustrated in Personality and Ability Testing." Manuscript in preparation.

Elashoff, J. D., Book Review, *Contemp. Psych.*, Vol. 15, pp. 102–104, 1970.

Endler, N. S., "Estimating Variance Components from Mean Squares for Random and Mixed Effects Analysis of Variance Models," *Percep. Motor Skills*, Vol. 22, pp. 559–570, 1966.

Endler, N. S., and Hunt, J. McV., "Sources of Behavioral Variance as Measured by the S–R Inventory of Anxiousness," *Psych. Bull.* Vol. 65, pp. 336–346, 1966.

Endler, N. S., and Hunt, J. McV., "S–R Inventories of Hostility and Comparisons of the Proportions of Variance from Persons, Responses, and Situations for Hostility and Anxiousness," *J. Person. Soc. Psych.*, Vol. 9, pp. 309–315, 1968.

Endler, N. S., and Hunt, J. McV., "Generalizability of Contributions from

Sources of Variance in the S–R Inventories of Anxiousness," *J. Person.*, Vol. 37, pp. 1–24, 1969.

Endler, N. S., Hunt, J. McV., and Rosenstein, A. J., "An S–R Inventory of Anxiousness," *Psych. Mono.*, Vol. 76, No. 17 (Whole No. 536), 1962.

Engelhart, M. D., "A Note on the Analysis of Gains and Posttest Scores," *Educ. Psych. Meas.*, Vol. 27, pp. 257–260, 1967.

Ferguson, R. L., *Computer-Assisted Criterion-Referenced Testing*, Technical Report. Learning Research and Development Center, University of Pittsburgh, Pittsburgh, Pa., 1970.

Fleiss, J. L., "Estimating the Reliability of Interview Data," *Psychometrika*, Vol. 35, pp. 143–162, 1970.

Finlayson, D. S., "The Reliability of Marking Essays," *Brit. J. Educ. Psych.*, Vol. 21, pp. 126–134, 1951.

Fisher, R. A., *Statistical Methods for Research Workers*. Oliver and Boyd, London, 1925.

Gleser, G. C., "Psychometric Contributions to the Assessment of Patients," in *Psychopharmacology, Review of Progress, 1957–1967*, pp. 1029–1037. Edited by D. H. Efron *et al*. Government Printing Office, Washington, 1968.

Gleser, G. C., Cronbach, L. J., and Rajaratnam, N., "Generalizability of Scores Influenced by Multiple Sources of Variance," *Psychometrika*, Vol. 30, pp. 395–418, 1965.

Gleser, G. C., and Ihilevich, D., "An Objective Instrument for Measuring Defense Mechanisms," *J. Consult. Clin. Psych*, Vol. 33, pp. 51–60, 1969.

Goodenough, F. L., "A Critical Note on the Use of the Term 'Reliability' in Mental Measurement," *J. Educ. Psych.*, Vol. 27, pp. 173–178, 1936.

Goodwin, D. L., *Training Teachers in Reinforcement Techniques to Increase Pupil Task-Oriented Behavior: An Experimental Evaluation*, Unpublished doctoral dissertation. Stanford University, 1966.

Graybill, F. A., *An Introduction to Linear Statistical Models: Vol. I.* McGraw-Hill, New York, 1961.

Gross, R. B., and Marsh, M., "An Instrument for Measuring Creativity in Young Children: The Gross Geometric Forms," *Develop. Psych.*, Vol. 3, p. 267, 1970.

Guilford, J. P., *Psychometric Methods*, McGraw-Hill, New York, 1954.

Gulliksen, H., "The Content Reliability of a Test," *Psychometrika*, Vol. 1, pp. 189–194, 1936.

Gulliksen, H. *Theory of Mental Tests*. John Wiley & Sons, Inc. New York, 1950.

Guttman, L., "A Special Review of Harold Gulliksen, *Theory of Mental Tests*," *Psychometrika*, Vol. 18, pp. 123–130, 1953.

Guttman, L., "What Lies Ahead for Factor Analysis?" *Educ. Psych. Meas.*, Vol. 18, pp. 497–515, 1958.

Harris, C. W. (Ed.), *Problems in Measuring Change*. University of Wisconsin Press, Madison, Wis., 1963.

Harris, H., "Development of Moral Attitudes in White and Negro Boys," *Develop. Psych.*, Vol. 2, pp. 376–383, 1970.

Hartley, H. O., and Rao, J. N. K., "Maximum-Likelihood Estimation for the Mixed Analysis of Variance Model," *Biometrika*, Vol. 54, pp. 93–108, 1967.

Hays, W. L., *Statistics for Psychologists*. Holt, Rinehart, and Winston, New York, 1964.

Hill, B. M., "Inferences About Variance Components in the One-Way Model," *J. Am. Stat. Assoc.*, Vol. 60, pp. 806–825, 1965.

Hill, B. M., "Some Contrasts Between Bayesian and Classical Inference in the Analysis of Variance and in the Testing of Models," in *Bayesian Statistics*, pp. 29–36. Edited by D. L. Meyers and R. O. Collier, Jr. Peacock, Itasca, Ill., 1970.

Hively, W., II, Patterson, H. L., and Page, S. H., "A Universe-Defined System of Arithmetic Achievement Tests." *J. Educ. Meas.*, Vol. 5, pp. 275–290, 1968.

Honzik, M. P. Book Review, *Merrill-Palmer Quarterly*, Vol. 11, p. 81, 1965.

Hooke, R. "Some Applications of Bipolykays to the Estimation of Variance Components and Their Moments," *Ann. Math. Stat.*, Vol. 27, pp. 80–98, 1956.

Horst, P., "Determination of Optimal Test Length to Maximize the Multiple Correlation," *Psychometrika*, Vol. 14, pp. 79–88, 1949.

Hunter, J. E., "Probabilistic Foundations for Coefficients of Generalizability," *Psychometrika*, Vol. 33, pp. 1–18, 1968.

Jackson, R. W. B., and Ferguson, G. A., *Studies on the Reliability of Tests*. University of Toronto, Toronto, 1941.

Kaiser, H. F., and Caffrey, J. "Alpha Factor Analysis," *Psychometrika*, Vol. 30, pp. 1–14, 1965.

Kaplan, A., *The Conduct of Inquiry*. Chandler, San Francisco, 1964.

Kelley, T. L., *Fundamentals of Statistics*. Harvard University Press, Cambridge, Mass, 1947.

Klotz, J. H., Milton, R. C., and Zacks, S., "Mean Square Efficiency of Estimators of Variance Components," *J. Am. Stat. Assoc.*, Vol. 64, pp. 1383–1402, 1969.

Knapp, T. R., "An Application of Balanced Incomplete Block Designs to the Estimation of Test Norms," *Educ. Psych. Meas.*, Vol. 28, pp. 265–272, 1968.

Krumboltz, J. D., and Goodwin, D. L., *Increasing Task-Oriented Behavior: An Experimental Evaluation of Training Teachers in Reinforcement Techniques*. Stanford University, Palo Alto, Calif., 1966.

Kuhn, T. S., "The Function of Measurement in Modern Physical Science," in *Quantification: A History of The Meaning of Measurement in the Natural and Social Sciences*, pp. 31–63. Edited by H. Woolf. Bobbs–Merrill, Indianapolis, Ind., 1961.

LaForge, R. "Components of Reliability," *Psychometrika*, Vol. 30, pp. 187–195, 1965.

Lehmann, H. E., "The Place and Purpose of Objective Methods in Psychopharmacology," in *Drugs and Behavior*, pp. 107–127. Edited by L. Uhr and J. G. Miller. John Wiley & Sons, Inc., New York, 1960.

Leler, H. O., *Mother-Child Interaction and Language Performance in Young Disadvantaged Negro Children*, Unpublished doctoral dissertation. Stanford University, Palo Alto, Calif., 1970.

Leone, F. C., and Nelson, L. S., "Sampling Distributions of Variance Components. I. Empirical Studies of Balanced Nested Designs," *Technometrics*, Vol. 8, pp. 457–468, 1966.

Lesser, G. S., Davis, F. B., and Nahemow, L., "The Identification of Gifted Elementary School Children with Exceptional Scientific Talent," *Educ. Psych. Meas.*, Vol. 22, pp. 349–364, 1962.

Levin, J. R., Rohwer, W. D., Jr., and Cleary, T. A., "Individual Differences in the Learning of Verbally and Pictorially Presented Paired Associates," *Am. Educ. Res. J.*, Vol. 8, pp. 11–26, 1971.

Lindley, D. V., *A Bayesian Estimate of True Scores That Incorporates Prior Information*, Research Bulletin 69–75. Educational Testing Service, Princeton, N.J., 1969.

Lindquist, E. F., *Design and Analysis of Experiments in Psychology and Education*. Houghton–Mifflin, Boston, 1953.

Loevinger, J., "Person and Population as Psychometric Concepts," *Psych. Rev.*, Vol. 72, pp. 143–155, 1965.

Lord, F. M., "Estimating Test Reliability," *Educ. Psych. Meas.*, Vol. 15, pp. 324–336, 1955. (a)

Lord, F. M., "Sampling Fluctuations Resulting from the Sampling of Test Items," *Psychometrika*, Vol. 20, pp. 1–22, 1955. (b)

Lord, F. M., "The Measurement of Growth," *Educ. Psych. Meas.*, Vol. 16, pp. 421–237, 1956. See also: Errata, *ibid.*, Vol. 17, p. 452, 1957.

Lord, F. M., "Further Problems in the Measurement of Growth," *Educ. Psych. Meas.*, Vol. 18, pp. 437–454, 1958.

Lord, F. M., "Large-Sample Covariance Analysis When the Control Variable is Fallible," *J. Am. Stat. Assoc.*, Vol. 55, pp. 309–321, 1960.

Lord, F. M., "Test Reliability: a Correction," *Educ Psych. Meas.*, Vol. 22, pp. 511–512, 1962.

Lord, F. M., "Elementary Models for Measuring Change," in *Problems in Measuring Change*, pp. 21–38. Edited by C. W. Harris. University of Wisconsin Press, Madison, Wis., 1963.

Lord, F. M., "Estimating True-Score Distributions in Psychological Testing (an Empirical Bayes Estimation Problem)," *Psychometrika*, Vol. 34, pp. 259–300, 1969.

Lord, F. M., and Novick, M., *Statistical Theories of Mental Test Scores*. Addison-Wesley, Reading, Mass., 1968.

Loveland, E. H., *Measurement of Factors Affecting Test-Retest Reliability*. Unpublished doctoral dissertation. University of Tennessee, 1952.

Mahalanobis, P. C., "Next Steps in Planning," *Sankhya*, Vol. 22, pp. 143–172, 1960.

Mahmoud, A. F. "Test Reliability in Terms of Factor Theory." *Brit. J. Stat. Psych.*, Vol. 8, pp. 119–135, 1955.

Mathur, R. K. and Kumar, P. "Errors in the Classification by Ability Groups in Two-Stage Sequential Decision Strategy," Indian Science Congress, Kharagpur, 1969.

Maxwell, A. E., "The Effect of Correlated Error on Estimates of Reliability Coefficients," *Educ. Psych. Meas.*, Vol. 28, pp. 803–811, 1968.

Maxwell, A. E., and Pilliner, A. E. G., "Deriving Coefficients of Reliability and Agreement for Ratings," *Brit. J. Math. Stat. Psych.*, Vol. 21, pp. 105–116, 1968.

McNemar, Q., *Psych. Stat.*, Fourth Edition. John Wiley and Sons, Inc., New York, 1969.

Medley, D. M., and Mitzel, H. E., "Measuring Classroom Behavior by Systematic Observation," in *Handbook of Research on Teaching*, pp. 247–328. Edited by N. L. Gage. Rand McNally, Chicago, Ill., 1963.

Medley, D. M., Mitzel, H. E., and Doi, A. N., "Analysis-of-Variance Models and Their Use in a Three-Way Design without Replication," *J. Exper. Educ.*, Vol. 24, pp. 221–229, 1956.

Meehl, P. E., "Nuisance Variables and the Ex Post Facto Design," in *Minnesota Studies in the Philosophy of Science, Vol. IV.* Edited by M. Radner and S. Winokur. University of Minnesota Press, Minneapolis, Minn., 1970. Pp. 373–402.

Meredith, W., "Canonical Correlations with Fallible Data," *Psychometrika*, Vol. 29, pp. 55–65, 1964.

Merrifield, P. R., "Parameter Factors of the Structure-of-Intellect Tests," Paper presented to the Society of Multivariate Experimental Psychology, 1970.

Miller, R. G., Jr., "Jackknifing Variances," *Ann. Math. Stat.*, Vol. 39, pp. 567–582, 1968.

Millman, J., and Glass, G. C., "Rules of Thumb for Writing the ANOVA Table," *J. Educ. Meas.*, Vol. 4, pp. 41–51, 1967.

Mischel, W., *Personality and Assessment.* John Wiley & Sons, Inc., New York, 1968.

Mosteller, F., and Tukey, J. W., "Data Analysis, Including Statistics," in *Handbook of Social Psychology: Vol. II*, pp. 80–203. Edited by G. Lindzey and E. Aronson. Addison-Wesley, Reading, Mass., 1968.

Nanda, H., *Factor Analytic Techniques for Interbattery Comparison and Their Application to Some Psychometic Problems*, Unpublished doctoral dissertation. Stanford University, Palo Alto, Calif., 1967.

Nash, L. K., Editor's remarks, "Plants and the Atmosphere, Case 5," in *Harvard Case Histories in Experimental Science: Vol. II.* Edited by J. B. Conant and L. K. Nash. Harvard University Press, Cambridge, Mass. 1957.

Nelder, J. A., "The Interpretation of Negative Components of Variance," *Biometrika*, Vol. 41, pp. 544–548, 1954.

Norman, W., "On Estimating Psychological Relationships: Social Desirability and Self-Report," *Psych. Bull.*, Vol. 67, pp. 273–293, 1967.

Novick, M. R., "Multiparameter Bayesian Inference Procedures (with discussion)," *J. Roy. Stat. Soc.: Ser. B*, Vol. 31, pp. 29–64, 1969.

Novick, M. R., "Bayesian Considerations in Educational Information Systems," in *Proceedings, Invitational Testing Conference*, pp. 77–88. Educational Testing Service, Princeton, N.J., 1971.

Novick, M. R., and Grizzle, J. E., "A Bayesian Indifference Procedure," *J. Am. Stat. Assoc.*, Vol. 60, pp. 1104–1117, 1965.

Novick, M. R., and Jackson, P. H., "Bayesian Guidance Technology," *Rev. Educ. Res.*, Vol. 40, pp. 459–494, 1970.

Novick, M. R., Jackson, P. H., and Thayer, D. T., "Bayesian Inference and the Classical Test Theory Model. Reliability and True Scores," *Psychometrika*, Vol. 36, 261–288, 1971.

Nunnally, J. C., Jr., *Introduction to Psychological Measurement*. McGraw-Hill, New York, 1971.

O'Connor, E. F., Jr., *Extending Classical Test Theory to the Measurement of Change*, Unpublished doctoral dissertation. University of California, Los Angeles, 1970.

Osburn, H. G., "Item Sampling for Achievement Testing," *Educ. Psych. Meas.*, Vol. 28, pp. 95–104, 1968.

Overall, J. E., "Estimating Individual Rater Reliabilities from Analysis of Treatment Effects," *Educ. Psych. Meas.*, Vol. 28, pp. 255–264, 1968.

Palmer, G. J., Jr., *A Method for Objective Measurement of Interpersonal Relations in Group Behavior*. Tulane University, New Orleans, La., 1960. (Mimeographed.) (a)

Palmer, G. J., Jr., *Tests of Interpersonal Knowledge: Some Development Considerations and Specifications for a Universe of Items*. Tulane University, New Orleans, La., 1960. (Mimeographed.) (b)

Pearson, K., "On the Mathematical Theory of Errors of Judgment," *Phil. Trans., Roy. Soc. London, Ser. A*, Vol. 198, pp. 235–299, 1902.

Pilliner, A. E. G., "The Application of Analysis of Variance to Problems of Correlation," *Brit. J. Psych., Stat. Sect.*, Vol. 5, pp. 31–38, 1952.

Pilliner, A. E. G., *The Application of Analysis of Variance Components in Psychometric Experimentation*. Unpublished doctoral dissertation. University of Edinburgh, Edinburgh, 1965.

Pilliner, A. E. G., Sutherland, J., and Taylor, E. G., "Zero Error in Moray House Verbal Reasoning Tests," *Brit. J. Educ. Psych.*, Vol. 30, pp. 53–62, 1960.

Porch, B. E., *Multidimensional Quantification of Gestural, Verbal, and Graphic Responses of Patients with Cerebral Pathology*. Unpublished doctoral dissertation. Stanford University, Palo Alto, Calif., 1966.

Porch, B. E., *The Porch Index of Communicative Ability*. Consulting Psychologists Press, Palo Alto, Calif., 1970.

Porter, A. C., *The Effects of Using Fallible Variables in the Analysis of Covariance*. Unpublished doctoral dissertation. University of Wisconsin, Madison, Wis., 1967.

Rabinowitz, W. and Eikeland, H. M., "Estimating the Reliability of Tests with Clustered Items." *Pedag. Forsk.*, Vol. 8, pp. 85–106, 1964.

Rajaratnam, N., "Reliability Formulas for Independent Decision Data When Reliability Data are Matched." *Psychometrika*, Vol. 25, pp. 261–271, 1960.

Rajaratnam, N., Cronbach, L. J., and Gleser, G. C., "Generalizability of Stratified–Parallel Tests," *Psychometrika*, Vol. 30, pp. 39–56, 1965.

Rao, C. R., "Familial Correlations or the Multivariate Generalisation of the Intraclass Correlation," *Curr. Sci.*, Vol. 14, pp. 66–67, 1945.

Rao, C. R., "Discriminant Functions for Genetic Differentiation and Selection," *Sankhya*, Vol. 12, pp. 229–246, 1953.

Rao, C. R., *Linear Statistical Inference and Its Applications*. John Wiley & Sons, Inc., New York, 1965.

Ross, J., and Lumsden, J., "Attribute and Reliability," *Brit. J. Math. Stat. Psych.*, Vol. 21, pp. 251–263, 1968.

Ross, J., and Smith, P., "Orthodox Experimental Designs," in *Methodology in Social Research,* pp. 333–389. Edited by H. M. Blalock, Jr. and A. B. Blalock. McGraw-Hill, New York, 1968.

Ross, R. T., and Morledge, J., "Comparison of the WISC and WAIS at Chronological Age Sixteen," *J. Consult. Psych.*, Vol. 31, pp. 331–332, 1967.

Rothenburg, B. B., "Children's Social Sensitivity and the Relationship to Interpersonal Competence, Intrapersonal Comfort, and Intellectual Level," *Develop. Psych.*, Vol. 2, pp. 335–350, 1970.

Rozeboom, W. W., *Foundations of the Theory of Prediction.* Dorsey Press, Homewood, Ill., 1966.

Scheffé, H., *The Analysis of Variance.* John Wiley & Sons, Inc., New York, 1959.

Siegel, P. M., and Hodge, R. W., "A Causal Approach to the Study of Measurement Error," in *Methodology in Social Research*, pp. 28–59. Edited by H. M. Blalock and A. B. Blalock. McGraw-Hill, New York, 1968.

Silverstein, A. B., and Fisher, G., "Estimated Variance Components in the S–R Inventory of Anxiousness," *Percep. Motor Skills*, Vol. 27, pp. 740–742, 1968.

Sirotnik, K., "An Analysis of Variance Framework for Matrix Sampling," *Educ. Psych. Meas.*, Vol. 30, pp. 891–908, 1970.

Snyder, F. W., *A Unique Variance Model for Three-Mode Factor Analysis.* Department of Psychology, University of Illinois, Urbana, Ill., 1968. (Mimeographed.)

Soskin, W. F., "Bias in Postdiction from Projective Tests," *J. Abnorm. Soc. Psych.*, Vol. 49, pp. 69–74, 1954.

Spearman, C., "The Proof and Measurement of Association between Two Things," *Amer. J. Psych.*, Vol. 15, pp. 72–101, 1904.

Spearman, C., "Correlation Calculated from Faulty Data," *Brit. J. Psych.*, Vol. 3, pp. 271–295, 1910.

Stanley, J. C., *Intraclass r as Related to Product-Moment r in Multiple Classification Designs*, University of Wisconsin, Madison, Wis., 1955. (Mimeographed.)

Stanley, J. C., "Analysis of Unreflected Three-Way Classifications with Applications to Rater Bias and Treatment Independence," *Psychometrika*, Vol. 26, pp. 205–219, 1961.

Stanley, J. C., "General and Special Formulas for Reliability of Differences," *J. Educ. Meas.*, Vol. 4, pp. 249–252, 1967.

Stanley, J. C., and Wiley, D. E., *Development and Analysis of Experimental Designs for Ratings.* University of Wisconsin, Madison, Wis., 1962. (Mimeographed.)

Stein, C. M., "Confidence Sets for the Mean of a Multivariate Normal Distribution (with discussion)," *J. Roy. Stat. Soc., Ser. B*, Vol. 24, pp. 265–296, 1962.

Sutcliffe, J. P., "A Probability Model for Errors of Classification. I. General Conditions," *Psychometrika*, Vol. 30, pp. 73–96, 1965.

Thorndike, R. L., *Research Problems and Techniques*, Report No. 3. AAF Aviation Psychology Program Research Reports. Washington: U.S. Government Printing Office, Washington, 1947.

Thorndike. R. L., "Reliability," In *Testing Problems in Perspective*, pp. 284–291. Edited by Anne Anastasi. American Council on Education, Washington, 1967. (Originally presented in 1963.)

Tiao, G. C., and Tan, W. Y., "Bayesian Analysis of Random-Effect Models in the Analysis of Variance. I. Posterior Distribution of Variance Components," *Biometrika*, Vol. 52, pp. 37–53, 1965.

Tippett, L. H. C., *Technological Applications of Statistics*. John Wiley & Sons, Inc., New York, 1950.

Tryon, R. C., "Reliability and Behavior Domain Validity: Reformulation and Historical Critique," *Psych. Bull.*, Vol. 54, pp. 229–249, 1957.

Tucker, L. R., "An Inter-Battery Method of Factor Analysis," *Psychometrika*, Vol. 23, pp. 111–136, 1958.

Tucker, L. R., "The Extension of Factor Analysis to Three-Dimensional Matrices," in *Contributions to Mathematical Psychology*, pp. 109–127. Edited by N. Frederiksen and H. Gulliksen. Holt, Rinehart, & Winston, New York, 1964.

Tucker, L. R., "Some Mathematical Notes on Three-Mode Factor Analysis," *Psychometrika*, Vol. 31, pp. 279–312, 1966.

Tukey, J. W., "Dyadic Anova, an Analysis of Variance for Vectors," *Human Biol.*, Vol. 21, pp. 65–110, 1949.

Tukey, J. W., "Components in Regression," *Biometrics*, Vol. 7, pp. 33–69, 1951.

Vaughn, G. M., and Corballis, M. C., "Beyond Tests of Significance: Estimating Strength of Effects in Selected Anova Designs," *Psych. Bull.*, Vol. 72, pp. 204–213, 1969.

Wang, Y. Y., "A Comparison of Several Variance Component Estimators," *Biometrika*, Vol. 54, pp. 301–305, 1967.

Wechsler, D., *Wechsler Preschool and Primary Scale of Intelligence*, *Manual*. Psychological Corporation, New York, 1967.

Welch, B. L., "On Linear Combination of Several Variances," *J. Am. Stat. Assoc.*, Vol. 51, pp. 132–148, 1956.

Westbrook, B. W., and Jones, C. I., "The Reliability and Validity of a Class-Constructed Measure of Achievement in Tests and Measurements," *Educ. Psych. Meas.*, Vol. 28, pp. 485–486, 1968.

Wiley, D. E., and Wiley, J. A., "The Estimation of Measurement Error in Panel Data," *Am. Soc. Rev.*, Vol. 35, pp. 112–117, 1970.

*The Evaluation of Instruction: Issues and Problems*. Wittrock, M. C., and Wiley, D. E. (Eds.). Holt, Rinehart, and Winston, New York, 1970.

# Index